# BANANA

## *Quantum Mechanics for Primates*


JEFFREY BUB

*Distinguished University Professor*
*Philosophy Department, Institute for Physical Science and Technology,*
*and Joint Center for Quantum Information and Computer Science*
*University of Maryland, College Park, USA*


*Illustrations by*

Tanya Bub, with a bow to John Tenniel

Great Clarendon Street, Oxford, OX2 6DP,
United Kingdom

Oxford University Press is a department of the University of Oxford.
It furthers the University's objective of excellence in research, scholarship,
and education by publishing worldwide. Oxford is a registered trade mark of
Oxford University Press in the UK and in certain other countries



First published 2016
First published in paperback 2018

Impression: 1



Published in the United States of America by Oxford University Press
198 Madison Avenue, New York, NY 10016, United States of America

British Library Cataloguing in Publication Data
Data available

Library of Congress Cataloging in Publication Data
Data available

ISBN 978–0–19–871853–6 (Hbk.)
ISBN 978–0–19–881784–0 (Pbk.)

Printed and bound by
CPI Group (UK) Ltd, Croydon, CR0 4YY

*For Gil and Tanya*

# FOREWORD

*Bananaworld. Bananaworld*?!? This surely can't be serious!

But it is. It is at the same time very serious scientifically and easily accessible.

Read it. You will enjoy it. And in the process you will really understand one of the most fascinating things in physics today—nonlocality.

When my friends ask me what I'm working on, I know I'm in trouble. Most scientists have it easy: "We thought we knew everything about volcanoes—what pushes the hot lava upwards, how strong the explosion will be, how to detect that a volcano is going to erupt. But this volcano that erupted last year broke all the rules; it was nothing like what we expected and sent us back to the drawing board. I'm now analysing various possible explanations....". Even the most complex area that is biology is quite comprehensible. "We wondered why, in people who have Alzheimer's, healthy brain cells (called neurons) die. We found that some other cells, the 'cleaners', eat them. Cleaners were supposed to eat only damaged neurons, but we discovered that they attack the healthy ones as well. We are now trying to find out what makes them do this ...".

So simple. But I'm a theoretical physicist. And theoretical physics is hard to explain. Even to state what the question is, let alone what the answers could be, is hard. This is because physics is a "vertical" science. Like in a tower, the upper layer rests on the one underneath, which rests on the one underneath, and so on. You can understand Chinese literature even if you don't know Spanish literature, and vice versa, but you cannot understand quantum mechanics if you don't understand classical mechanics, which you can't understand if you don't understand advanced mathematics, which you cannot understand ... I can't tell my friends what I'm working on right now because there is so much they need to know to start understanding what the question I'm trying to answer is.

Faced with such a daunting task, most of the popular presentations of the fascinating issues that physics deals with these days struggle. The tool they tend to use is making analogies. "Imagine an atom like a miniature solar system, with electrons moving around the nucleus like the planets around the sun ...". No, the atom is nothing like that. And any effort made to understand the behaviour of atoms from such an analogy is doomed to fail. Just try to use that analogy to understand why when two hydrogen atoms—solar systems with a single planet—come close to each other they attract each other to form a molecule, while two helium ones—solar systems with two planets—don't. In both cases the planets would just collide and all will be a mess. You get the idea. So the reader is left frustrated. The more the reader tries to really understand, the more frustrated the reader becomes. It's hard to write a popular physics book well.

Yet, from time to time, a real jewel emerges that gives you the real stuff. This book is one of them.

That such a book can exist at all is utterly surprising. After all, I did my best to convince you that theoretical physics cannot be explained simply. And that is true: most things in theoretical physics are basically incomprehensible for the general public. But here and there are some isolated islands of simplicity. Finding them is the trick, and this book succeeds.

*Bananaworld* focuses on one of the most exciting and deep issues in present day physics: nonlocality. Nonlocality and the associated phenomenon of entanglement are very much in the news these days since they are at the basis of fascinating applications such as teleportation and quantum computation. But it is far more than that. It is now becoming recognised that nonlocality is one of the, if not *the,* most important aspects of nature.

And what a dramatic and paradoxical effect nonlocality is. All we see in everyday life is local: what you do here only affects things here; things far away are affected only via perturbations that propagate from neighbouring region to neighbouring region. A stone goes from here to there by crossing via points in between. But once you go down to the microscopic level where quantum mechanics applies, you find that in addition to this local behaviour there are also subtle effects which manifest themselves in strange correlations that occur instantaneously between remote locations, with seemingly nothing going in between. It doesn't seem to make sense. It even seems to violate high authority—didn't Einstein prove that nothing moves faster than the speed of light? And yet nature is like this, and everyday life will change dramatically when we will start using this effect. And "no", it doesn't violate what Einstein said, but it pushes it to the limit.

But the book does much more than merely explain what nonlocality is. That would have been more than enough; however the book goes far beyond. That nonlocality can exist at all, given the constraints imposed by relativistic causality, is an extraordinary fact. Yet, it was recently discovered that the nonlocal correlations predicted by quantum mechanics are not as nonlocal as Einstein's relativity permits. Nonlocal correlations stronger than the quantum mechanical ones, yet consistent with relativity, are theoretically possible. Whether or not such correlations exist in nature is an open experimental question. We have not seen anything like that yet, but it is quite possible that we will see them in the future. If they exist, quantum mechanics is wrong and has to be replaced—that would be a revolution in physics. But if they do not exist—why not? What is the fundamental physical principle—something different from relativity—that forbids them? We don't know yet and the quest to find it is one of the most important and challenging quests in modern physics. The book describes the different ways that are taken as we speak to try and solve this problem.

And again, why bananas? That is the great skill of the author. Using magic bananas and simple actions—peeling and eating them—you can REALLY UNDERSTAND what nonlocality is in simple terms. The more I think of the bananas, the more I like their use: The two peeling actions are complementary and cannot both be implemented on the same banana—once it's peeled it's peeled, exactly like two measurements that cannot be simultaneously performed on microscopic particles. And there is only one answer—once it's eaten it's eaten. Again, exactly like in quantum mechanics. The magic bananas are a perfect model for what is going on. So much better than the abstract models the physicists use ... I just love it. I'm confident you will too.

Sandu Popescu
Professsor of Physics
University of Bristol

# PREFACE

The first thing to clear up is that this is not another "quantum mechanics for dummies" book. The aim of the book is really quite serious. There's a new game in town in the foundations of quantum mechanics. I want to say something about this in a way that's accessible and interesting to a wide range of readers, not just the experts. The reader I have in mind is someone with a certain sophistication about science, perhaps a non-physicist scientist like a biologist, or a computer scientist, or an economist, who has read a bit about quantum entanglement and nonlocality, and would like to understand what's so weird about quantum mechanics, and why a Nobel physicist like Richard Feynman could say that nobody understands it. Physics is a broad field, and I also have in mind physicists who have only a passing familiarity with quantum foundations and would like to learn something about what's going on in this area, as well as non-scientists who know about Schrödinger's cat and quantum teleportation and browse the science shelves in bookstores for something new. Finally, the book is also addressed to philosophers of physics, who worry about the measurement problem of quantum mechanics, and of course to students.

So why the bananas?

It's quite common nowadays to see papers in major physics journals that are not about actual physical systems or actual experiments, but about abstractly defined "boxes," with inputs and outputs, and some hypothetical probabilistic correlation between inputs and outputs. The idea is to explore what you can and can't do with these boxes, or the extent to which you can simulate the correlation of a box if you are limited by certain constraints. It's a novel way of exploring the terrain of possible quantum-like theories, and the approach has been enormously fruitful in posing new questions and in providing some precise answers.

The Popescu–Rohrlich box or PR box is perhaps the most familiar of these boxes. It was proposed by Sandu Popescu and Daniel Rohrlich in the context of asking whether you could have nonlocal correlations stronger than the correlations of entangled quantum states, without violating a "no signaling" principle. A PR box has an input and output for Alice, and an input and output for Bob. Inputs and outputs can be 0 or 1, and the correlation is simply that the outputs are the same, except when both inputs are 1. The box is nonlocal in the sense that you can pull the Alice part and the Bob part of the box as far apart as you like without changing the correlation. There's something extremely strange about this correlation (which I talk about in Chapter 3, "Bananaworld"), but the correlation is possible and not inconsistent, and PR boxes have proved to be an extremely powerful theoretical tool.

I replace the boxes with bananas. The two possible inputs to a PR box correspond to two ways of peeling a banana, either from the stem end or the top end, and the two possible outputs of the box correspond to a banana tasting *o*rdinary ("o" for 0) or *i*ntense, *i*ncredible, *i*ndescribably delicious ("i" for 1). That makes things more concrete, but I admit to liking the apparent absurdity of doing quantum foundations with bananas. It's more straightforward to imagine Alice and Bob separating their bananas by any distance without altering the correlation than to think of the Alice part of a box and the Bob part of a box being pulled or stretched apart. Apart from Popescu–Rohrlich correlations, you can define all sorts of interesting correlations in Bananaworld for bananas that grow in bunches of two, three, or more bananas. Some of the correlations aren't observed in our quantum world, while some are actual quantum correlations, but it's easy to imagine bananas with peelings and tastes correlated in various ways. The point is to show that a correlation can or can't be simulated with certain resources. If Alice and Bob can't simulate a correlation with local resources, but can succeed or do better with entangled quantum states, that's a fact about our world, not just about Bananaworld.

The nice thing about all this as far as readers are concerned is that a large part of the discussion is about bananas with hypothetical correlations between peelings and tastes, and this can all be understood without any familiarity with quantum mechanics. It would be really nice if I could just talk about bananas and avoid bringing in quantum states and observables and the mathematical machinery of quantum mechanics at all. Sophisticated readers familiar with the new work on quantum information and quantum foundations would see the connection with quantum mechanics, but most of the readers I have in mind would not. So I've had to relate the banana discussion to qubits, the simplest quantum systems, using polarized photons as the standard example. I try to keep things simple, but I've not found it possible to avoid a technical discussion at some points. There's an introduction to photon polarization and the mathematical language of quantum mechanics in a brief chapter, "Qubits," before getting to Bananaworld correlations. It's meant to be self-contained and ought to be readable by anyone with some feeling for mathematical concepts. There's more in a Supplement, "Some Mathematical Machinery," at the end of the book, and I occasionally refer to sections in this Supplement at various places in the book.

Another advantage in tackling these issues with bananas is that the puzzling features of quantum mechanics noted by Feynman apply equally to quantum-like theories that are more nonlocal than quantum mechanics—in particular, they apply to PR boxes or Popescu–Rohrlich bananas. You could imagine discovering an island, Bananaworld, with banana trees that have bunches of bananas with strange correlations. So as far as the conceptual problems of the theory that philosophers worry about are concerned, you might as well talk about bananas rather than quantum states.

Some chapters have a final section called "More." The idea came from my son, Gil, who is a physiologist and one of my target readers. The "More" sections are a little more challenging than the rest of the discussion, and I've tried to keep the mathematical arguments to these sections. You can skip the "More" sections and still get a lot out of the book, but there's more if you want it. If you're feeling lost, look for the summary box in each section outlining the main points, a suggestion from my daughter-in-law, Caroline.

My illustrator, who is also my daughter Tanya, came up with the idea of using elements from the Tenniel drawings for *Alice in Wonderland* in some of the illustrations. These add a whimsical touch, but they've also turned out to be wonderfully effective in conveying information. My wife, Robin, always my muse, inspired some of the chapter titles. So the book has turned out to be something of a family affair.

The new work on quantum foundations was sparked by John Bell's nonlocality theorem, which amounts to showing that Alice and Bob, restricted to local resources, can't simulate the correlations of Popescu–Rohrlich bananas with a probability of success greater than 3/4, but they can do better with entangled qubits (see the "Bananaworld" chapter). Two figures stand out for me in conveying the scope and richness of this research: Sandu Popescu and Nicolas Gisin. The book reflects what I have learned from them, but of course they are not responsible for any foolishness on my part. Readers interested in going to the source should read Sandu Popescu's review article 'Nonlocality beyond quantum mechanics' in *Nature Physics* 10, 264–270 (2014) and Nicolas Gisin's popular book *Quantum Chance: Nonlocality, Teleportation and Other Quantum Marvels* (Copernicus, Göttingen, 2014).

Several people read versions of some of the chapters and have given me useful feedback that resulted in quite extensive revisions. I am very grateful for all their input. My son Gil has been really helpful in gently steering me away from my usual academic writing style with lots of useful suggestions and requests for clarity. Leah Henderson, who is both a philosopher and a physicist with expertise in quantum information, has been a perceptive critic and pushed me to sharpen some of the arguments. My cousin, Stanley Barkan, who is a mathematician, and his son Shaul, were amazingly diligent in sending me detailed critiques of sections that were confusing or poorly written. Michael Dascal, my graduate student, carefully read through the entire manuscript and caught numerous mistakes, omissions, and passages that were less than clear. The final version reflects his suggestions for editorial improvements.

Finally, I should mention my colleague, Allen Stairs. We've worked together for years and our discussions in Washington, DC coffee bars while the book was taking shape have been a major influence on my thinking.

This book took a while to write. For part of the time, I enjoyed the support of a grant from the John Templeton Foundation, which freed me from teaching. The opinions expressed in the book are my own and don't necessarily reflect the views of the John Templeton Foundation.

# PREFACE TO THE REVISED PAPERBACK EDITION

Apart from minor revisions, the changes in this edition are mostly to "The information-Theoretic Interpretation," the final section 4 of the last chapter, "Making Sense of It All." The changes were prompted by a workshop on *Bananaworld* organized by Lucas Dunlap and Michael Cuffaro at Western University, London, Ontario, June 11–12, 2016. The official title was "Information-Theoretic Interpretations of Quantum Mechanics," but the workshop quickly became known as "Bananarama," a perhaps more apt title conferred by Wayne Myrvold.

I am very grateful to the organizers, to the participants for stimulating comments, and particularly to the commentators, Leah Henderson, Allen Stairs, and Matthew Leifer. "The Information-Theoretic Interpretation" now focuses on the Frauchiger-Renner argument that any "one-world" interpretation of quantum mechanics is inconsistent, which appeared after the publication of the book. I've dropped the discussion of Hepp's toy model of quantum measurement, which Matthew Leifer pointed out was flawed. Special thanks to Matthew Leifer for clarifying discussions, and to Michel Janssen, Allen Stairs, Michael Dascal, and Michael Cuffaro for comments on the revised section.

# CONTENTS

# 1

# Nobody Understands Quantum Mechanics

## 1.1 QUANTUM BLUES

Richard Feynman, who won the Nobel prize in physics in 1965, is often quoted as saying that nobody understands quantum mechanics:[1]

> There was a time when the newspapers said that only twelve men understood the theory of relativity. I do not believe there ever was such a time. There might have been a time when only one man did, because he was the only guy who caught on, before he wrote his paper. But after people read the paper, a lot of people understood the theory of relativity in some way or other, certainly more than twelve. On the other hand, I think I can safely say that nobody understands quantum mechanics. ... I am going to tell you what nature behaves like. If you will simply admit that maybe she does behave like this, you will find her a delightful entrancing thing. Do not keep saying to yourself, if you can possibly avoid it, "But how can it be like that?" because you will get "down the drain," into a blind alley from which nobody has yet escaped. Nobody knows how it can be like that.

What is the theory of relativity about? Even non-physicists are likely to say that it's about space and time. The notion of space as a sort of universal three-dimensional arena in which events take place, and time as the ticking of a universal clock, turns out to be wrong. According to Einstein's theory of special relativity,[2] space and time depend on the state of motion of a system and so are different for Alice on a flight to Rome and Bob on a train to New York, and quite a lot different for neutrinos moving at light speed. Hermann Minkowski, Einstein's former mathematics professor, showed that space and time in special relativity can be represented by a four-dimensional non-Euclidean geometry and predicted that "space by itself, and time by itself, are doomed to fade away into mere shadows, and only a kind of union of the two will preserve an independent reality."[3] General relativity goes further and treats gravity as the bending of space-time, so objects falling under the influence of gravity move by following the curvature of space-time.

One might think that the idea of space and time fading away into "mere shadows" relative to some sort of merging of the two notions is pretty wild and not that easy to get one's head around. How is it that "nobody understands quantum mechanics," but there isn't a similar difficulty in making sense of the theory of relativity? There's an introduction to special relativity in the "More" section at the end of this chapter. It's not hard to see how Alice and Bob, equipped with identical synchronized clocks, could disagree about how much time passes between two events, or about the distance between them, given the two central assumptions of the theory.

If relativity is about the geometrical structure of space-time, what is quantum mechanics about? There are a surprising variety of answers to this question: that quantum mechanics is about energy being quantized in discrete lumps or quanta, or about particles being wave-like, or about the universe continually splitting into countless coexisting quasi-classical universes, with many copies of ourselves, and so on. A rather more mundane answer, with quite remarkable implications, has emerged over the past thirty years or so from the study of the difference between classical information and quantum information: quantum mechanics is about new sorts of probabilistic correlations in nature, so *about the structure of information*, insofar as a theory of information in the sense relevant to physics is essentially a theory of probabilistic correlations.

Here's a very short history of the birth of quantum mechanics. A hot solid body glows, which means that it emits light. If you pass the light through a prism, you see a continuous band of colors like the spectrum of a rainbow, corresponding to light waves of varying frequencies. For a hot gas, what you see is an "emission spectrum" of discrete colored lines with dark gaps between the different colors. A cool gas surrounding a hotter object like a star produces the opposite effect: a continuous "absorption spectrum" with dark lines at the discrete frequencies of the emission spectrum of the gas. Different gases have different spectra, with the lines in different places. These facts were known in the 19th century. The problem was that no one could figure out how to explain the discrete spectra in the conceptual framework of classical physics.

In 1913, the Danish physicist Niels Bohr proposed a new theory of atomic structure[4] as an explanation. In Bohr's theory, negatively charged electrons orbit a positively charged nucleus, like planets orbiting a star. There's a difference between the behavior of an orbiting planet, described by Newton's theory of gravity, and the way a charged particle moves in an electromagnetic field, described by Maxwell's theory of classical electrodynamics. The details are irrelevant to what follows, but the significant difference here is that, according to classical electrodynamics, a charged particle emits electromagnetic radiation when it accelerates and so loses energy, unlike an accelerating Newtonian body. An electron orbiting the nucleus of an atom is constantly accelerating (because its velocity is constantly changing direction), and so should radiate energy in the form of light and spiral into the nucleus as it loses energy. That doesn't happen, so Bohr's theory stipulates that the energy of an orbiting electron is "quantized"—there's a discrete set of allowed orbits associated with different energies that an electron can occupy, with lower energy orbits closer to the nucleus. The theory also stipulates that an atom radiates or absorbs energy only when an electron jumps from one of these quantized orbits to another orbit, with the frequency of the radiation depending on the energy gap between the two orbits. An emission spectrum is a picture of the radiation emitted when electrons jump from higher to lower energy orbits in the gas atoms. An absorption spectrum represents electron jumps from lower to higher energy orbits, when the atoms absorb energy radiated by a hotter body.

Bohr's theory explains the distribution of spectral lines for a gas in terms of its atomic structure, but the ad hoc "quantum" rules for orbiting electrons conflict with classical electrodynamics, and also with classical mechanics, Newton's theory of motion. In 1925, Heisenberg published a breakthrough paper[5] in the journal *Zeitschrift für Physik* and shortly afterwards developed the idea into an early version of quantum mechanics in collaboration with Max Born and Pascual Jordan. The title of Heisenberg's paper in English is "On the quantum-theoretical re-interpretation [Umdeutung] of kinematical and mechanical relations." The "Umdeutung" changes the ball game. The thought was that the discrete orbits were an artificial theoretical fix that "saves the appearances,"[6] but this was not the right way to think about the structural features of atoms responsible for the spectral lines. Heisenberg proposed to *reinterpret* classical mechanical quantities, like position, momentum, energy, angular momentum, as operations, later represented by operators that act on and transform the states of quantum systems. The aim was to explain the discrete frequencies of light emitted by atoms *without appealing to electron orbits*, and later to explain other phenomena that couldn't be explained by classical physics. In a 1925 letter to Wolfgang Pauli, Heisenberg wrote:[7]

> All of my meagre efforts go toward killing off and suitably replacing the concept of the orbital paths that one cannot observe.

In Heisenberg's theory, the effect of applying an operation $A$ followed by an operation $B$ can differ from applying $B$ followed by $A$ for certain operations, which is to say that quantities $A$, $B$ represented by operators needn't commute with each other: you can have $AB \neq BA$ in terms of the effect on a quantum state. If that's so, then it turns out that quantum systems can't have definite values for all these quantities simultaneously—in particular, an electron can't have definite position and momentum values and so can't have a well-defined orbit in an atom. If a quantity has a definite value in a quantum state, then certain other quantities are indefinite, and what you find if you measure a quantity with an indefinite value is intrinsically random, in the sense that the outcome is independent of any information available before the measurement, as I'll show in Chapter 4, "Really Random."

Indefiniteness or intrinsic randomness is related to a feature of the theory that Heisenberg later formulated as an "uncertainty" or "indeterminacy" principle, but it's a much more radical departure from classical or commonsense ways of thinking about physical systems than the uncertainty principle. As derived in quantum mechanics, the uncertainty principle is simply a statement about a reciprocal relation between two noncommuting quantities, or "observables" in the jargon of quantum mechanics. Quantum states assign probabilities to measurement outcomes. If an observable is most likely to have a value in a certain range of values when measured, the likely outcomes of measuring a noncommuting observable can't be pinned down more precisely than a range of values that is inversely related to the first range: when one range of values is small, the other is correspondingly large. In particular, the uncertainty principle says that it's impossible to prepare a system in a quantum state in which two noncommuting observables both have definite values, so the observables are said to be "incompatible."

The uncertainty principle is open to various interpretations, and Heisenberg himself explained the relationship in the 1927 paper in which he introduced the principle as the result of irreducible measurement disturbances. He argued that the procedure for measuring an observable necessarily disturbs the value of an incompatible or noncommuting observable—which presupposes that there are definite values there to be measured or disturbed in the first place. An uncertainty principle in this sense would be a feature of any theory with noncommuting observables, but the indefiniteness of observables, and the intrinsic randomness of the value revealed when an observable of a quantum system is measured, is a feature of the particular way in which commuting and noncommuting observables are related in quantum mechanics.

Schrödinger published a wave mechanical version of the theory in 1926[8] that kept the orbits and explained their quantization as a wave phenomenon. Shortly afterwards, he proved the formal equivalence of Heisenberg's noncommutative mechanics and his own wave mechanics. Not surprisingly, physicists found wave mechanics more intuitively appealing as a picture of reality at the subatomic level than the abstract notion of a noncommutative mechanics, but the intuitive appeal is misleading. As Schrödinger pointed out in a lecture to the Royal Institution in London in March, 1928, the wave associated with a quantum system evolves in an abstract, multidimensional representation space, not real physical space:[9]

> The statement that what *really* happens is correctly described by describing a wave-motion does not necessarily mean exactly the same thing as: what *really* exists is a wave-motion. We shall see later on that in generalizing to an *arbitrary* mechanical system we are led to describe what really happens in such a system by a wave-motion in the generalized space of its co-ordinates ($q$-space). Though the latter has quite a definite physical meaning, it cannot very well be said to "exist"; hence a wave-motion in this space cannot be said to "exist" in the ordinary sense of the word either. It is merely an adequate mathematical description of what happens. It may be that also in the case of a single mass-point, with which we are now dealing, the wave-motion must not be taken to "exist" in *too* literal a sense, although the configuration space happens to coincide with ordinary space in this particular simple case.

That's the end of the short history. The point is not to suggest that "Heisenberg saw it all in 1925," or that Heisenberg had anything like an information-theoretic interpretation in mind. It's rather that the "Umdeutung" contained the germ of a radically new way of thinking about physical systems that developed as Heisenberg's idea was applied to more complex systems, while Schrödinger's wave mechanics evoked a very different structural picture that has turned out to be misleading in many ways and the source of a lot of confused thinking—not necessarily Schrödinger's fault.

The idea of a wave as a representation of quantum reality continues to shape contemporary discussions of conceptual issues in the foundations of quantum mechanics. But, as

Schrödinger pointed out, it is "merely an adequate mathematical description of what happens." From the perspective adopted here, the later formalization of quantum mechanics by Paul Dirac[10] in 1930 and John von Neumann[11] in 1932 as a theory of observables represented by operators on a space of quantum states is fundamentally an elaboration of Heisenberg's "Umdeutung" rather than a wave theory. Operators needn't commute—the order in which they are applied generally makes a difference—and the really significant thing about a noncommutative mechanics is the novel possibility of correlated events that are *intrinsically random*, not merely apparently random like coin tosses, where the outcome of a toss is determined by the way a coin is tossed, and the probabilities of "heads" and "tails" represent an averaging over differences among individual coin tosses that we don't keep track of for practical reasons. This intrinsic randomness allows *new sorts of nonlocal probabilistic correlations* for "entangled" quantum states of separated systems, where the probabilities are, as von Neumann put it, "perfectly new and *sui generis* properties of physical reality."[12] Schrödinger, who coined the term, referred to entanglement ("Verschränkung" in German) as "*the* characteristic trait of quantum mechanics, the one that enforces its entire departure from classical lines of thought."[13] So the deep significance of Heisenberg's "Umdeutung" is that quantum mechanics, as a noncommutative modification of classical mechanics, is a theory about a *structurally different sort of information* than classical information.

Quantum entanglement plays a major role throughout the book. At this point I haven't said what it is, other than that entanglement is somehow associated with "new sorts of nonlocal probabilistic correlations." I'll say more about entanglement, and about quantum states and operators, when I talk about polarization in Chapter 2, "Qubits."

What do correlations have to do with information? The classical theory of information was initially developed by Claude Shannon to deal with certain problems in the communication of messages as electromagnetic signals along a channel such as a telephone wire. A communication setup involves a transmitter or source of information, a communication channel, and a receiver. An information source produces messages composed of sequences of symbols from an alphabet, with certain probabilities for the different symbols. The fundamental question for Shannon was how to quantify the minimal physical resources required to represent messages produced by a source, so that they could be communicated via a channel and reconstructed by a receiver. For this problem, and related communication problems, the meaning of the message is irrelevant.

As Shannon remarked in his seminal paper "A mathematical theory of communication":[14]

> The fundamental problem of communication is that of reproducing at one point either exactly or approximately a message selected at another point. Frequently the messages have *meaning*; that is they refer to or are correlated according to some system with certain physical or conceptual entities. These semantic aspects of communication are irrelevant to the engineering problem. The significant aspect is that the actual message is one *selected from a set* of possible messages. The system must be designed to operate for each possible selection, not just the one which will actually be chosen since this is unknown at the time of design.

So a theory of information for Shannon is about the "engineering problem" of communicating messages over a channel efficiently. In this sense, the concept of information has nothing to with anyone's knowledge and everything to do with the stochastic or probabilistic process that generates the messages. The standard unit of information is the bit, short for "binary digit." An information source that produces sequences of 0's and 1's from a two-symbol alphabet, with equal probability for each symbol, is said to produce one bit of information per symbol. Shannon showed that it's possible to compress the information required to communicate a message—to reduce the average number of bits per symbol—up to a certain optimal compression, if the probabilities of the different symbols produced by an information source are not all equal.

In modern formulations, information theory is about random variables—variables that take values from a set of values with certain probabilities—and correlations between random variables. As such, information theory is a branch of the mathematical theory of probability, and the physical communication of messages is just one application. In proposing that quantum mechanics is about the structure of information, I mean that the theory deals with new sorts of probabilistic correlations that are structurally different from correlations that arise in classical theories (and, of course, the theory is also able to handle standard classical correlations). What we have discovered is that the "engineering" possibilities for representing, manipulating, and communicating information in a quantum world are different than we thought, irrespective of what the information is about.

John Wheeler's slogan "it from bit"[15] and Rolf Landauer's influential comment that "information is physical"[16] are often taken to suggest that information is primary and that "stuff"—what physics is usually understood to be about—is in some sense derived from information. Vlatko Vedral, for example, in his book *Decoding Reality* explicitly endorses the view that "our reality is ultimately made up of information" and that "the laws of Nature are information about information."[17]

The idea that the basic building blocks of reality might be information is dizzyingly intriguing, but I have no idea how to make sense of this. It can't be information in Shannon's sense. To say that quantum mechanics is about the structure of information is not to say that "stuff"—particles, fields, planets, people—is somehow made of information. The theory of relativity rests on the recognition that events have a spatiotemporal structure, and that this structure is not what Newton thought it was. This is not usually taken to mean that "stuff" is made of space and time (although there have been such suggestions, notably by Wheeler). Similarly, the conceptual revolution in the transition from classical to quantum physics should be understood as resting on the recognition that there is an *information-theoretic structure* to the mosaic of events, and this structure is not what Shannon thought it was.[18]

Of course, this is not a description of the actual historical development of quantum mechanics—Shannon published "A mathematical theory of communication" in 1948 and Heisenberg's "Umdeutung" paper appeared in 1925. Rather, this book develops the idea that what is revolutionary about quantum mechanics is analogous to what is revolutionary about the theory of relativity: a fundamental structural change in the way we represent how events fit together, where the change involves spatiotemporal structure in the case of relativity, and the structure of information in the case of quantum mechanics.

The claim that quantum mechanics is about the structure of information is often met with John Bell's dismissive response: "*Information? Whose* information? Information about *what*?"[19] But we don't ask these questions about a USB flash drive. A 64 GB drive is an information storage device with a certain capacity, and whose information or information about what is irrelevant.

 **The bottom line**

- Quantum mechanics is fundamentally a theory about the structure of information, insofar as a theory of information in the physical sense is essentially a theory of probabilistic correlations.
- This is largely implicit in Heisenberg's "Umdeutung" paper, in which classical quantities like position and momentum are "reinterpreted" as operations, later represented by operators that act on and transform the states of quantum systems. Operations needn't commute—the order in which they are applied to a quantum state can make a difference—and a noncommutative mechanics allows the novel possibility of intrinsically random events associated with new sorts of nonlocal probabilistic correlations for so-called "entangled" quantum states of separated systems.

## 1.2 WHY BANANAWORLD?

What is it about quantum correlations that leads prominent physicists like Feynman to say that "nobody understands quantum mechanics"? The usual way of approaching this question is from the familiar perspective of classical physics, or from commonsense intuitions about correlations, but that's a limited perspective that risks introducing implicit assumptions and prejudices. The idea of Bananaworld is to look at quantum correlations "from the outside." Bananaworld is an imaginary world in which there are classical and quantum correlations, but also superquantum correlations between separated systems that are even more nonclassical than the correlations of entangled quantum states. The conceptual puzzles of quantum correlations arise without the distraction of the mathematical formalism of quantum mechanics, and you can see what is at stake—where the clash lies with the usual presuppositions about the physical world.

In the contemporary quantum information literature, you'll find references to "Boxworld."[20] A "box" is an imaginary device, with an input and output port on one side for Alice and an input and output port on the other side for Bob. Inputs can be 0 or 1, and outputs can be 0 or 1 (and, more generally, there could be more than two inputs, or more than two outputs). A box is defined by a particular correlation between inputs and outputs. So a box is really just an abstract device that produces a correlation, classical, quantum, or superquantum, without any specification of an internal mechanism that could produce the correlation. A nonlocal box can be stretched by an arbitrary amount, or perhaps separated into two parts, an Alice part and a Bob part, by any distance, without affecting the correlation. A box can be used only once: all it does is produce an output following an input at each side (whether or not there is an input at the other side), and once that happens the box is done for that side—you need a new box for a new input. Boxworld has turned out to be a powerful conceptual tool in exploring nonclassical features of quantum mechanics. In effect, Bananaworld is Boxworld, with bananas and peelings and tastes instead of boxes with 0, 1 inputs and outputs.

The protagonists of information-theoretic scenarios are Alice and Bob (and sometimes Charlie or Clio as well), who are in different locations and can each choose to perform one of at least two alternative actions—say Alice can choose to measure one of two noncommuting or incompatible observables labeled $A$ and $A'$ on a quantum system in her possession, and Bob can choose to measure $B$ or $B'$ on his system—with at least two possible outcomes in each case, say 0 or 1. Here $A$ and $A'$ are just labels for Alice's choices, and $B$ and $B'$ are labels for Bob's choices, where $A$ and $B$ might represent the same measurement choice by Alice and by Bob, but could also represent different choices (and similarly for $A'$ and $B'$). Once Alice or Bob measures a particular observable, the quantum system is changed in an intrinsically random way and is no longer in the same quantum state, so no longer available in the original state for the alternative choice. The fact that a choice is free (in a precise sense clarified in Chapter 4, "Really Random") turns out to be essential here. So there must be at least two actions, with at least two possible outcomes for each action, because otherwise there would be no variation and so no correlation. The choice of action could be random, so Alice and Bob could be replaced by random number generators that produce 0's and 1's with equal probability, corresponding to the two possible choices, 0 for $A$ and 1 for $A'$ for Alice, and 0 for $B$ and 1 for $B'$ for Bob.

Correlations are correlations, irrespective of the nature of the systems that manifest the correlations, and the discussion needn't be confined to measurement correlations. So imagine discovering Bananaworld, an island covered with banana trees. A banana tree is really a giant herb, and bananas grow in bunches pointing up from the stem, so the non-stem end is actually the top end. There are two, and only two, ways to peel one of these bananas (two possible actions that can be performed). Some primates on the island peel a banana from the stem end ($S$), while other primates prefer peeling from the top end ($T$). A Bananaworld banana simply can't be peeled any other way—suppose, if you like, that a banana is simply inedible if you try to get at the fruit without peeling from the stem end or the top end. Once a banana is peeled a certain way, of

**Figure 1.1** Alice and Bob peel two bananas from the same two-banana bunch in Bananaworld, where tastes and peelings are correlated in a particular way. Alice peels her banana from the top end (*T*) and Bob peels his banana from the stem end (*S*). In this case, both bananas taste the same: *i*ncredible, *i*ntense, *i*ndescribably delicious (1). The correlation could be an instance of the Popescu–Rohrlich correlation described in Chapter 3, "Bananaworld."

course, it's a peeled banana—the alternative peeling is no longer available—and it tastes just like an *o*rdinary banana ("o" or 0), or the flavor is *i*ntense, *i*ncredible, *i*ndescribably delicious ("i" or 1). See Figure 1.1. Whether the taste is 0 or 1 is an objective fact, not a subjective matter of opinion, and anyone who checks will agree that the taste is definitely ordinary or definitely intense. Various correlations between the tastes of bananas (0 or 1) and alternative peelings (*S* or *T*), which persist when bananas from the same bunch are separated by any distance, can be considered, with features that correspond to classical, quantum, or even superquantum correlations. The same conceptual problems that arise for quantum systems like electrons or photons arise for bananas in Bananaworld.

If you're not used to thinking about Boxworld, it's less confusing to talk about peeling a banana in one of two ways, *S* or *T*, with two possible tastes, ordinary (0) or intense (1) in each case, than to talk about two inputs, 0 or 1, to a box, with two possible outputs, also represented by 0 and 1, for each input. To compare with measurement correlations: a particular peeling, *S* or *T*, by Alice or Bob corresponds to the measurement of a particular quantum observable, $A$ or $A'$ for Alice, or $B$ or $B'$ for Bob, and the two possible tastes correspond to the two possible measurement outcomes. Alice's two observables are noncommuting or incompatible, as are Bob's, so measuring one observable precludes measuring the other, just as peeling *S* or *T* precludes the alternative peeling. I'll say more about the observables in question as I go along (in the "Bananaworld" chapter, peeling *S* or *T* corresponds to measuring the polarization of a photon in one of two directions 45° apart, or measuring the spin of an electron in one of two orthogonal directions), but all that matters here is that the two ways of peeling a banana correspond to measuring two noncommuting or incompatible quantum observables.

The idea is to get at what's puzzling about quantum correlations by considering the extent to which Alice and Bob, limited to certain resources, can simulate various correlations in Bananaworld. A simulation can be thought of as a game between Alice and Bob, and a moderator, played over several rounds. Think of a simulation game as like a TV game show. At the beginning of each round of a simulation game, the moderator, who can communicate with Alice and Bob separately, gives each player one of two prompts, *S* or *T*, and the player is supposed to respond to his or her prompt with one of two responses, 0 or 1. They win a round if the responses and the prompts are correlated in the right way. Alice and Bob are allowed to discuss strategy before the first round, and they are then sent to separate locations (say, separate soundproof booths from which neither player is visible to the other, each separately linked by telephone to the moderator) and are not allowed to communicate with each other during the game. The game is played over many rounds, and at the end of the game they win a prize, where the value of the prize depends on the number of rounds they win. So the aim is to figure out an optimal strategy: a strategy that will enable them to win the maximum number of rounds.

Of course, Alice and Bob could win a round of the game, or several consecutive rounds, purely by chance even if they respond randomly without any strategy. The relevant question is whether there is a winning strategy for the simulation game, assuming Alice and Bob have access to certain resources. For example, Alice and Bob might be allowed to take pencil and paper or calculators with them to perform calculations when they are separated, or written instructions for responding to a given prompt that they prepare during the strategy session before they are separated. Or they might be allowed to base their responses on the outcomes of measurements on shared pairs of entangled quantum particles they prepare during the strategy session, and so on.

Some correlations can be perfectly simulated with local resources available to Alice and Bob separately. These could be local lists of instructions that tell Alice and Bob separately how to respond to prompts ("local" in the sense of a list of instructions for Alice independent of Bob's instructions, and a list of instructions for Bob independent of Alice's instructions). Other correlations can't be simulated with local resources, but can be perfectly simulated with shared nonlocal entangled quantum states. Still other correlations can't be perfectly simulated with quantum resources, but a simulation with shared entangled quantum states does a better job than the optimal simulation with local resources. So the plan will be to set things up in terms of a correlation in Bananaworld that is counterintuitive in some way, and the punchline will then be to show that the correlation can be simulated with nonlocal quantum resources but not with local resources, or that a quantum simulation is better than the best possible simulation restricted to local resources. If Alice and Bob can't simulate a correlation with local resources, but can succeed or do better with entangled quantum states, *that's a fact about our world*, not just about Bananaworld. The point is to use Bananaworld in this way to say something significant about our quantum world.

Local quantum resources provide no advantage over local classical resources, so I'll sometimes refer to correlations that can be perfectly simulated with local resources as classical correlations. For classical correlations, there is a winning strategy for the simulation game that does not involve access to a nonlocal resource like entangled quantum states and measuring instruments capable of measuring quantum observables. Specifically, local lists of instructions that could include shared lists of random numbers generated before the start of the game ("shared randomness") are allowed as classical resources, but not shared copies of a pair of particles in an entangled quantum state and appropriate quantum measuring instruments, the quantum analogue of classical shared randomness. If there is no winning strategy, the interesting question is whether there is an optimal strategy, and what that would be. I'll give a precise characterization of the difference between classical and quantum correlations in Chapter 5, "The Big Picture."

Correlations between events at different places cry out for explanation, as John Bell put it,[21] and in classical physics or, for that matter, in everyday life, there are just two sorts of explanation. Either there is a direct causal connection between the events, a physical signal that takes a certain amount of time to travel continuously from one event to the other and transmit information between them, or there is a common cause: some event in the common past of the correlated events that is responsible for the correlation, like a flash of lightning that is the common cause of Alice and Bob both hearing thunder at more or less the same time if they are located at different places in the vicinity of a thunderstorm. Nothing moves between Alice and Bob. Rather, a sound wave, the common cause of the vibration in their eardrums, moves from the disturbance in the atmosphere to Alice and Bob.

In Bananaworld there are correlations between the tastes of bananas from the same bunch peeled in various ways by Alice and Bob, even if they peel their bananas in separate locations on the island, where *both types of explanation are excluded*. As I'll show in the "Bananaworld" chapter, bananas correlated in this way can't have definite or predetermined tastes before they are peeled, and in Chapter 4, "Really Random," I'll show that in this case a banana tasting 0 or 1 when it's peeled is an intrinsically random event. Feynman's comment that nobody understands quantum mechanics applies with equal force to Bananaworld, but here it's clear that the mystery has to do with the weirdness of the nonclassical correlations rather than the properties of microsystems. The bananas have ordinary properties in the objective sense—what's extraordinary and counterintuitive is the nature of the correlations between peelings and tastes.

### The bottom line

- Correlations can be defined for two possible actions performed by Alice and by Bob, with two possible outcomes for each action. Bananaworld makes this strategy concrete. The two possible actions correspond to two ways of peeling a banana, from the stem end ($S$) or the top end ($T$), and the two possible outcomes correspond to two possible tastes, ordinary (0) and intense (1). Specific correlations are considered in Chapter 3, "Bananaworld," and subsequent chapters.
- Correlations between events at different places cry out for explanation. In classical physics, there are just two sorts of explanation: either there is a direct causal connection between the events, or there is a common cause: some event in the common past of the correlated events that is responsible for the correlation.
- In Bananaworld there are correlations between the tastes of bananas from the same bunch peeled in various ways by Alice and Bob, even if they peel their bananas in separate locations on the island, where *both types of explanation are excluded.*

## 1.3 YES! WE HAVE NO BANANAS, BUT ...

Bananaworld is a possible world. You could imagine discovering an island like Bananaworld, but of course there are no banana trees in our world like the superquantum banana trees in Bananaworld.[22] The really amazing thing about our world is that there are correlations that are closer to these Bananaworld correlations than classical correlations. What's mind-boggling is the discovery that we live in a world in which there are nonlocal correlations, where a direct causal influence can be excluded as an explanation, that can't have a common cause explanation either.

As in Bananaworld, observables correlated in this way can't have definite or predetermined values before they are measured, so the property we attribute to one of the correlated quantum systems after a measurement couldn't have been there before the measurement. A quantum observable taking a particular value with a probability between 0 and 1 when measured is an intrinsically random event, not only for correlated observables in entangled quantum states, but in general for the values of observables in any quantum state (see Section 4.3, "Really Random Qubits," in Chapter 4, "Really Random"). This is a structural feature of quantum information, related to the way in which observables corresponding to properties of a system are related to other observables.

The weirdness of quantum correlations shows up in photon polarization measurements, where the outcomes can be "horizontal" or "vertical" for linear polarization in some direction, or electron spin measurements, where the outcomes can be "up" or "down" for spin in some direction. For definiteness, I'll stick to photon polarization rather than electron spin in the following, since that's likely to be more familiar to most readers. Choosing to measure the polarization of a photon in one of two directions 45° apart is like choosing to peel a banana by the stem end ($S$) or the top end ($T$), and the two possible tastes correspond to horizontal or vertical polarization in the respective directions.

The only really crucial thing you need to know about quantum mechanics to understand most of the discussion in this book is that quantum probabilities depend on the angle between the directions of polarization. Suppose you have a photon that is in a state of horizontal polarization in a certain direction, and you measure the polarization in a different direction at an angle $\theta$ to the first direction. The probability of finding the photon horizontally polarized is $\cos^2\theta$, and the probability of finding the photon vertically polarized is $1 - \cos^2\theta = \sin^2\theta$. The smaller the angle $\theta$, the closer the second direction is to the first direction, and the closer the probability of finding the photon to be horizontally polarized is to 1 ($\cos^2 0$), which is what you'd expect.

For two separated photons in an entangled state called a Bell state, which plays an important role in the following chapters, the two possible outcomes of measuring the polarization of one photon in some direction are equally probable. If you measure the polarizations of both photons in the same direction, you get the same outcome: either both photons are horizontally polarized in the direction of the polarization measurement, or they are both vertically polarized in a direction orthogonal to the direction of the polarization measurement, and the probability of each of these possibilities is $1/2$. Take this as a fact about photon polarization, a particular case of a general rule for quantum probabilities. As Feynman says: "If you will simply admit that maybe [nature] does behave like this, you will find her a delightful entrancing thing."

If you measure the polarizations of the two photons, $A$ and $B$, in directions an angle $\theta$ apart, the probability of getting the same outcome for both measurements is $\cos^2\theta$ (with equal probability that both photons are horizontally polarized in the directions of their respective measurements, or that they are both vertically polarized), and the probability of getting different outcomes is $\sin^2\theta$ (again with equal probability that $A$ is horizontally polarized and $B$ is vertically polarized, or that $A$ is vertically polarized and $B$ is horizontally polarized).

That's it. There's an introductory account of polarization and spin in Chapter 2, "Qubits," and more about how probabilities are calculated in the Supplement, "Some Mathematical Machinery," at the end of the book, but the point of "Bananaworld" is to get at the heart of the conceptual revolution involved in the transition from classical to quantum physics by a more direct route than working through the formalism. What's really at issue is the nature of quantum correlations, and how these correlations can be exploited in information-theoretic tasks that are impossible with classical correlations. All this can be understood without getting bogged down in the mathematical machinery of quantum mechanics.

 **The bottom line**

- The connection between Bananaworld correlations and quantum correlations is through photon polarization measurements, where Alice and Bob can each choose to measure linear polarization in one of two directions, with possible outcomes "horizontal" or "vertical." Quantum probabilities depend on the angle between the directions of polarization. If you have a photon that is in a state of horizontal polarization in a certain direction, and you measure the polarization in a direction at an angle $\theta$ to the first direction, the probability of finding the photon horizontally polarized is $\cos^2\theta$, and the probability of finding the photon vertically polarized is $1 - \cos^2\theta = \sin^2\theta$.
- For a perfectly correlated entangled state of two photons called a Bell state that plays an important role in the following chapters, if you measure the polarizations of both photons in the same direction, you get the same outcome: either both photons are horizontally polarized in the direction of the polarization measurement, or they are both vertically polarized in a direction orthogonal to the direction of the polarization measurement, with equal probability of $1/2$ for these two alternatives.
- If you measure the polarizations of the photons in this Bell state in different directions an angle $\theta$ apart, the probability of getting the same outcome for both measurements is $\cos^2\theta$ (with equal probability that both photons are horizontally polarized in the directions of their respective measurements, or that they are both vertically polarized), and the probability of getting different outcomes is $\sin^2\theta$ (again with equal probability for the two alternatives).

## 1.4 MORE

### 1.4.1 Special Relativity: The Basics

To amplify Feynman's remark that quantum mechanics is deeply puzzling in a way that the theory of relativity isn't, here's a brief account of special relativity, following Hermann Bondi's formulation in terms of his *k*-calculus.[23]

Don't be intimidated by the term. Bondi says:

> I have discovered that what I call the *k*-calculus is still unknown in educated mathematical circles. In uneducated ones, of course, it is well-known. And I propose, therefore, to talk a little bit about it here, partly because I enjoy it, partly because I hope to put those of you who

> don't know it into a state where you can also easily derive the consequences of relativity. You will then see where it leads you and see that Einstein's principle of relativity gives perfectly reasonable results capable of being tested by experiment and observation.

The core insight of Bondi's formulation is the significance of the difference between the Doppler effect for sound and the Doppler effect for light. The Doppler effect for sound is the familiar increase in the pitch of a train's siren as it approaches a station, and the decrease in pitch as it recedes. Sound involves the vibration of air, the medium in which the sound waves propagate. The *k*-factor for a bystander at the station is the ratio of the interval of reception $\Delta_r$ of the sound signals, corresponding to successive pressure peaks of the sound wave, to the interval of transmission $\Delta_t$ by the moving train: $k = \Delta_r / \Delta_t$.

Figure 1.2 is a space–time diagram showing the relation between the two intervals. In these diagrams, time is in the vertical direction, and space in the horizontal direction. The thick vertical line illustrates a temporal sequence of events for the bystander at the same spatial point in the station. The thick sloping line illustrates a sequence of events in space and time for the train, which is moving towards the station, so the interval of reception is shorter than the interval of transmission. The dotted lines represent sound signals, moving from the train to the bystander. It should be clear from the diagram that the signals are moving faster than the train: as time goes by in the vertical direction, the signals cover more ground than the train in the horizontal spatial direction towards the bystander.

The same point can be made, without involving the physics of wave motion, by considering the Doppler effect when the signals are large objects, like cars moving along a highway, where the highway corresponds to the air as the stationary medium.

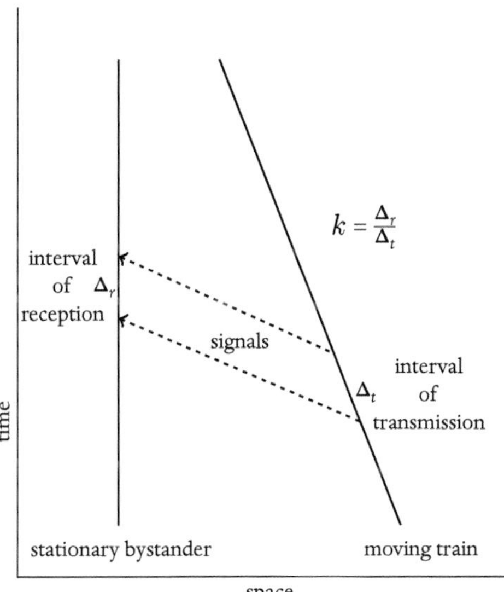


**Figure 1.2** The *k*-factor: $k = \Delta_r / \Delta_t$. Adapted, with changes, from a diagram in H. Bondi, *Assumption and Myth in Physical Theory* (Cambridge University Press, Cambridge, 1967), p. 35.

Consider a truck moving at 30 miles per hour leaving a stationary parking garage at 12 noon. Suppose 10 cars leave the parking garage at intervals of 6 minutes, traveling at 60 miles per hour towards the slower moving truck. The cars can be regarded as signals, with a signal velocity relative to the highway that is twice the velocity of the truck. The interval of transmission between car signals is 6 minutes. What is the interval of reception of these signals by the truck? If the last car leaves the parking garage at 1 pm when the truck has traveled 30 miles, it will reach the truck at 2 pm, when the truck has traveled an additional 30 miles and the car has traveled 60 miles in the hour between 1 pm and 2 pm. Since 10 cars pass the truck at equal time intervals over 120 minutes (the first car leaves at 12:06 pm and the last car leaves at 1 pm), the interval of reception of the car signals by the truck is 12 minutes. So $k_{rs} = \frac{\Delta_r}{\Delta_t} = 12/6 = 2$, where the $r$ here indicates that the moving truck is *receding* from the stationary parking garage, and the $s$ indicates that the signal source is *stationary*. See Figure 1.3.

Now suppose the truck approaches the parking garage, beginning 30 miles away at noon, and the car signals leave the parking garage and travel towards the moving truck as it approaches the parking garage. If the first car leaves the parking garage at noon traveling at 60 miles per hour, it will reach the truck at the 20 mile mark, when the slower-moving truck has traveled 10 miles, so at 12:20 pm. Since 10 cars pass the truck in the 40 minutes between 12:20 pm and 1:00 pm when the truck arrives at the parking garage, the interval of reception is 4 minutes, and $k_{as} = 4/6 = 2/3$, where the $a$ here indicates that the moving truck is *approaching* the stationary parking garage. See Figure 1.4.

In both these cases, $k_{rs}$ and $k_{as}$, the source of the car signals is the *stationary* parking garage. There are two other cases to consider, $k_{rm}$ and $k_{am}$, where the source is the *moving* truck, either receding from or approaching the parking garage.

**Figure 1.3** The Doppler effect for cars. The truck is receding from the parking garage, the stationary source of the car signals. The truck moves at 30 mph, the cars at 60 mph, beginning at noon. The illustration shows the situation at 1 pm. Here, $k_{rs}$ = interval of reception/interval of transmission = 12/6 = 2.

**Figure 1.4** The Doppler effect for cars. The truck is approaching the parking garage, the stationary source of the car signals. The truck moves at 30 mph, beginning at the 30 mile mark at noon. The cars move at 60 mph. The illustration shows the situation at 12:20 pm. Here, $k_{as}$ = interval of reception/interval of transmission = 4/6 = 2/3.

For the *receding* case, suppose the truck leaves the parking garage at noon traveling at 30 miles per hour. Suppose 10 cars are parked along the highway at 3 mile intervals from the parking garage. Every 6 minutes the moving truck passes a car, and as it passes, the car begins traveling towards the parking garage at 60 miles per hour. The cars can be regarded as signals from the moving truck (like the sound signals from a moving train's siren, which produces a wave vibration in the stationary air as the train passes). If the last car begins traveling towards the parking garage at 1 pm, after the truck has traveled 30 miles, it will take half an hour to cover the 30 miles to the parking garage, so it will arrive at the parking garage at 1:30 pm. Since 10 cars arrive at equal time intervals over 90 minutes, the interval of reception of the car signals at the parking garage is 9 minutes, and $k_{rm} = 9/6 = 3/2$. See Figure 1.5.

For the *approaching* case, suppose the truck approaches the parking garage, beginning 30 miles away at noon. If 10 cars are parked along the highway at 3 mile intervals from the parking garage, and the first car begins traveling towards the parking garage at noon at 60 miles per hour, it will reach the parking garage after half an hour, at 12:30 pm. Since 10 cars arrive in the 30 minutes between 12:30 pm and 1:00 pm when the truck arrives at the parking garage, the interval of reception is 3 minutes, and $k_{am} = 3/6 = 1/2$. See Figure 1.6.

So there are four $k$-factors for signals passing between a stationary system and a moving system: two $k$-factors when the moving system is receding from the stationary system, depending on whether the signal source is moving ($k_{rm} = 3/2$) or stationary ($k_{rs} = 2$), and two $k$-factors when the moving system approaches the stationary system, depending on whether the signal source is moving ($k_{am} = 1/2$) or stationary ($k_{as} = 2/3$). (The values of $k$ also depend on the ratio of the speed of the moving system to the

**Figure 1.5** The Doppler effect for cars. The truck is receding from the parking garage, and the car signals begin to move towards the parking garage as the truck passes, so the source of the signals is the moving truck. The truck moves at 30 mph, the cars at 60 mph, beginning at noon. The illustration depicts the situation at 1:00 pm. Here, $k_{rm}$ = interval of reception/interval of transmission = $9/6 = 3/2$.

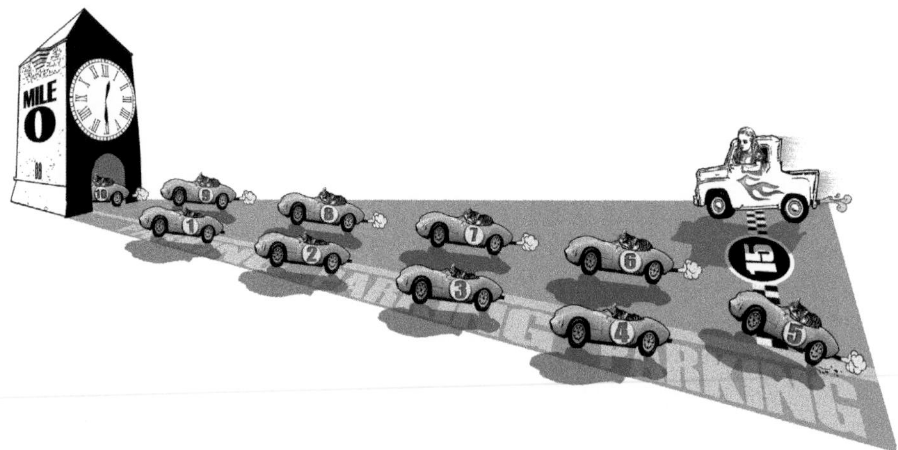


**Figure 1.6** The Doppler effect for cars. The truck is approaching the parking garage, and the car signals begin to move towards the parking garage as the truck passes, so the source of the signals is the moving truck. The truck moves at 30 mph, beginning at the 30 mile mark at noon. The cars move at 60 mph. The illustration depicts the situation at 12:30 pm. Here, $k_{am}$ = interval of reception/interval of transmission = $3/6 = 1/2$.

signal speed, so these values would change for different truck speeds and car speeds.) Notice that $k_{am} = 1/k_{rs}$ and $k_{as} = 1/k_{rm}$, which is true in general for a moving system approaching or receding from a stationary system at the same speed.

Now here's the punchline: *for light, there is no stationary medium for the transmission of light signals corresponding to the stationary highway with respect to which rest and motion are defined for the stationary parking garage, the car signals, and the moving truck.* For sound signals, the air corresponds to the highway as the stationary medium in which sound waves propagate at around 760 miles per hour at sea level. The theory of special relativity drops the idea of a "luminiferous ether" as the medium of propagation for light signals—an idea proposed by Lorentz in his rival theory.

Aristotle's physics distinguished a state of rest from other states of motion, but Galileo and Newton realized that a particular set of states of motion, the "inertial" states of motion, are all equivalent and distinguished from other states of motion. Inertial systems are systems moving with constant velocity relative to each other, as distinct from accelerating systems. Simply put: velocity is relative, but acceleration is absolute.

The theory of special relativity rests on two principles: the light postulate, that there is "no overtaking of light by light,"[24] as Bondi puts it, and the relativity principle, which Bondi sums up as the slogan "velocity doesn't matter."[25] The light postulate says that the velocity of light in empty space is constant, independent of the motion of the source, and the same for all inertial observers. So "no overtaking of light by light" means, in particular, no overtaking of light by anything else: the relative velocity of light is the same for differently moving inertial observers, quite unlike the relative velocity of sound or cars. The relativity principle is an extension of the Galilean or Newtonian principle of relativity from mechanical phenomena to all phenomena, in particular to electromagnetic phenomena such as light. The statement is simply that physics is the same for all inertial systems, so nothing in the way physical systems behave could tell you how you are moving if you are in an enclosed space like a spaceship, provided the spaceship is not accelerating or rotating.

It follows that for light signals passing between two systems moving away from each other with constant relative velocity, $k_{rm} = k_{rs} = k_r$, because there is no fact of the matter about which system is "really" at rest and which system is moving: all we have is that the two systems are receding from each other at a certain constant relative velocity. Similarly, $k_{am} = k_{as} = k_a$.

## The bottom line

- In Bondi's *k*-calculus, the *k*-factor is the ratio of the interval of reception of signals to the interval of transmission.
- In a classical (Newtonian) theory of space and time, where it makes sense to think of motion relative to a state of rest (relative to the earth for sound signals in the air, or relative to a hypothetical "luminiferous ether" for light signals), there are four *k*-factors for signals passing between a stationary system and a moving system: two *k*-factors when the moving system is receding from the stationary system, depending on whether the signal source is moving ($k_{rm}$) or

(*continued*)

 **The bottom line** *(continued)*

stationary ($k_{rs}$), and two $k$-factors when the moving system approaches the stationary system, depending on whether the signal source is moving ($k_{am}$) or stationary ($k_{as}$). For the car and truck example in the text, $k_{rm} = 3/2$, $k_{rs} = 2$, $k_{am} = 1/2$, and $k_{as} = 2/3$.

- Special relativity rests on two principles: the light postulate, "no overtaking of light by light," and the relativity principle, "velocity doesn't matter," for both mechanical phenomena and electromagnetic phenomena involving light. There's no state of motion relative to a state of absolute rest, only relative velocities between systems.
- It follows that for light signals passing between two systems moving away from each other with constant relative velocity, $k_{rm} = k_{rs} = k_r$ and $k_{am} = k_{as} = k_a$, because there is no fact of the matter about which system is "really" at rest and which system is moving. So there are only two $k$-factors, $k_r$ and $k_a$. That makes all the difference.

### 1.4.2 The Lorentz Transformation

You might suspect that the $k$-factor for two systems approaching each other at a certain speed, $k_a$, is the reciprocal of the $k$-factor for two systems receding from each other with the same speed, $k_r$, and this is so. Here's the argument.

Two inertial systems are associated with a specific $k$-value that is a function of their relative velocity. In particular, by the relativity principle, the same $k$-factor applies if Alice is the source of signals to Bob, or if Bob is the source of signals to Alice, and the velocity of light is the same from Alice to Bob as from Bob to Alice. Now suppose that Alice, Bob, and Clio are three inertial observers, equipped with identical clocks, moving in a straight line with different relative velocities so that the $k$-factor is $k_{AB}$ between Alice and Bob, and $k_{BC}$ between Bob and Clio. See Figure 1.7. If Alice sends light signals to Bob at regular intervals $T$ by her clock, the signals will be received by Bob at intervals $k_{AB}T$ by his clock. Suppose that when Bob receives Alice's signals, he immediately sends light signals to Clio. Bob's signals will be received by Clio at intervals $k_{BC}(k_{AB}T)$ by Clio's clock. If Alice's signals continue traveling towards Clio, they will be received by Clio simultaneously with Bob's signals, since there is no overtaking of light by light. It follows that if the $k$-factor between Alice and Clio is $k_{AC}$, then $k_{AC}T = k_{BC}k_{AB}T$, and so $k_{AC} = k_{AB}k_{BC}$. If Alice and Clio are stationary with respect to each other, and Bob moves from Alice towards Clio, then $k_{AC} = 1$ (the interval of reception is the same as the interval of transmission), and so $k_{BC} = 1/k_{AB}$.

Suppose that Alice and Bob are inertial observers moving relative to each other and that $k_{AB} = 3/2$. Suppose Bob passes Alice at noon, receding from Alice in a certain

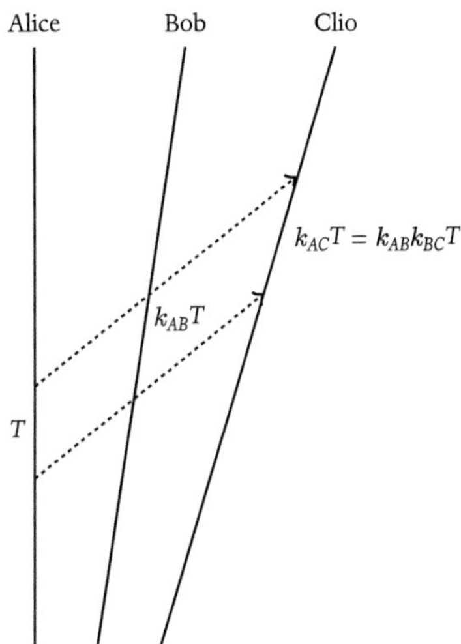


**Figure 1.7** Composition of *k*-factors. As in Figure 1.2, only one direction of space is shown. The time axis is in the vertical direction and the space axis is in the horizontal direction. Adapted, with changes, from a diagram in H. Bondi, *Assumption and Myth in Physical Theory* (Cambridge University Press, Cambridge, 1967), p. 40.

direction, and that they synchronize their clocks as they pass at noon. (See Figure 1.8.) Bob sends Alice ten light signals six minutes apart. Alice receives the signals at intervals $k_{AB} \cdot 6 = 9$ minutes, so the last signal reaches her at 1:30 pm by her clock. At 1 pm by Bob's clock, Clio passes Bob moving towards Alice at the same speed that Bob recedes from Alice. Bob and Clio synchronize their clocks as they pass at 1 pm. Clio now sends Alice ten light signals six minutes apart. Since $k_{CA} = 1/k_{AB} = 2/3$, Alice receives the signals at intervals of four minutes, so the last signal arrives 40 minutes after 1:30 pm at 2:10 pm by her clock—just as Clio passes and registers 2:00 pm by her clock! Astonishingly, the time measured by Alice's clock between two events (meeting Bob and meeting Clio) is a longer time interval than the combined times measured by Bob's clock and Clio's clock, even though all three clocks are identical and synchronized. This phenomenon is known as time dilation: clocks moving relative to Alice run slow relative to Alice's clock.

Where did the extra ten minutes come from? As puzzling as this may seem at first sight, the time difference is an immediate consequence of the assumptions about relativistic *k*-factors, which follow from the light postulate (no overtaking of light by light) and the relativity principle (the state of motion doesn't matter for different inertial systems: the physics is the same). You don't get the difference in time measurements with cars as signals. In that case there are four *k*-factors and, for receding Bob and approaching Clio who are the source of signals to Alice, the relevant *k*-factors are $k_{rm} = 3/2$ for Alice and Bob, and $k_{am} = 1/2$ for Clio and Alice. Then Alice receives Bob's signals at intervals of nine minutes as before, so that the last signal arrives at 1:30 pm, but she receives Clio's signals at intervals of three minutes, so the last signal arrives at 2 pm, just as Clio passes.

The essential point here is that there is no universal highway for signals in our universe. We can suppose that Alice, Bob, and Clio are equipped with identical clocks, and we can

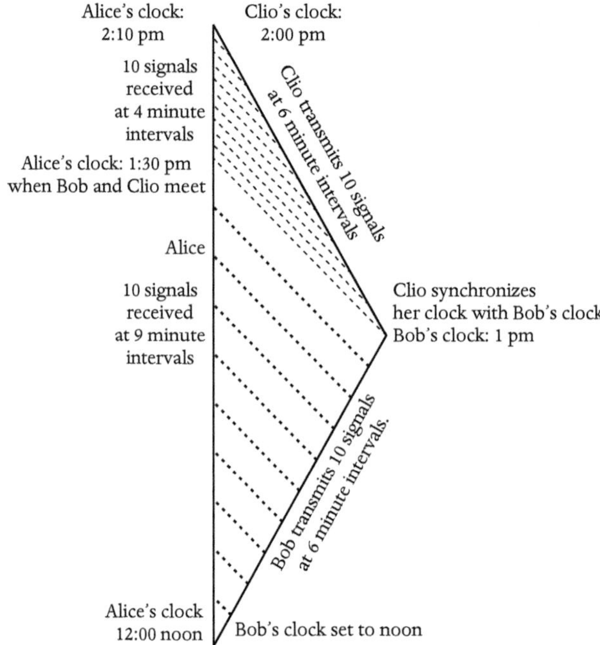


**Figure 1.8** Time dilation for Alice relative to Bob and Clio. As in the previous space–time diagrams, only one direction of space is shown. The time axis is in the vertical direction and the space axis is in the horizontal direction. Adapted, with changes, from a diagram in H. Bondi, *Relativity and Common Sense* (Heinemann Educational Books, London, 1965), p. 84.

suppose that clocks can be synchronized when two physical systems are momentarily in the same place. But we can't assume that clocks run at the same rate when they are in different parts of space moving relative to each other. The only way to relate the times of moving clocks in separate parts of space is for Alice, Bob, and Clio to communicate by signaling to each other, and the only general way to do this is via light signals, or electromagnetic radiation, for which there is no medium or universal highway that could define an absolute state of rest.

The relation between the space and time coordinates of different inertial observers in special relativity is known as the Lorentz transformation. It's straightforward to derive the Lorentz transformation in Bondi's $k$-calculus. Time dilation, length contraction, and the relativity of simultaneity—features of Minkowski space-time—all follow from the Lorentz transformation.

First, what's the relation between the $k$-factor and relative velocity? Suppose, for definiteness, that Alice and Bob are receding from each other along a straight line. For convenience, you can think of Alice at rest and Bob as moving, but you could equally well think of Bob at rest and Alice as moving away from Bob. Suppose Alice and Bob synchronize their clocks at zero as they pass each other. At a time $T$ later, Alice sends Bob a

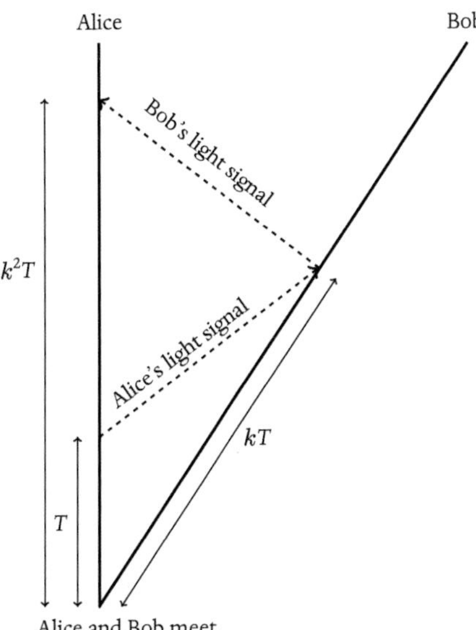


**Figure 1.9** The relation between the $k$-factor and relative velocity. As in the previous space–time diagrams, only one direction of space is shown. The time axis is in the vertical direction and the space axis is in the horizontal direction. Adapted, with changes, from a diagram in H. Bondi, *Assumption and Myth in Physical Theory* (Cambridge University Press, Cambridge, 1967), p. 38.

light signal, which is received by Bob at the time $kT$ by his clock. Bob immediately reflects Alice's light signal back to her. Alice receives the returning signal at the time $k(kT)$ by her clock and calculates the round-trip time taken by the signal to cover the distance from her to Bob and from Bob back to her as $k^2T - T = (k^2 - 1)T$ by her clock. See Figure 1.9.

To figure out the distance between her and Bob at the moment her light signal is reflected by Bob, Alice multiplies the time taken for the light to travel this distance by the speed of light. She calculates the time as half the round-trip time, $1/2(k^2 - 1)T$. The speed of light is about 300, 000 kilometers per second. It's convenient to take the unit of distance as the distance light covers in a second, the unit of time. Then the speed of light is 1 (one unit of distance per one unit of time). In these units, the distance is $1/2(k^2 - 1)T$. Alice has no way of measuring the time of this remote event, and no direct access to the time registered by Bob's clock for this event, so she assigns the time $T + 1/2(k^2 - 1)T = 1/2(k^2 + 1)T$ to the event, halfway between the time she transmitted her signal and the time she received Bob's signal. This is the time taken by Bob to cover the distance $1/2(k^2 - 1)T$, so Bob's velocity is:

$$v = \frac{k^2 - 1}{k^2 + 1}$$

which is always less than 1, the speed of light, or

$$k = \sqrt{\frac{1 + v}{1 - v}}.$$

If Alice is walking in a moving train, the velocity of Alice relative to the earth is the velocity of Alice relative to the train plus the velocity of the train relative to the earth, according to the classical or Galilean law of composition of velocities:

$$v_{AC} = v_{AB} + v_{BC},$$

where $A$ here refers to Alice, $B$ to the train, and $C$ to the earth. The corresponding relativistic law follows from $k_{AC} = k_{AB}k_{BC}$ by substituting the appropriate expressions for the relative velocities in the $k$-factors:

$$v_{AC} = \frac{v_{AB} + v_{BC}}{1 + v_{AB}v_{BC}}.$$

Even if $v_{AB}$ and $v_{BC}$ are both very close to 1, the velocity of light, $v_{AC} < 1$, and if $v_{AB} = v_{BC} = 1$, $v_{AC} = 1$ (no overtaking of light by light).

To derive the Lorentz transformation, first note that if Alice transmits a light signal at time $t - x$ by her clock and the light signal is reflected by some event $E$ and reaches her at time $t + x$ by her clock, she will assign space and time coordinates $x, t$ to the event $E$, as she did in the derivation of the relation between $k$ and relative velocity. See Figure 1.10. (That's because the time taken by Alice's light signal to travel from Alice to $E$ and back to Alice is $(t+x)-(t-x) = 2x$. So Alice assigns a distance $x$ to the event $E$, taking the velocity of light as 1 and the time it takes the signal to travel to $E$ as half the round-trip time, $x$. She assigns a time $t = (t - x) + x$ to $E$, which is halfway between the transmission time and the reception time.) Suppose Bob is moving away from Alice and they synchronize clocks at zero as they pass each other. If Bob transmits a light signal towards the event $E$

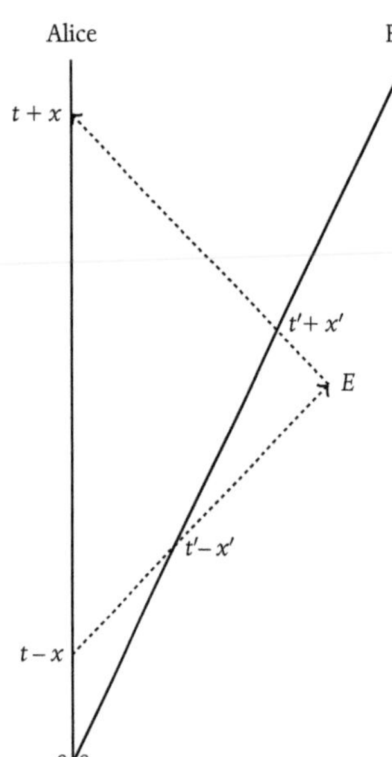


**Figure 1.10** The Lorentz transformation. As in the previous space–time diagrams, only one direction of space is shown. The time axis is in the vertical direction and the space axis is in the horizontal direction. Adapted, with changes, from a diagram in H. Bondi, *Relativity and Common Sense* (Heinemann Educational Books, London, 1965), p. 117.

just as Alice's signal passes him at time $t' - x'$ by his clock, and the signal is reflected back to Bob by $E$ together with Alice's signal and reaches him at time $t' + x'$ by his clock, Bob will assign coordinates $x', t'$ to $E$.

Now, $t' - x' = k(t - x)$ and $t + x = k(t' + x')$. Substitute $\sqrt{\frac{1+v}{1-v}}$ for $k$ and the two equations simplify to the Lorentz transformation:

$$x' = \frac{x - vt}{\sqrt{1 - v^2}},$$
$$t' = \frac{t - vx}{\sqrt{1 - v^2}}.$$

Length contraction and time dilation follow from the Lorentz transformation. If Bob is moving relative to Alice and assigns the space coordinates $x' = 0$ and $x' = 1$ to the beginning and end of a ruler at rest in his frame, Alice will assign the beginning and end of the ruler the coordinates $x = 0$ and $x = \sqrt{1 - v^2}$ at time $t = 0$ by her clock, and so a shorter length $\sqrt{1 - v^2}$. Similarly, if a clock at position $x = 0$ in Alice's frame ticks at $t = 0$ and then again at $t = 1$, the ticks will occur at times $t' = 0$ and $t' = 1/\sqrt{1 - v^2}$ for Bob's clock. So the time between the ticks will be 'dilated' by $1/\sqrt{1 - v^2}$ relative to Alice's clock, and Bob's clock, moving relative to Alice, will appear to Alice to run slow.

The relativity of simultaneity also follows from the Lorentz transformation. Suppose Alice assigns a time $t$ by her clock to two events at different places, $x_1$ and $x_2$. Bob will assign different times $t'_1 = \frac{t - vx_1}{\sqrt{1-v^2}}$ and $t'_2 = \frac{t - vx_2}{\sqrt{1-v^2}}$ to the events by his clock. So there is a time difference of $t'_2 - t'_1 = \frac{v(x_1 - x_2)}{\sqrt{1-v^2}}$ for Bob between two events that are simultaneous for Alice. If $x_2 > x_1$ and $v$ is positive, $t'_2 - t'_1$ is a negative number, so $t'_1 > t'_2$, which means that Bob sees event $E_1$ occurring *after* $E_2$, while Alice sees the two events occurring at the same time.

Events in relativistic (Minkowski) space-time can be divided into three sets. See Figure 1.11. Events that can be connected to an event $O$ by a light ray are said to be lightlike separated from $O$. These are the events that lie on the "light cone" in four-dimensional space-time (here only one dimension of space is shown), represented by the diagonal lines at a 45° angle (because the velocity of light is conventionally taken as 1, which means that light covers one unit of distance in the horizontal direction in one unit of time in the vertical direction). Events that can be connected to $O$ by a signal traveling slower than light are said to be timelike separated from $O$. These are the events represented by the points in the top and bottom regions of the light cone. A line drawn from $O$ to any event in these regions will have a slope of more than 45°, indicating a velocity of less than 1 (in one unit of time in the vertical direction, less than one unit of distance is covered in the horizontal direction). Events that can't be connected to $O$ by a signal traveling at or less than the speed of light are said to be spacelike separated from $O$. These are the events represented by the points in the left and right regions of the diagram, outside the light cone. A line drawn from $O$ to any event in these regions will have a slope less than 45°, indicating a velocity of more than 1 (in one unit of time in the vertical direction, more than one unit of distance is covered in the horizontal direction).

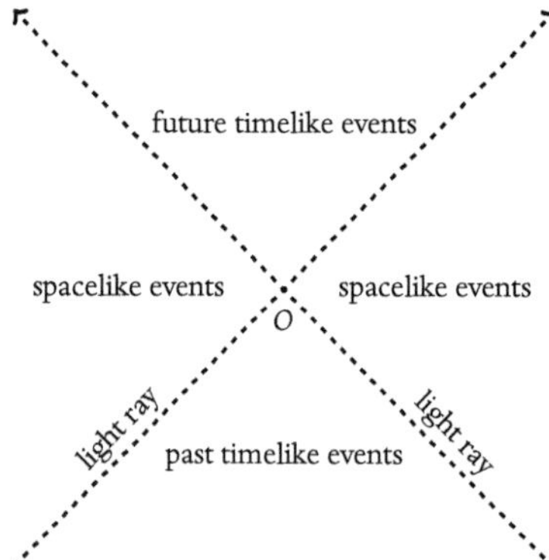


**Figure 1.11** The light cone defined by light rays divides events into three sets. Events represented by points on the light cone are lightlike separated from $O$: they can be connected to $O$ by a light ray. Events represented by points in the top and bottom regions of the light cone are timelike separated from $O$: they can be connected to $O$ by a signal traveling slower than light. Events represented by the points in the left and right regions of the diagram, outside the light cone, are spacelike separated from $O$: they can't be connected to $O$ by a signal traveling at or less than the speed of light. As in the previous space–time diagrams, only one direction of space is shown. The time axis is in the vertical direction and the space axis is in the horizontal direction.

If two events are spacelike separated, the time order can be different for different inertial observers. From the Lorentz transformation,

$$t_2' - t_1' = \frac{(t_2 - t_1) - v(x_2 - x_1)}{\sqrt{1 - v^2}}.$$

For spacelike separated events $E_1$ and $E_2$, the time difference between the two events is smaller than the space difference (in units in which the velocity of light is 1). The relative velocity of two observers, $v$, is less than or equal to 1, the velocity of light. For sufficiently small values of $v$, $t_2 - t_1$ will be greater than $v(x_2 - x_1)$, but for values of $v$ sufficiently close to 1, $t_2 - t_1$ will be smaller than $v(x_2 - x_1)$. So you could have $t_2 - t_1$ positive but $t_2 - t_1 - v(x_2 - x_1)$ negative, which means that $t_2' - t_1'$ could be negative while $t_2 - t_1$ is positive. In other words, $E_2$ could occur later than $E_1$ for Alice, but earlier than $E_1$ for Bob.

For timelike separated events, the time difference between two events is always greater than the space difference, so you can't have $t_2 - t_1 < v(x_2 - x_1)$ for any value of $v$ less than or equal to 1, and the time order is the same for all inertial observers.

The aim of this subsection was to show that, even if the nature of space and time remains mysterious in some deep metaphysical sense, the puzzle about how space and time could be different for Alice and Bob is resolved once we see that time dilation and length contraction follow from the two physically motivated principles of Einstein's theory: the light postulate and the principle of relativity. "Nobody understands quantum mechanics" because there's no agreement among physicists on a similar analysis for quantum phenomena. We don't see that quantum phenomena must be the way they are because of some basic physical principles.

Some of the concepts introduced in this subsection are relevant to later sections, in particular the notion of spacelike separated events, for which the time order can be different for different inertial observers.

**The bottom line**

- Inertial systems are systems moving with constant velocity relative to each other, as distinct from accelerating systems. The relation between the space and time coordinates of different inertial observers in special relativity follows from the *k*-calculus and is known as the Lorentz transformation.
- Length contraction, time dilation, and the relativity of simultaneity—features of relativistic (Minkowski) space-time—all follow from the Lorentz transformation.
- Events in relativistic space-time can be divided into three sets: lightlike (events that can be connected to an event *O* by a light ray), timelike (events that can be connected to *O* by a signal traveling slower than light, so events inside the forward or backward light cones of *O*), and spacelike (events that can't be connected by a signal traveling at or less than the speed of light, so events outside the light cone of *O*).
- For timelike separated events, the time order is the same for all inertial observers, but for spacelike separated events, the time order can be different for different inertial observers.

## Notes

1. Richard P. Feynman, Julian Schwinger, and Sin-Itiro Tomonago were joint winners of the 1965 Nobel Prize "for their fundamental work in quantum electrodynamics, with deep-ploughing consequences for the physics of elementary particles." The quotation "nobody understands quantum mechanics" is from Richard Feynman, *The Character of Physical Law* (MIT Press, Cambridge, 1967), p. 129. The book is a transcript of the Messenger Lectures that Feynman gave at Cornell University in 1964.
2. The title of Einstein's historic paper on the theory of special relativity is "On the electrodynamics of moving bodies" ("Zur Elektrodynamik bewegter Körper"), *Annalen der Physik*, 17, 891–921 (1905). Einstein was awarded the Nobel Prize in 1922 for his discovery of the photoelectric effect and "for his services to Theoretical Physics." Since there was no award for physics in 1921, Einstein's award in 1922 was officially the Nobel prize in physics for 1921. Curiously, the Nobel committee did not regard the theory of relativity as worth a Nobel prize.
3. Minkowski's comment that "space by itself, and time by itself, are doomed to fade away into mere shadows" is from Minkowski's article "Space and time," in W. Perrett and G. Jefferey (eds.), *The Principle of Relativity* (Dover, New York, 1952), p. 75.

4. Niels Bohr's theory of the atom was published in a two-part paper "On the constitution of atoms and molecules," in *Philosophical Magazine* 26, 1–24, 476–502 (1913). Arnold Sommerfeld published an improvement and extension of the theory in a two-part paper "Zur Quantentheorie der Spektrallinien" in *Annalen der Physik* 51, 1–94, 125–167 (1916). The full theory is referred to as the Bohr–Sommerfeld theory.
5. Werner Heisenberg's 1925 breakthrough paper, "Über Quantentheoretischer Umdeutung kinematischer und mechanischer Beziehungen," was published in *Zeitschrift für Physik* 33, 879–893 (1925). In the same year, Max Born and Pascual Jordan published the first part of a two-part paper "Zur Quantenmechanik" in *Zeitschrift für Physik* 34, 858–888 (1925). Part II of this paper, referred to by historians as the "three-man paper," was coauthored with Heisenberg and published in *Zeitschrift für Physik* 35, 557–615 (1926).
6. The expression "saving the appearances" has its origin in Osiander's preface to Copernicus's 1543 heliocentric theory *De Revolutionibus*, proposed as a rival to Ptolemy's geocentric theory. To avoid censure by the Catholic Church, Osiander presented Copernicus's theory as a mathematical model that "saves the appearances," in the sense that it was empirically adequate in representing astronomical phenomena—in fact, superior in some ways to Ptolemy's theory in this respect—without purporting to present a true picture of the relationship between the earth, the sun, and the planets. Osiander's position is now known as "instrumentalism," the view that a scientific theory is an instrument for prediction, with the sole aim of being empirically adequate to the phenomena, not necessarily a true account of the way things are.
7. Heisenberg's comment about all his efforts going toward killing off and replacing the electron orbits is in a letter Heisenberg wrote to Wolfgang Pauli dated July 9, 1925. The remark is quoted in David C. Cassidy, *Uncertainty: The Life and Science of Werner Heienberg* (W.H. Freeman, New York, 1992), p. 197.
8. Erwin Schrödinger's papers on wave mechanics were published as "Quantisierung als Eigenwertproblem," *Annalen der Physik* 79, 361–376 (1926) and "An undulatory theory of the mechanics of atoms and molecules," *Physical Review* 28, 1049–1070 (1926). His proof of the equivalence of his wave mechanics and Heisenberg's matrix mechanics was published as "Über das Verhältnis der Heisenberg–Born–Jordanschen Quantenmechanik zu der meinen," *Annalen der Physik* 79, 734–756 (1926).
9. Schrödinger gave his Royal Institution (London) lecture in 1928. It was published as "Four lectures on wave mechanics," in *Collected Papers on Wave Mechanics* (Chelsea Publishing Company, New York, 1982). The quotation about a wave as a representation of quantum reality being "merely an adequate mathematical description of what happens" is on p. 160.
10. Paul Dirac's formalization of quantum mechanics is developed in his book *The Principles of Quantum Mechanics* (Clarendon Press, Oxford, 1958).
11. Von Neumann's more rigorous 1932 formalization is set out in his book *Mathematical Foundations of Quantum Mechanics* (Princeton University Press, Princeton, 1955).
12. Von Neumann's comment on probabilities in quantum mechanics being "perfectly new and *sui generis* aspects of physical reality" is from "Quantum logics: strict- and probability-logics," a 1937 unfinished manuscript published in A. H. Taub (ed.), *Collected Works of John von Neumann IV* (Pergamon Press, Oxford and New York, 1961), pp. 195–197.
13. Erwin Schrödinger's comment about entanglement being "*the* characteristic trait of quantum mechanics, the one that enforces its entire departure from classical lines of thought" is from his article "Discussion of probability relations between separated systems," *Proceedings of the Cambridge Philosophical Society*, 31, 555–563 (1935). The quotation is on p. 555.

14. Claude Shannon's seminal two-part paper on information theory: "A mathematical theory of communication," *The Bell System Technical Journal* 27, 379–423, 623–656 (1948). The statement about the semantic aspects of communication being irrelevant to the engineering problem is on p. 379.
15. John Wheeler's slogan "it from bit" is from J. A. Wheeler, "Information, physics, quantum: the search for links," in W. Zurek (ed.), *Complexity, Entropy and the Physics of Information* (Addison-Wesley, Redwood City, CA, 1990), pp. 3–28.
16. Rolf Landauer's comment that "information is physical" is from his article "Information is physical," *Physics Today* 44, 23–29 (1991). See also R. Landauer, "The physical nature of information," *Physics Letters* A 217, 188–193 (1996), and "Information is a physical entity," *Physica* A 263, 63–67 (1999). From Landauer's writings, there's no reason to think that "information is physical" should be understood as the claim that information is primary and that "stuff" is in some sense derived from information, although Landauer's slogan has been cited to support such claims.
17. The quotation from Vlatko Vedral's book *Decoding Reality* (Oxford University Press, Oxford, 2010) that "our reality is ultimately made up of information" is on p. 13. On p. 215, he writes: "This book has argued that everything in our reality is made up of information." The statement that "the laws of Nature are information about information" is on p. 218.
18. For a rival view, diametrically opposed to Vedral's, but also at odds with the information-theoretic interpretation of quantum mechanics I put forward in this book, see Christopher Timpson's book, *Quantum Information Theory and the Foundations of Quantum Mechanics* (Oxford University Press, Oxford, 2013).
19. John Bell's comment "*Information*? *Whose* information? Information about *what?*" is from his article "Against measurement," in *Physics World* 8, 33–40 (1990). The comment is on p. 34. It is reprinted in A. Miller (ed.) *Sixty-Two Years of Uncertainty: Historical, Philosophical and Physical Inquiries into the Foundations of Quantum Mechanics* (Plenum, New York, 1990), pp. 17–31.
20. Sandu Popescu and Daniel Rohrlich introduced the nonlocal box now referred to as a PR box in their paper "Quantum nonlocality as an axiom," *Foundations of Physics* 24, 379 (1994). The origin of Boxworld can perhaps be traced to this paper.
21. John Bell's remark that "correlations cry out for explanation" is from his paper "Bertlmann's socks and the nature of reality," originally published in *Journal de Physique*, Colloque C2, suppl. au numero 3, Tome 42, 4161 (1981), and reprinted in J. S. Bell, *Speakable and Unspeakable in Quantum Mechanics* (Cambridge University Press, Cambridge, 1987), pp. 139–158. The quotation is on p. 55 in the original article, and on p. 152 in the collection.
22. 'Yes! We Have No Bananas' is a song from the 1922 Broadway revue *Make It Snappy*, by Frank Silver and Irving Cohn. Eddie Cantor sang the song in the revue, and it has been recorded by hundreds of artists since then. Apparently, there was a banana blight and a shortage of bananas at the time.
23. Bondi's *k*-calculus: Hermann Bondi, *Relativity and Common Sense: A New Approach to Einstein* (Heinemann Educational Books, London, 1965), and *Assumption and Myth in Physical Theory* (Cambridge University Press, Cambridge, 1967).
24. The formulation of the light postulate as "no overtaking of light by light" is from *Assumption and Myth in Physical Theory*, p. 27.
25. The formulation of the relativity principle as "velocity doesn't matter" is from *Relativity and Common Sense*, pp. 4, 58, 147.

# 2
# Qubits

To see the connection between quantum correlations and Bananaworld correlations, you'll need to have some understanding of quantum correlations, in particular the correlations of entangled quantum states. This short chapter is an introduction to the simplest quantum systems, qubits, using polarized photons—particles of light—as the standard example. It contains pretty much all you need to know about quantum mechanics to follow the "Bananaworld" chapter and most of the discussions in the book, so it may be worth reading through a couple of times before moving on. There's more about polarization and entangled states in the Supplement "Some Mathematical Machinery" at the end of the book that gets applied in some chapters, especially the "More" sections. I'll get to the bananas in the next chapter, "Bananaworld."

## 2.1 ABOUT LIGHT

A "bit" is the basic unit of classical (Shannon) information, but the term is also used to refer to an elementary classical system that can be in one of two states. For example, a fair coin is a bit if all that's relevant about the coin is whether it lands heads or tails when tossed, regardless of any other feature of the coin, such as its weight or its size. The two states are labeled 0 and 1, and a measurement always reveals one of the two values. The corresponding elementary quantum system is referred to as a qubit, short for "quantum bit." If you measure a two-valued observable of a qubit, you also get one of two values. But, unlike a classical bit, a qubit is associated with an infinite set of noncommuting two-valued observables that can't all have definite values simultaneously. In this sense, noncommuting observables are "incompatible": measuring one observable and finding a definite value precludes measuring a second noncommuting observable at the same time. (Popular statements that a bit can be either 0 or 1, but a qubit can "somehow" be both at the same time are simply confused.)

Here's a concrete example. Classical optics treats light as an electromagnetic wave produced by the oscillation of electric and magnetic fields in the plane orthogonal to the direction in which the wave moves. The polarization of a light wave is associated with the direction in which the electric field oscillates. Light is said to be linearly polarized if the direction of the oscillation, represented by the electric field vector vector $\vec{E}$, is back and

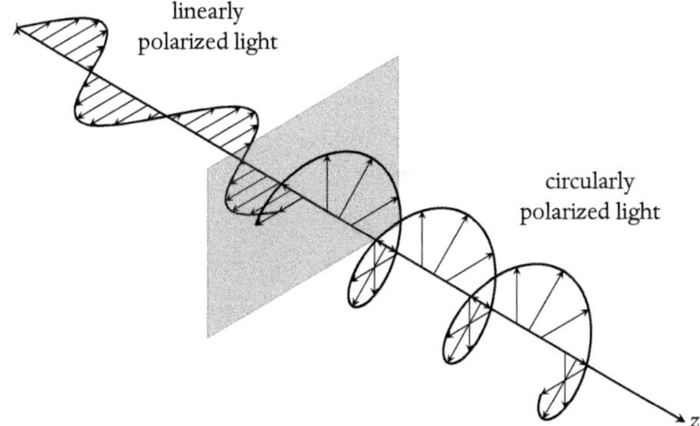


**Figure 2.1** Polarized light showing the electric field $\vec{E}$ represented by the arrow oscillating in a plane orthogonal to $z$, the direction in which the light wave is moving. Linearly polarized light is transformed to circularly polarized light after passing through a crystal represented by the gray rectangle. (Adapted, with changes, from a diagram by Cyril Langlois, under the Creative Commons Condition http://creativecommons.org/licenses/by/2.5/.)

forth along a straight line in a fixed direction. The vector $\vec{E}$ can also rotate as it oscillates, with the tip of $\vec{E}$ tracing out an ellipse in the plane orthogonal to the direction of the light wave. If $\vec{E}$ traces out a circle, the light is said to be right-circularly polarized or left-circularly polarized, depending on whether the rotation is clockwise or counterclockwise about the direction in which the wave is moving. See Figure 2.1.

Optical devices of various sorts (certain polymers, sheets of glass seeded with particles of silver or copper, so-called "birefringent" crystals like calcite or mica) can be used to rotate the direction of linear polarization, to transform linear to circular polarization or circular to linear polarization, to transmit light in a particular polarization state, to split a light beam into two orthogonally polarized beams, and in general to manipulate the polarization state of light.

A polarizer is a sheet of material with a preferred optical axis. If a linearly polarized light wave moving in the $z$ direction of a Cartesian coordinate system, with $\vec{E}$ oscillating along the $x$ direction in the $x, y$ plane, hits a polarizing filter, called an analyzer in this context, with its optical axis in a direction at an angle $\theta$ to the $x$ axis, the analyzer transmits only the component of $\vec{E}$ oscillating in the $\theta$ direction, $\vec{E}_\theta = \vec{E}\cos\theta$, and blocks the orthogonal component, $\vec{E}\sin\theta$. The transmitted beam is now linearly polarized in the $\theta$ direction. According to Malus's law, the intensity of the transmitted beam is reduced to a fraction $\cos^2\theta$ of the incoming beam intensity. So as the angle $\theta$ goes from $0°$ to $90°$, the intensity of the transmitted beam varies smoothly from 100% of the incoming beam to 0%. Take this as simply an experimental fact about polarized light. If you've worn a pair of polarizing sunglasses, you'll be familiar with the phenomenon: the lenses act as polarizing filters.

Einstein received the Nobel prize in physics in 1922 for his contribution to the early quantum theory—surprisingly, not for the theory of relativity—in particular, for proposing that a light beam is composed of quantum particles, now called photons. Today, the picture of an oscillating electric field associated with the wave model of light is regarded as a classical approximation that works for a beam of many photons. The quantum theory treats polarization as an observable associated with some internal "degree of freedom" of a photon, revealed through the behavior of single photons—some physical quantity characterizing a photon that can take one of two possible values. With respect to its polarization, a photon is a qubit (ignoring other dynamical properties of the photon, such as position, or momentum, or energy). It's possible to prepare a single photon in a particular state of polarization by passing the photon through a polarizing filter, to transform the polarization state of a photon with polarizers or polarizing beamsplitters, and to detect single photons by photon detectors.

A beamsplitter is a device with an optical axis that splits an incoming beam of light into two outgoing beams with orthogonal directions of linear polarization. One beam, the "horizontally polarized" beam, is linearly polarized in the direction of the optical axis of the beamsplitter. The other beam, the "vertically polarized" beam, is linearly polarized in the orthogonal direction. If a single photon is sent through a beamsplitter, it can be detected by photon detectors placed in the horizontal and vertical outgoing paths of the beamsplitter as being in one of the paths, and so as having a new linear polarization, horizontal or vertical, associated with that path. Instead of Malus's law for a light beam entering a beamsplitter, $\cos^2\theta$ and $\sin^2\theta$ now represent the probabilities that a photon will take one or other of the two paths in a beamsplitter. In the case of a linearly polarized photon hitting an analyzer, with its optical axis at an angle $\theta$ to the direction of the photon's linear polarization, $\cos^2\theta$ is the probability that the photon will be transmitted through the analyzer and $\sin^2\theta$ the probability that it will be blocked. If a photon gets through the analyzer, it has a new polarization state in the direction of the optical axis of the analyzer. In this sense, an analyzer is said to perform a "yes–no" measurement for a particular polarization state. This is a particular application of the Born rule for quantum probabilities, first proposed by Max Born in a paper published in *Zeitschrift für Physik* in 1926. (For more on quantum probabilities, see the section "The Born Rule" in the Supplement "Some Mathematical Machinery" at the end of the book.)

What the Born rule specifies is the probability that a photon in a certain initial polarization state, prepared by passing the photon through a preparation polarizer, will be detected by a photon detector after being directed to pass through an analyzer. In effect, the photon is faced with a choice when it meets the analyzer: to be transmitted or blocked. In the case of a beamsplitter, the choice is to take one of two paths. There is a binary choice with associated probabilities for linear polarization in any direction, so there are an infinite number of two-valued polarization observables.

Of course, the photon isn't choosing in the sense of "making up its mind." Rather, the photon is put in a physical situation where there are two possible outcomes, only one of which can occur, and which outcome actually occurs is selected randomly. In Chapter 4, "Really Random," I'll show that this "choice" is an intrinsically random event.

A word about terminology: the term "observable" in quantum mechanics is used to refer to what would be called a physical quantity in classical physics—something like energy or temperature that can be measured as having a definite value, or a value that lies in a certain range of values. If you measure the temperature of water in a glass and find that the temperature is 1°C, or that the temperature lies between 0°C and 2°C, so that the water has a temperature in a range of values that you would call cold, then you could say that the water has the "property" of being cold, or that it has the property of having a temperature of 1°C. To say that a system has a certain property is to say that an observable of the system, in the sense of a physical quantity, has a certain value, or that the value of the observable lies in a certain range of values. For any physical property, a classical system either has that property or it doesn't, so every property can be associated with a "yes–no" or two-valued observable that takes the value "yes" or 1 when the system has the property, and the value "no" or 0 when it doesn't.

A classical system can be in a physical state in which it has a particular property with a certain probability. Such a state is understood to be incomplete, in the sense that some information is left out that would specify whether the system has the property or not. In quantum mechanics, the Born rule assigns probabilities to the values of observables of quantum systems, depending on the quantum state, and the question of how these probabilities are to be understood has always been a central foundational problem in the theory. Bohr took the view that quantum mechanics is a complete theory, so nothing is left out even though the description is probabilistic. Einstein argued that the theory is incomplete.

With respect to polarization, a photon is a two-state quantum system or qubit because each type of polarization—linear or circular or elliptical—is associated with a binary alternative. For a classical system, you could combine the values for all of these binary alternatives into a cumulative list of the system's properties to completely define the state of the system. In the case of a playing card, for example, the state could be specified by the color of a card, and given the color, by whether the card is a club or a spade, or a diamond or a heart. A more precise specification could represent the value of the card, or its position in the deck.

The bottom line here is that such a list of properties associated with all possible polarizations doesn't apply to a qubit. If a photon is in a particular state of linear polarization, it has a probability 1 of being transmitted by an analyzer with its axis in the direction of the photon's polarization and a probability 0 of being transmitted by an analyzer with its axis in the orthogonal direction. The photon also has a probability $\cos^2\theta$ of being transmitted by a polarizer with its axis in a direction $\theta$ to the photon's polarization, and if the photon passes through the $\theta$-polarizer, it will be in a new state of linear polarization, in the $\theta$ direction.

So a photon that has zero probability of passing through two crossed analyzers, with their axes at right angles, has a non-zero probability of being transmitted if a third analyzer, with its axis at an angle $\theta$ to the axis of the first analyzer, is placed between the two crossed analyzers. If the angles between the first analyzer and the middle analyzer,

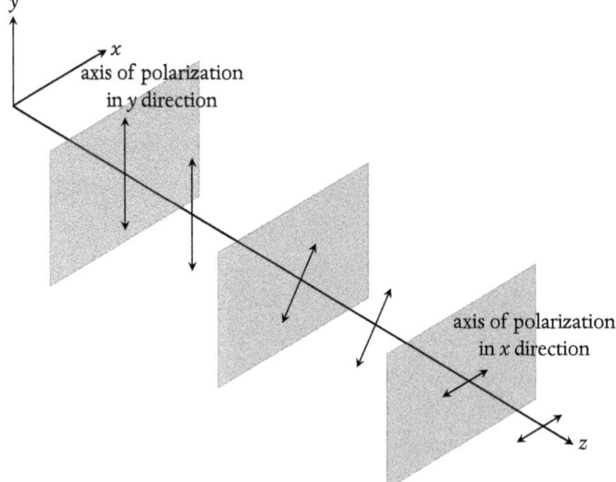


**Figure 2.2** A photon blocked by the first and third crossed analyzers (with orthogonal axes of polarization) has a probability 1/4 of being transmitted if a third analyzer is placed in between the crossed analyzers, with its axis of polarization at a 45° angle to the axes of polarization of the crossed analyzers. (Adapted, with changes, from a diagram by Cyril Langlois, under the Creative Commons Condition http://creativecommons.org/licenses/by/2.5/.)

and the middle analyzer and the final analyzer, are both 45°, the probability that a photon that gets through the first analyzer also gets through the middle and final analyzer is $\cos^2 45° \cdot \cos^2 45° = (\frac{1}{\sqrt{2}})^2 \cdot (\frac{1}{\sqrt{2}})^2 = 1/4$. You can easily check this phenomenon yourself, not with single photons of course, but with a beam of light from a window or other light source and three polaroid polymer sheets or polarized sunglass lenses, which act as analyzers for linear polarization. See Figure 2.2.

There is no physical property of a photon that determines whether or not a photon will be transmitted through an analyzer with its axis of polarization in some arbitrary direction. It's possible to construct a "hidden variable" model for a single qubit, where values of the hidden variables label complete lists of definite polarization properties, and the quantum probabilities given by the Born rule can be derived from classical probability distributions over the possible values of the hidden variables, but this is a quirk of a single qubit. Other sorts of quantum systems are associated with three-valued alternatives (a "qutrit"), or in general *d*-valued alternatives (a "qudit"), or even a continuous infinity of alternatives in the case of observables with possible values that vary continuously. For a qutrit, or any sort of quantum system other than a qubit, the Kochen–Specker theorem rules out "noncontextual" hidden variable models. I'll explain what this is all about in Chapter 6, "Quantum Magic." For two qubits, Bell's theorem shows that no local hidden variable model can recover the statistics of entangled qubit pairs. So there is no list of properties that defines the "real state" of a qubit in the classical sense in a universe with more than one qubit.

It's perhaps a bit odd to refer to the operation of detecting a photon, after it has, in effect, randomly chosen an alternative imposed by placing an analyzer in its path, as a "measurement" of the photon's polarization. The usual sense of the term refers to some operation by which you find out the value of a certain quantity associated with a feature of a physical system, or equivalently whether or not a system has a certain property, and there's an implicit presupposition here that the properties form a Boolean algebra. There's an account of Boolean algebras in subsection 3.5.3, "Boolean Algebras," in the section "More" at the end of the chapter. Basically, a Boolean algebra is the structure that everyone takes for granted when you talk about the properties of objects. The properties of an object are assumed to be there and to "fit together" in a way that makes it possible to use "and," "or," and "not" in the usual way when you talk about them. In the case of a photon, the space of polarization states is not associated with a Boolean algebra of properties—there's no *there* there, as Gertrude Stein said of Oakland, California,[1] in the sense of a predefined list of definite polarization properties for all possible polarizations. A polarization measurement is a procedure for putting the photon in a physical situation where the photon is forced to make an intrinsically random transition to one of two alternative polarizations: two orthogonal directions of linear polarization, or right-circular versus left-circular polarization, or a similar alternative for elliptical polarization.

For a photon, there is no measurement in the classical Boolean sense. You can choose an orthogonal pair of polarization states and perform an operation that amounts to asking the photon: are you in this state or that state? The photon undergoes a random transition, recorded in a measurement device, that is interpreted as an answer to the question. So it would be more appropriate to think of the measurement outcome probabilities as transition probabilities, where the transition is to a new polarization state.

A photon is a qubit with respect to its state of polarization. Another example of a qubit is an electron (or a proton or a neutron), with respect to its spin state. The "spin" of an electron is a technical term referring to a two-valued dynamical variable with some formal properties associated with rotation, but an electron shouldn't be pictured as a tiny spinning top. It would be better to think of electron spin, purely abstractly, as a two-valued quantum observable associated with every direction in three-dimensional space. Each direction is associated with two possible values of spin, conventionally referred to by the labels "up" and "down," analogous to horizontal and vertical polarization in a particular direction for a photon. Yet another example of a qubit is an electron in an atom, where the two states are energy states of the electron (usually the lowest energy state, called the ground state, and the first "excited" or higher-energy state).

If $x, y, z$ represent three orthogonal directions in space, and the "up" state of spin in the direction $z$ is represented by a vector from the center of a unit sphere to the north pole, the "down" state of spin in the direction $z$ would be represented by a vector pointing in the opposite direction, from the center of the sphere to the south pole. Relative to this representation of the spin states for $z$-spin, the spin states for the $x$ and $y$ directions would then be represented by two orthogonal lines in the plane of the equator, with the "up" and "down" spin states represented by vectors pointing in opposite directions along these lines, from the center to the equator. This representation of spin states is referred to as the Bloch sphere representation. A similar representation, called the Poincaré sphere representation, works for polarization, but it's a little more complicated and plays no role in the following discussion.

The difference between spin and polarization is that the *half-angle* is relevant for spin, rather than the angle. An electron in a spin state "up" in some direction has a probability $\cos^2(\theta/2)$ of being found to have a spin state "up" in a measurement of the spin in a direction at an angle $\theta$ to the first direction, and a probability $\sin^2(\theta/2)$ of being found to have a spin state "down." This means that the "up" and "down" states of spin in a particular direction, the two states that are the possible outcomes of a measurement of this spin observable, are not associated with two orthogonal directions in space, as in the case of the "horizontal" and "vertical" polarization states that are the possible outcomes of a linear polarization measurement, but rather with opposite directions, 180° or $\pi$ radians apart. So if spin in a particular direction is "up," the probability of finding it "down" in a measurement is $\cos^2(\pi/2) = 0$, as you would expect.

Electron spin is the standard example of a qubit in discussions of Bell's theorem, which I'll get to in the "Bananaworld" chapter. It's sometimes simpler to talk about qubit correlations in terms of electron spin rather than photon polarization. For an electron, you can talk about spin observables $X, Y, Z$ associated with the spin of the electron in three orthogonal directions, $x, y$, and $z$. For a photon, the corresponding observables would be linear polarization in some direction $z$ associated with the observable $Z$, "diagonal" polarization in a direction 45° to $z$ associated with the observable $X$, and right-circular versus left-circular polarization associated with the observable $Y$. I'll stay with photon polarization states as an example of qubit states, because polarization measurements with beamsplitters and polarization filters are more familiar than spin measurements on electrons, which involve devices like Stern–Gerlach magnets and inhomogeneous magnetic fields. You can actually do polarization measurements yourself with beams of photons and a calcite crystal or polaroid filter.

In the "Bananaworld" chapter, peeling $S$ corresponds to measuring the $Z$ observable of a photon (polarization in the direction $z$), and peeling $T$ corresponds to measuring the $X$ observable (diagonal polarization at an angle 45° to the original $z$ direction). In later chapters the correspondence between peeling $S$ or $T$ and the observables measured might be different. The tastes 0 and 1 correspond to the two possible values ("horizontal" and "vertical") of $Z$ and $X$.

## The bottom line

- The simplest quantum system is referred to as a qubit, short for "quantum bit." A "bit" is the basic unit of classical (Shannon) information, but the term is also used to refer to an elementary classical system that can be in one of two states.

*(continued)*

**The bottom line** *(continued)*

- The standard example of a qubit in this book is a photon, or light particle. A photon is a qubit with respect to its state of polarization. A linearly polarized photon can be either horizontally or vertically polarized with respect to some direction in the plane orthogonal to the direction in which the photon is moving. In the classical wave theory of light, the direction of polarization is the direction in which the electric field oscillates. A photon can also be right-circularly polarized or left-circularly polarized, depending on whether the electric field rotates clockwise or counterclockwise about the direction in which the wave is moving.
- The Born rule specifies the probability that a photon in a certain initial polarization state, prepared by passing the photon through a preparation polarizer, will be detected by a photon detector after being directed to pass through an "analyzer." There is a binary alternative with associated probabilities for linear polarization in any direction, and for circular polarization, so there are infinitely many two-valued polarization observables.
- In the "Bananaworld" chapter, peeling a banana from the stem end ($S$) corresponds to measuring the observable $Z$, linear polarization in some direction $z$, and peeling from the top end ($T$) corresponds to measuring the observable $X$, diagonal polarization at an angle 45° to the $z$ direction. The tastes 0 and 1 correspond to the two possible values of $Z$ and $X$, horizontal or vertical.

## 2.2 QUANTUM STATES

I'm now going to get a bit technical about how quantum states are represented. Dirac introduced an ingenious notation for representing quantum states as vectors, and observables or physical quantities (like the polarization of a photon or the spin of an electron) as operators that transform these states to new states.[2] In the Dirac notation, a quantum state is denoted by the symbol $|\cdots\rangle$, where $\cdots$ is a label for the state. So the two states of horizontal and vertical linear polarization in some direction might be represented by $|H\rangle$ and $|V\rangle$, the two states of diagonal (45°) polarization by $|\nearrow\rangle$ and $|\searrow\rangle$, and the two states of right circular and left circular polarization by $|\circlearrowright\rangle$ and $|\circlearrowleft\rangle$. (The vertical bar and the wedge are notational devices that turn out to be very useful. I don't make use of this feature of the notation in this chapter, but it plays a significant role in the discussion of quantum states and operators in the Supplement, "Some Mathematical Machinery," at the end of the book.) You might want to read the following paragraphs until the end of this section a few times to get the terminology straight, and to understand how to read the notation.

For a general qubit, it's usual to pick two states labeled $|0\rangle$ and $|1\rangle$ associated with the two possible values of some two-valued observable $Z$ as the standard or "computational" states—the reference states for computations. Don't confuse the states $|0\rangle$ and $|1\rangle$ with the numbers 0 and 1. The "0" and "1" in the quantum states are labels for the two states associated with the 0 and 1 values of $Z$. For a photon, these states are usually the horizontal and vertical states of linear polarization, $|H\rangle$ and $|V\rangle$, in some standard direction $z$. The two states associated with the two possible values of $X$ are usually represented as $|+\rangle$ and $|-\rangle$. For photons, $|+\rangle$ and $|-\rangle$ correspond to the two possible states of diagonal polarization along a direction 45° to $z$. I won't need the observable $Y$ in the "Bananaworld" chapter, but it does play a role in Chapter 6, "Quantum Magic." See Figure 2.3.

The states $|0\rangle, |1\rangle, |+\rangle, |-\rangle$ are "pure" quantum states. Although these states assign probabilities to the possible outcomes of polarization measurements, as in the passage of a photon along one of two outgoing paths in a beamsplitter, they cannot be represented as mixtures or probability distributions of other quantum states. So the probabilities don't represent ignorance of the actual quantum state. A photon can also be in a "mixed" quantum state: a probability distribution or mixture of pure states, say the output of a source that produces the states $|0\rangle$ and $|1\rangle$ with probabilities $p_0$ and $p_1$. (See the section "Mixed States" in the Supplement "Some Mathematical Machinery" at the end of the book for more on mixed states.)

The qubit states $|0\rangle$ and $|1\rangle$ are represented by orthogonal vectors of unit length in a two-dimensional space of quantum states, called a Hilbert space (although the term is more properly used for a state space of infinite dimensions, associated with observables like position and momentum that have a continuous range of values). The unit vectors $|0\rangle$ and $|1\rangle$ define a coordinate system in the state space, just as the unit vectors $\vec{x}$ and $\vec{y}$ in the directions of the $x$ and $y$ coordinates of a Cartesian coordinate system define the two orthogonal directions of a coordinate system in the Cartesian plane. Any unit length vector $\vec{v}$, say at an angle $\theta$ to the $x$ axis, can be represented as a linear combination or "superposition" of its projections in the $x$ and $y$ directions: $\vec{v} = \cos\theta\,\vec{x} + \sin\theta\,\vec{y}$.

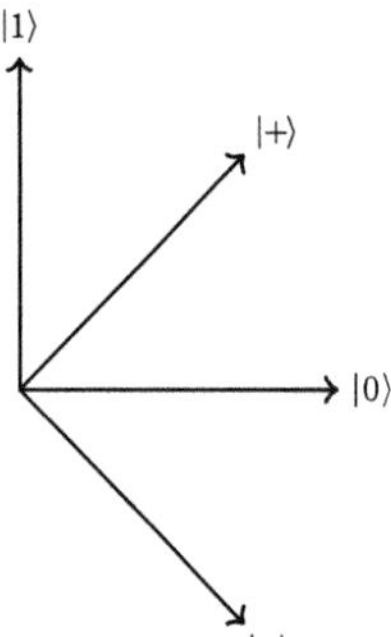


**Figure 2.3** Two bases in the quantum state space: the standard or computational basis $|0\rangle, |1\rangle$, and the diagonal basis $|+\rangle, |-\rangle$.

The coefficients $\cos\theta$ and $\sin\theta$ multiplying the unit vectors $\vec{x}, \vec{y}$ represent the lengths of the projections of $\vec{v}$ onto the coordinate directions. (A linear combination or "superposition" of vectors is just a sum of vectors, where each vector is multiplied by a numerical coefficient, here $\cos\theta$ and $\sin\theta$.) For a qutrit, the state space is three-dimensional and a coordinate system is defined by three quantum states represented by three mutually orthogonal unit vectors associated with the three possible values of some three-valued observable.

A coordinate system in a quantum state space is referred to as a "basis," and the basis defined by the qubit states $|0\rangle$ and $|1\rangle$ is referred to as the "computational basis."

You can have qubit states like $|0\rangle$ or $|1\rangle$, or a qubit in a state that is a superposition of $|0\rangle$ and $|1\rangle$. For example, the states $|+\rangle = \frac{1}{\sqrt{2}}|0\rangle + \frac{1}{\sqrt{2}}|1\rangle$ and $|-\rangle = \frac{1}{\sqrt{2}}|0\rangle - \frac{1}{\sqrt{2}}|1\rangle$ associated with the two possible values of $X$ representing horizontal and vertical polarization in the diagonal direction are linear superpositions of $|0\rangle$ and $|1\rangle$. The coefficients of the superposition needn't be $1/\sqrt{2}$, and needn't be the same. They could be any two numbers, even complex numbers, so long as the squares of the absolute values of the coefficients add up to 1. The vectors representing quantum states are unit vectors, so if you consider the orthogonal triangle defined by a state vector and its two orthogonal projections onto $|0\rangle$ and $|1\rangle$ then, by Pythagoras's theorem, the square of the length of the state vector, which is 1, should be equal to the sum of the squares of the lengths of its orthogonal projections.

If you're unfamiliar with complex numbers and puzzled by the reference to "absolute values" here, there's a discussion of complex numbers in the section "Dirac's Ingenious Idea" in the Supplement, "Some Mathematical Machinery." The absolute value of a complex number corresponds to the length of the vector. You won't need complex numbers to understand correlations in the "Bananaworld" chapter, which are all associated with quantum states with real coefficients, but they do come up in some of the simulations in the "More" sections of the "Quantum Magic" chapter.

It's easier to read a quantum state represented by a superposition of unit vectors with equal coefficients by dropping the coefficients: $|0\rangle + |1\rangle$ instead of $\frac{1}{\sqrt{2}}|0\rangle + \frac{1}{\sqrt{2}}|1\rangle$. In that case, the state is said to be "unnormalized," which is a technical way of saying that the state vector does not have unit length (if the two coefficients are both 1, the length of the state vector is $\sqrt{1+1} = \sqrt{2}$, by Pythagoras's theorem). I'll routinely leave out the coefficients when they are all equal and it's clear how to put them back in to "normalize" the state to unit length. If there are two terms in the superposition, the two equal coefficients are $\frac{1}{\sqrt{2}}$. If there are three terms in the superposition, the three equal coefficients are $\frac{1}{\sqrt{3}}$, and so on.

The unit length is associated with probability: the squares of the absolute values of the coefficients (both $1/2$ for the state $|+\rangle$) are interpreted as representing the probabilities of finding the qubit in one of the two computational basis states $|0\rangle$ or $|1\rangle$ in a measurement of the observable associated with these two states. So they should sum to 1 for the probability interpretation to make sense. For more on this, see the section

"The Born Rule" in the Supplement, "Some Mathematical Machinery," at the end of the book.

**The bottom line**

- Two states, labeled $|0\rangle$ and $|1\rangle$, associated with the two possible values of a two-valued observable Z, are chosen as the standard or "computational" states, represented by orthogonal unit vectors. The two state vectors define a coordinate system called a "basis" in the state space of a qubit. Other qubit states can be represented as linear combinations or "superpositions" of these basis states.
- In this chapter, Z represents linear polarization in the $z$ direction for a photon, and $|0\rangle$ and $|1\rangle$ represent states of horizontal and vertical polarization in the $z$ direction. The observable $X$ represents diagonal polarization in a direction at a 45° angle to $z$, and $|+\rangle$ and $|-\rangle$ represent states of horizontal and vertical polarization in this direction.
- The states $|+\rangle$ and $|-\rangle$ are represented as orthogonal unit vectors at a 45° angle to $|0\rangle$ and $|1\rangle$ in the qubit state space, so as linear superpositions of the computational states: $|+\rangle = \frac{1}{\sqrt{2}}|0\rangle + \frac{1}{\sqrt{2}}|1\rangle$ and $|-\rangle = \frac{1}{\sqrt{2}}|0\rangle - \frac{1}{\sqrt{2}}|1\rangle$. For ease of reading, the coefficients $\frac{1}{\sqrt{2}}$ are usually dropped.
- According to the Born rule, if a photon in the state $|+\rangle = |0\rangle + |1\rangle$ (equal coefficients $\frac{1}{\sqrt{2}}$ omitted for ease of reading) is passed through a beamsplitter oriented in the $z$ direction, it has a probability $\cos^2 45 = 1/2$ of emerging horizontally polarized in the state $|0\rangle$ and a probability $\sin^2 45 = 1/2$ of emerging vertically polarized in the state $|1\rangle$.
- This is referred to as a "measurement" of Z, but it's not a measurement in the sense of revealing a preexisting value of polarization in the $z$ direction.
- Similarly, a photon in the state $|+\rangle$ has a probability $\cos^2 45 = 1/2$ of being transmitted by an analyzer with its axis of polarization in the $z$ direction and a probability $\sin^2 45 = 1/2$ of being blocked. This is referred to as a "yes–no" measurement of Z, but a "yes" ("transmitted") value doesn't reveal a preexisting value of Z.

## 2.3 ENTANGLED PHOTONS

If you have two qubits, the first qubit in a state $|\psi\rangle$ and the second qubit in a state $|\phi\rangle$, the pair of qubits is in a product state, represented as $|\psi\rangle \otimes |\phi\rangle$ or just $|\psi\rangle|\phi\rangle$ for short, where the symbol $\otimes$ denotes a product. You can also have a pair of qubits in a state that is a superposition of product states, like

$$|\phi^+\rangle = |0\rangle|0\rangle + |1\rangle|1\rangle.$$

A state of this form is said to be "entangled" because it cannot be represented as a product state of two qubits $|\psi\rangle$, $|\phi\rangle$, for any states $|\psi\rangle$, $|\phi\rangle$. Entangled polarization states of a photon can be prepared by passing a beam of laser light through a nonlinear crystal. Most of the light, called the "pump," goes straight through the crystal, but sometimes, in a process called "spontaneous parametric downconversion" (don't worry about the terminology), the interaction with the crystal creates a pair of photons, called the "signal" and the "idler," in an entangled state of polarization. The signal and idler photons can be detected on different paths. For more on entangled states, see the section "Entanglement" in the Supplement, "Some Mathematical Machinery," at the end of the book.

What's relevant here is that if you measure the $Z$ observable, polarization in the $z$ direction, on a pair of separated photons in the state $|\phi^+\rangle$, in whatever order, by passing the photons through beamsplitters with their optical axes in the $z$ direction, the probability is $1/2$ (the square of the missing coefficient $1/\sqrt{2}$) of finding both photons horizontally polarized in the $z$ direction, or both photons vertically polarized in the $z$ direction. It turns out that the state $|\phi^+\rangle$ takes exactly the same form for diagonal polarization 45° to the $z$ direction, represented by the observable $X$ (and, in fact, for linear polarization in any direction):

$$|\phi^+\rangle = |0\rangle|0\rangle + |1\rangle|1\rangle = |+\rangle|+\rangle + |-\rangle|-\rangle.$$

So the probabilities are also $1/2$ of finding both photons horizontally polarized in the diagonal direction, or both photons vertically polarized in the diagonal direction. In other words, the photons emerge in the states $|0\rangle|0\rangle$ or $|1\rangle|1\rangle$ with equal probability $1/2$ if beamsplitters with optical axes in the $z$ direction are placed in the paths of the two photons, and they emerge in the states $|+\rangle|+\rangle$ or $|-\rangle|-\rangle$ with equal probability $1/2$ if beamsplitters with optical axes in the diagonal direction are placed in the paths of the two photons. This is an example of a correlation that comes up in the "Bananaworld" chapter.

What if you measure the polarization observable $Z$ on one photon and polarization in some direction $\theta$ to the $z$ direction on the second photon? In that case, the $Z$-measurement leaves the first photon in the state $|0\rangle$ (corresponding to horizontal polarization in the $z$ direction) with probability $1/2$, and the second photon ends up horizontally polarized in the $\theta$-direction with probability $\cos^2\theta$, or vertically polarized in the $\theta$-direction with probability $\sin^2\theta$. Alternatively, the $Z$-measurement leaves the first photon in the state $|1\rangle$ (corresponding to vertical polarization in the $z$ direction) with probability $1/2$, and the second photon ends up horizontally polarized in the $\theta$-direction with probability $\sin^2\theta$, or vertically polarized in the $\theta$-direction with probability $\cos^2\theta$. If the polarization observable $X$ is measured on the second photon, $\theta = 45°$

and $\cos^2\theta = \sin^2\theta = 1/2$, so there are four possible outcomes for this pair of measurements, each with probability $1/2 \cdot 1/2 = 1/4$: horizontal in the $z$ direction and horizontal in the diagonal direction, or horizontal in the $z$ direction and vertical in the diagonal direction, or vertical in the $z$ direction and horizontal in the diagonal direction, or vertical in the $z$ direction and vertical in the diagonal direction.

These properties of the entangled quantum state $|\phi^+\rangle$ are relevant for the correlations in the next chapter, "Bananaworld": the Einstein–Podolsky–Rosen correlation, and the optimal quantum simulation of the Popescu–Rohrlich correlation.

**The bottom line**

- A pair of qubits can be in a state that is a superposition of product states called an "entangled state," like $|\phi^+\rangle = |0\rangle|0\rangle + |1\rangle|1\rangle$ (unnormalized: the coefficients $\frac{1}{\sqrt{2}}$ are omitted here for ease of reading). This state is equivalent to the entangled state $|+\rangle|+\rangle + |-\rangle|-\rangle$ for $Z$ and $X$ measurements.
- Whether you measure $Z$ or $X$ on both these photons, the probability is $1/2$ of finding both photons horizontally polarized (in the state $|0\rangle|0\rangle$ if you measure $Z$ or in the state $|+\rangle|+\rangle$ if you measure $X$), and $1/2$ of finding both photons vertically polarized (in the state $|1\rangle|1\rangle$ if you measure $Z$ or in the state $|-\rangle|-\rangle$ if you measure $X$).
- If you measure $Z$ on one photon and $X$ on the other, the Born rule gives equal probability of $1/4$ to the four possible outcomes $|0\rangle|+\rangle$, $|0\rangle|-\rangle$, $|1\rangle|+\rangle$, $|1\rangle|-\rangle$.

## Notes

1. Gertrude Stein's comment about Oakland, California is from *Everybody's Autobiography* (New York: Random House, 1937, p. 289):

   > She took us to see her granddaughter who was teaching in the Dominican convent in San Raphael, we went across the bay on a ferry, that had not changed but Goat Island might just as well not have been there, anyway what was the use of my having come from Oakland it was not natural to have come from there yes write about it if I like or anything if I like but not there, there is no there there.

2. The Dirac notation is introduced in P. A. M. Dirac, *The Principles of Quantum Mechanics* (Oxford University Press, Oxford, 1958).

# 3

# Bananaworld

The really remarkable thing about our quantum world is the existence of nonlocal correlations—correlations between events at separate locations—that can't be explained by either of the two sorts of explanation we are familiar with in classical physics or in everyday life: a direct causal connection in which information is transmitted from one event to the other by some physical system moving continuously at finite speed between the correlated events, or a common cause that is the source of the same information transmitted to the correlated events.

The plan in this chapter is to consider correlations in Bananaworld, an imaginary island covered with banana trees, with a variety of peculiar correlations between the tastes of the bananas and how they are peeled. To remind you: a Bananaworld banana can only be peeled in one of two ways, either by peeling from the stem end ($S$) or the top end ($T$), and, once peeled, a Bananaworld banana tastes just like an *o*rdinary banana (0), or it tastes *i*ntense, *i*ncredible, *i*ndescribably delicious (1). Whether the taste is 0 or 1 is an objective fact, not a subjective matter of opinion. So there are two things you can do to a banana in Bananaworld (corresponding to a binary choice of quantum measurements), and two possible outcomes (corresponding to two possible measurement outcomes).

I'll begin with the Einstein–Podolsky–Rosen correlation and show how to simulate this correlation with a shared random local resource, which is equivalent to showing that the correlation has a common cause explanation. Then I'll consider a superquantum correlation, the Popescu–Rohrlich correlation, and I'll show that this correlation can't be perfectly simulated with local resources, but a quantum simulation with entangled quantum states does better than the optimal simulation with local resources. So there's no common cause explanation of the Popescu–Rohrlich correlation. Since the correlated events can be arbitrarily far apart, there can't be a direct causal explanation either, because information would have to be transmitted faster than light between the correlated events. As I'll show below, this is a version of a nonlocality result first proved by John Bell for the correlations of entangled quantum states.

## 3.1 EINSTEIN–PODOLSKY–ROSEN BANANAS

In May, 1935, Albert Einstein, Boris Podolsky, and Nathan Rosen published an article in the *Physical Review* with the title "Can quantum-mechanical description of physical reality be considered complete?"[1] They argued that the only way to make sense of quantum mechanics is to suppose that the state descriptions of the theory are incomplete: something has been left out of the story. The article created quite a stir at the time, with pre-publication headlines in the science section of *The New York Times* of May 4, 1935, based on an interview with Boris Podolsky, announcing:

> Einstein Attacks Quantum Theory. Scientist and Two Colleagues Find It Is Not "Complete" Even Though "Correct."

Einstein was upset by the interview. In a terse statement in the May 7 issue of *The New York Times* he pointed out that the interview was given without his authority, adding:

> It is my invariable practice to discuss scientific matters only in the appropriate forum and I deprecate advance publication of any announcement in regard to such matters in the secular press.

The response came in an article with the same title by Niels Bohr in the October issue of the same year in the *Physical Review*.[2] At the time, Bohr's analysis was regarded as sufficiently authoritative that the issue was treated as more or less settled in favor of the orthodox position. Most physicists took the view "if it's good enough for Bohr, it's good enough for me" or, if there was some lingering worry, would have endorsed Feynman's remark about quantum mechanics:[3]

> It has not yet become obvious to me that there's no real problem. I cannot define the real problem, therefore I suspect there's no real problem, but I'm not sure there's no real problem.

In 1952, David Bohm published a "hidden variable" extension of quantum mechanics in a two-part article in the *Physical Review* that reproduced all the empirical predictions of quantum mechanics.[4] In the previous year, Bohm had published a widely acclaimed textbook on orthodox quantum mechanics but was dissatisfied with his own discussion of the conceptual problems of the theory. The hidden variable theory is a deterministic theory, in which quantum phenomena arise from the behavior of particles that move on definite trajectories guided by Schrödinger's wave function as a field in the many-dimensional space of particle configurations. Bohm's theory is empirically equivalent to quantum mechanics if you assume that a quantum state at a particular time describes a whole ensemble of Bohmian particles distributed in real space according to the probabilities defined by the Born rule. It's an ingenious feature of the theory that if the Bohmian particles are distributed in this way at any time, Bohm's equation of motion for the particle trajectories guarantees that this continues to be the case as the quantum state evolves in time. One could say that even if the positions of the Bohmian particles were distributed in some wildly non-quantum way in the early history of the universe, the distribution must have become quantum at some point in time, because that's the way they are distributed now. So if you assume that the Bohmian particles have settled down to an "equilibrium distribution" in space that fits the probabilities defined by the quantum state of the universe, then the particle positions are hidden variables that remain hidden because they can't be pinned down more precisely than quantum mechanics allows.

I first came across the Einstein–Podolsky–Rosen paper and Bohr's reply as an undergraduate at the University of Cape Town and was quite shocked to discover that physicists like Einstein and Bohr could disagree about the foundations of physics. I thought reasonable people could disagree about art or literature or philosophy, but only cranks disagreed about the basic theories of physics. At the same time, I was introduced to Bohm's hidden variable theory, and I eventually became Bohm's graduate student. I worked on a different sort of hidden variable theory for my dissertation, where the linear quantum dynamics is modified by an additional nonlinear term involving hidden variables that kicks in during a measurement process and produces the notorious "collapse" of the wave function (see Section 10.2, "The Measurement Problem," in Chapter 10, "Making Sense of It All"). The theory is empirically distinguishable, in principle, from quantum mechanics and more or less met its demise in a 1967 experiment that failed to confirm a conjecture that the statistics of measurement outcomes in sufficiently rapid sequences of measurements would differ from quantum probabilities.[5] Today there are much more sophisticated "dynamical collapse" theories, notably the Ghirardi–Rimini–Weber theory and its variants.[6] These theories also conflict with quantum mechanics, but under conditions that are very difficult to produce with current technology.

One might have thought that Bohm's theory was at least a first step to the sort of completion of quantum mechanics that the Einstein–Podolsky–Rosen paper called for, but in correspondence with Max Born dated May 12, 1952, Einstein dismissed Bohm's theory as "too cheap for me."[7] Quite possibly, what Einstein objected to in Bohm's theory was its nonlocality, roughly that what happens *over here* instantaneously affects what happens *over there* (which conflicts with special relativity, even though the conflict is unobservable if the particle positions, the hidden variables in Bohm's theory, are in the equilibrium Born distribution). The behavior of a Bohmian particle—where its trajectory will go at a particular time—depends on where all the other Bohmian particles are located. So a change in the position of a particle is instantaneously transmitted nonlocally to remote particles via the quantum wave function as a guiding field, which means that whether a photon goes one way or the other in a beamsplitter instantaneously affects the outcome of a measurement on an entangled remote photon, no matter where it is in the universe.

In 1966, John Bell published a critical review of several results about the impossibility of reconstructing quantum mechanics as a hidden variable theory, given various assumptions about the hidden variables. The article concluded with a discussion of Bohm's theory in which Bell pointed out that the theory evades these "no go" results because the equations of motion are nonlocal, so that "an explicit causal mechanism exists whereby the disposition of one piece of apparatus affects the results obtained with a distant piece." He added that "the Einstein–Podolsky–Rosen paradox is resolved in the way in which

Einstein would have liked least," and asked whether one could prove "that *any* hidden variable account of quantum mechanics *must* have this extraordinary character."[8]

Bell answered this question by showing that adding hidden variables to quantum mechanics to complete the story is a non-starter if the hidden variables are required to be "local," in the sense that they provide a common cause explanation for quantum correlations when a direct causal connection isn't possible. (See Subsection 3.5.2, "Correlations," in the "More" section at the end of the chapter for the relation between a common cause and Bell's locality condition.) Bell showed that any such theory would have to satisfy a certain inequality, which is violated by the correlations of entangled quantum states for certain combinations of measurements that Einstein, Podolsky, and Rosen didn't consider. He concluded:[9]

> In a theory in which parameters are added to quantum mechanics to determine the results of individual measurements, without changing the statistical predictions, there must be a mechanism whereby the setting of one device can influence the reading of another instrument, however remote. Moreover, the signal involved must propagate instantaneously, so that such a theory could not be Lorentz invariant [so consistent with special relativity].

Bell's theorem, as it is now called, appeared in 1964—two years before the long-delayed publication of the article in which Bell posed the question about Bohm's theory.

The original 1935 formulation of the Einstein–Podolsky–Rosen argument in the *Physical Review* article was apparently mostly due to Podolsky, and Einstein thought it was rather convoluted. As he put it in a 1948 paper in the journal *Dialectica*,[10] the argument rests on two assumptions: a separability assumption, that physical objects in different parts of space have their own independent state of existence or "being-thus" ("So-sein" in the German original), and a locality assumption, that if $A$ and $B$ are separated, then you can't directly, or instantaneously, affect the "being-thus" of $B$ by doing something to $A$, and conversely—in particular, as he wrote in a letter to Max Born, you can't directly affect $B$ by measuring $A$:[11]

> That which really exists in $B$ should therefore not depend on what kind of measurement is carried out in part of space $A$; it should also be independent of whether or not any measurement at all is carried on in space $A$.

Einstein, Podolsky, and Rosen formulated their argument for correlations between position and momentum measurements on two separated quantum particles in an entangled state. Position and momentum are observables with continuous ranges of values, and a rigorous version of the argument would involve a rather sophisticated mathematical treatment. Instead of position and momentum, most modern discussions follow a version proposed by Bohm and consider correlations between the $X$ and $Z$ observables of a qubit (defined in the previous chapter) for a particular entangled state.

Suppose that two separated photons, $A$ and $B$, are in the entangled state $|\phi^+\rangle = |0\rangle|0\rangle + |1\rangle|1\rangle = |+\rangle|+\rangle + |-\rangle|-\rangle$ considered at the end of the previous chapter. (Remember: the equal coefficients $1/\sqrt{2}$ for each of the product terms in the linear superposition are left out for ease of reading.) If $Z$ stands for polarization in the $z$ direction, and $X$ for polarization in a diagonal direction at a 45° angle to the $z$ direction, then $|0\rangle$ and $|1\rangle$ represent states of horizontal and vertical polarization in the $z$ direction, and $|+\rangle$ and $|-\rangle$ represent states of horizontal and vertical polarization in the diagonal direction. If you measure either $Z$ or $X$ on photon $A$, you can predict photon $B$'s polarization with certainty, because the outcomes of a $Z$-measurement on $A$ and $B$ are perfectly correlated for the two photons in this entangled quantum state, and similarly the outcomes of an $X$-measurement on the two photons are perfectly correlated.

On Einstein's locality assumption, a $Z$-measurement of photon $A$ can't change what "really exists" in the remote region where photon $B$ is situated. So it seems that you would have to say that photon $B$ has a particular polarization in the $z$ direction, either horizontal or vertical, even before $B$'s polarization is measured. Or you would have to say that $B$ has a definite value of some variable that determines a particular value for polarization in the $z$ direction if one were to measure the polarization, in the sense that the values of this variable provide the photon with an "instruction set" on the basis of which the photon makes a transition to a state of horizontal or vertical polarization when the polarization is measured in a particular direction. The particular polarization values, or the instruction sets, correspond to a common cause of the correlation. Similarly, since a measurement of diagonal polarization corresponding to the observable $X$ on photon $A$ reveals $B$'s diagonal polarization with certainty, one would have to say that $B$ has a particular polarization, either horizontal or vertical, in a direction 45° to $z$ as well, or an instruction set for this polarization. But the quantum state $|\phi^+\rangle$ is *inconsistent* with the assumption that photon $B$ has a definite polarization in both polarization directions before the measurements. The observables $Z$ and $X$ don't commute, so there is no quantum state in which they both have definite values. (See the section "Noncommutativity and Uncertainty" in the Supplement "Some Mathematical Machinery" at the end of the book for more on this.) Einstein concluded that the state descriptions of quantum mechanics are incomplete: they leave something out of the full description.

### The Einstein–Podolsky–Rosen argument for photons

- If you have two separated photons, $A$ and $B$, in the entangled state $|\phi^+\rangle = |0\rangle|0\rangle + |1\rangle|1\rangle = |+\rangle|+\rangle + |-\rangle|-\rangle$, then the outcomes of a $Z$-measurement (linear polarization in some direction $z$) on the two photons are perfectly correlated, and the outcomes of an $X$-measurement (diagonal polarization at an angle 45° to $z$) are perfectly correlated. So you can predict photon $B$'s polarization, either $Z$ or $X$, with certainty by measuring the corresponding polarization of photon $A$.

*(continued)*

 **The Einstein–Podolsky–Rosen argument for photons** *(continued)*

- On Einstein's locality assumption, a measurement of photon *A* can't change what "really exists" in the remote region where photon *B* is situated.
- It seems that you would have to say that photon *B* has a particular polarization even before *B*'s polarization is measured. Or you would have to say that *B* has a definite value of some variable that provides the photon with an "instruction set" on the basis of which the photon makes a transition to a state of horizontal or vertical polarization when the polarization in a particular direction is measured. The instruction set, with a corresponding instruction set for *A*'s correlated measurement outcome, corresponds to a common cause for the correlation.
- The observables *Z* and *X* don't commute, so there is no quantum state in which they both have definite values. Einstein, Podolsky, and Rosen concluded that the state descriptions of quantum mechanics are incomplete: they leave something out of the full description.

Actually, as Einstein pointed out, you don't need to consider noncommuting observables, and you could make the argument with just one observable, say Z. The conclusion that *B*'s polarization in the *z* direction, either horizontal or vertical, is predetermined by an instruction set, even before it is measured, already shows that the state description by the quantum state $|\phi^+\rangle$ is incomplete, because the quantum description doesn't include information about this predetermined polarization value.

The Einstein–Podolsky–Rosen argument starts with an entangled quantum state, and it was this argument that brought entanglement to the attention of the physics community as a surprising and possibly problematic quantum phenomenon. But the argument itself does not really depend on anything particularly quantum about the *Z* and *X* correlations they consider. What Einstein, Podolsky, and Rosen do with an entangled state amounts to saying: "Look, the correlations between measurement outcomes for two particular noncommuting observables are exactly like the classical correlations you get when you have two systems that start out correlated from a common source, like two billiard balls moving in different directions after a collision, where the positions and momenta are correlated. You can predict with certainty the outcome of a measurement on one billiard ball from a measurement on the other, for either position or momentum as you choose." For two entangled photons in the state $|\phi^+\rangle$, the outcome of a *Z*-measurement on one photon is uncorrelated with the outcome of an *X*-measurement on the second photon, but the outcomes of *Z*-measurements on the two photons are perfectly correlated, as well as the outcomes of *X*-measurements. That's like the relation between position and momentum in the original Einstein–Podolsky–Rosen paper. So if Alice and Bob are separated and share two photons in the entangled quantum state $|\phi^+\rangle$, and they each have a choice between $Z$ and $X$ measurements, then the outcomes are perfectly correlated if they measure the same observables, but uncorrelated if they measure different observables. This is a classical correlation that can be explained quite simply by a common cause, as I'll show below. That's the point of the Einstein–Podolsky–Rosen argument: there must be "elements of reality" for such a correlation, to use their terminology (Einstein's "being-thus"), which are the common cause of the correlations, and quantum mechanics is incomplete because it doesn't include these elements of reality in the quantum state description.

Putting it a little differently, the Einstein–Podolsky–Rosen argument is that entangled states in quantum mechanics involve classical correlations that can be simulated with local resources, and this means that something has been left out of the quantum description—the theory is incomplete. For two photons $A$ and $B$ in the entangled state $|\phi^+\rangle$, the sort of correlation Einstein, Podolsky, and Rosen are talking about is between an $A$-observable, either $A_1$ or $A_2$, and a $B$-observable, either $B_1$ or $B_2$, where $A_1, A_2$ are the same observables for $A$ as $B_1, B_2$ are for $B$. The possible outcomes of measuring the observables can be labeled 0 and 1, and the quantum state $|\phi^+\rangle$ encodes correlations between $A$ and $B$ measurement outcomes—which persist as the photons separate—of the following sort:

- if the same observable is measured on $A$ and on $B$, the outcomes are the same, with equal probability for 00 and 11;
- if different observables are measured on $A$ and on $B$, the outcomes are uncorrelated, with equal probability for 00, 01, 10, 11;
- the "marginal" probabilities—the probabilities for each possible outcome, 0 or 1, when either observable is measured on $A$, or when either observable is measured on $B$—are $1/2$, irrespective of what observable is measured on the paired photon, or whether any observable is measured at all.

These are just the probabilities for the possible outcomes of measurements of the observables $Z$ ($A_1$ or $B_1$) or $X$ ($A_2$ or $B_2$) on two photons in the entangled state $|\phi^+\rangle$.

### The Einstein–Podolsky–Rosen correlation

- For two photons $A$ and $B$ in the entangled state $|\phi^+\rangle$ and measurements of observables $Z$ and $X$, the Einstein–Podolsky–Rosen correlation is:
  (i) if the same observable is measured on $A$ and on $B$, the outcomes are the same, with equal probability of $1/2$ for 00 and 11, and if different observables are measured, the outcomes are uncorrelated, with equal probability of $1/4$ for 00, 01, 10, 11;

*(continued)*

 **The Einstein–Podolsky–Rosen correlation** *(continued)*

(ii) the "marginal" probabilities—the probabilities for each possible outcome, 0 or 1, when either observable is measured on $A$, or when either observable is measured on $B$—are $1/2$, irrespective of what observable is measured on the paired photon, or whether any observable is measured at all.

- The analogous correlation in Bananaworld is:
  (i) if the peelings are the same ($SS$ or $TT$), the tastes are the same, with equal probability of $1/2$ for 00 and 11, and if the peelings are different ($ST$ or $TS$), the tastes are uncorrelated, with equal probability of $1/4$ for 00, 01, 10, 11;
  (ii) the marginal probabilities for the tastes 0 or 1 if a banana is peeled $S$ or $T$ are $1/2$, irrespective of how the paired banana is peeled, or whether or not the paired banana is peeled.

The term "marginal probability" is used when probabilities are defined for two or more random variables, $A, B, \ldots$, and you want to consider just one random variable, say $A$, or, more generally, some subset of all the random variables. (A random variable is just a variable whose values are associated with certain probabilities.) The *marginal probabilities* of $A$ are the probabilities that $A$ has certain values, irrespective of the values of the other random variables. The *joint probabilities* of $A, B, \ldots$ are the probabilities of sequences of values for $A, B, \ldots$. These notions apply to the probabilities of measurement outcomes of quantum observables.

In Bohm's version of the Einstein–Podolsky–Rosen argument, the entangled state is $|\psi^-\rangle = |0\rangle|1\rangle - |1\rangle|0\rangle$ and the observables are *oppositely correlated* rather than correlated, so the outcomes are *different* if the same observable is measured on the two qubits, rather than the same. The Einstein–Podolsky–Rosen argument is the same for the two versions: you could convert one version to the other just by having Alice (but not Bob) flip her outcomes, exchanging 0 for 1 and 1 for 0.

In Bananaworld, there are Einstein–Podolsky–Rosen banana trees with bunches of just two bananas, where peelings and tastes are correlated as in the Einstein–Podolsky–Rosen argument, with correlations that persist when the bananas are separated by an arbitrary distance, as follows (see Figure 3.1):

- if the peelings are the same ($SS$ or $TT$), the tastes are the same, with equal probability for 00 and 11;
- if the peelings are different ($ST$ or $TS$), the tastes are uncorrelated, with equal probability for 00, 01, 10, 11;
- the marginal probabilities for the tastes 0 or 1 if a banana is peeled $S$ or $T$ are $1/2$, irrespective of whether or not the paired banana is peeled.

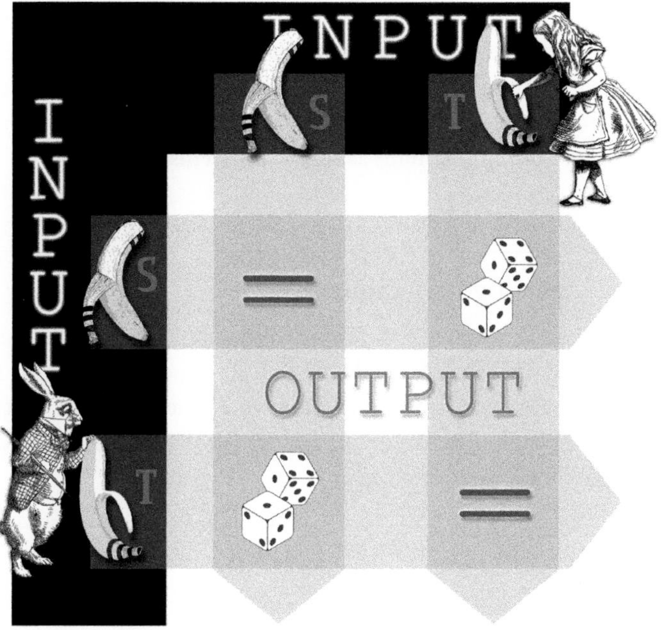


**Figure 3.1** The Einstein–Podolsky–Rosen correlation in Bananaworld. The inputs are the possible choices of how to peel a banana. The outputs are the possible tastes of a peeled banana. The = sign indicates that the tastes are the same. The dice symbol indicates that the tastes are uncorrelated, so that each possible pair of outcomes occurs with the same probability.

The symbol *ST* is short for "Alice peels *S* and Bob peels *T*," and similarly 01 is short for "Alice's banana tastes 0 and Bob's banana tastes 1," with Alice's peeling or taste represented by the first member of the pair, and Bob's peeling or taste represented by the second member of the pair. The stipulation that the marginal probabilities are all 1/2—the probabilities that a banana tastes 0 or 1 when peeled a certain way—guarantees that the correlations satisfy a "no-signaling" principle. Expressed in Bananaworld terms, the principle is that the taste of a banana is independent of how a remote banana is peeled, or whether or not a remote banana is peeled (which doesn't rule out correlations between the *tastes*). More generally, if you are talking about measurements, no information should be available in region **A** about alternative choices made by Bob in region **B**—Alice in region **A** shouldn't be able to tell what Bob measured in region **B**, or whether Bob performed any measurement at all, by looking at the statistics of her measurement outcomes, and conversely.

As Nicolas Gisin puts it, the no-signaling principle is "no communication without transmission," where transmission involves a physical signal that moves along a continuous trajectory at a certain finite speed through the space-time continuum.[12] Violating the no-signaling principle would mean the possibility of superluminal, or even instantaneous, signaling. But the no-signaling principle is not specifically a relativistic constraint on the motion of a physical entity through space-time—there's no reference to the velocity of light, even implicitly. It's a much more fundamental constraint. What's excluded is that something happening *here* has an effect over *there*, without delay and without anything physical passing from here to there. You might say that the principle is part of what's meant by treating two physical systems as separate systems.

### The "no-signaling" principle

- The "no-signaling" principle: no information should be available in region **A** about alternative choices made by Bob in region **B**—Alice in region **A** shouldn't be able to tell what Bob measured in region **B**, or whether Bob performed any measurement at all, by looking at the statistics of her measurement outcomes, and conversely.
- Expressed in Bananaworld terms, the principle is that the taste of a banana is independent of how a remote banana is peeled, or whether or not a remote banana is peeled.

Before considering how Alice and Bob, limited to local resources, could simulate the correlations of Einstein–Podolsky–Rosen banana pairs, consider a simpler problem. Suppose the correlation is that if the peelings are the same, the tastes are the same, and if the peelings are different, the tastes are different. If you think about it for a moment, there's a simple way Alice and Bob could win the simulation game for this correlation. (To remind you—see Section 1.2, "Why Bananaworld?" in Chapter 1—the moderator, who can communicate with Alice and Bob separately, gives each player one of two prompts, $S$ or $T$, at the beginning of each round of the game, and the players, who are not allowed to communicate with each other during the game, are supposed to respond to their prompts with one of two responses, 0 or 1. They win a round if the responses and the prompts are correlated in the right way. The game is played over many rounds, and at the end of the game they win a prize, where the value of the prize depends on the number of rounds they win. So the aim is to figure out a strategy that will enable them to win the maximum number of rounds.) For this game, a winning strategy is to respond 0 if the moderator's prompt is $S$ and 1 if the prompt is $T$. Then Alice and Bob both respond 0 or both respond 1 if the prompts are the same, and their responses will be different if the prompts are different.

That's easy, but now suppose the correlation is a little more complicated. Suppose the response for a given prompt has to be random, so for the prompt $S$ the response can't always be 0. Both possible responses, 0 and 1, should come up with equal probability over many rounds of the game, and similarly for the prompt $T$. Can Alice and Bob win this game? They can if they are allowed to consult the same list of random numbers during the game. You might want to think how they could do this before I give you the solution.

Here's what they could do. Alice and Bob generate a long list of random bits, 0's and 1's, during the strategy session before the game starts, perhaps by tossing a fair coin, with 0 for "heads" and 1 for "tails," with at least as many bits in the list as there are rounds in the simulation game. They each make a copy of the list of bits, which they take with them to consult during the game. The strategy is to respond with the bit on the list if the prompt is *S*, in order down the list for successive rounds, and to flip the bit on the list if the prompt is *T*. Then if Alice and Bob receive the same prompt, they both respond 0 or both respond 1, but the response will be independent of the prompt, and random, because the list is random. If they receive different prompts, one of them responds with the bit on the list, and the other with the flipped bit, so the responses are different, but 0's and 1's come up with equal probability.

Now consider how Alice and Bob, limited to local resources, could simulate the correlations of Einstein–Podolsky–Rosen banana pairs. They could adopt a modification of the previous strategy and generate two long lists of random 0's and 1's by tossing two fair coins in the strategy session before the start of the game. So Alice and Bob each have copies of two random lists of 0's and 1's that I'll call the *S*-list and the *T*-list, with at least as many bits in each list as there are rounds in the simulation game. At each round, they adopt the following strategy:

- if the prompt is *S*, the response (for both Alice and Bob) is a bit from the *S*-list;
- if the prompt is *T*, the response (for both Alice and Bob) is a bit from the *T*-list.

If the prompts are the same, they respond with the same random bit, 0 or 1, because they consult the same list of shared bits, and they read off the random bits in order for successive rounds. If the prompts are different, they respond with bits from different lists, which are randomly related to each other. The marginal probabilities of 0 and 1 for Alice and Bob separately are $1/2$ for either prompt, irrespective of the prompt to the remote player. Figure 3.2 illustrates the idea.

What does this have to do with the Einstein–Podolsky–Rosen argument for the incompleteness of quantum mechanics? The local simulation of Einstein–Podolsky–Rosen bananas shows that it is possible for two systems to exhibit Einstein–Podolsky–Rosen correlations if they both have their own "being-thus," but the separate "being-thuses" originate from a common source as the common cause of the correlation. In the case of the bananas, this will be the case for two bananas on an Einstein–Podolsky–Rosen bunch that grow as a pair on a single banana tree, each with a shared value of some banana variable $\lambda$, the "being-thus" of the bananas. The banana variable $\lambda$ corresponds to the shared simulation random variable, represented by a pair of random bits from the *S*-list and the *T*-list. The tastes for all possible pairs of peelings are determined by the shared value of $\lambda$, and the distribution of shared $\lambda$ values for many Einstein–Podolsky–Rosen pairs of bananas provides an explanation of the correlation.

**Figure 3.2** The Einstein–Podolsky–Rosen simulation game. Alice and Bob can win the game if they respond on the basis of two shared random lists of bits that they consult in order for each round of the game, an *S*-list for *S* prompts and a *T*-list for *T* prompts. This is Round 4, the prompts are both *S*, so according to the shared *S*-list, Alice and Bob both respond with a 0.

The correlations are just like the correlations you would get if you took playing cards from a deck of cards, cut each card in half and mailed the two halves to separate addresses. Opening an envelope containing a half-card would reveal the color of the half-card and, instantaneously, the color of the half-card at the distant address. In this case, the "being-thus" of a half-card would be specified by a shared color variable, with two possible values, red or black. Here, each of the subsystems, a banana in an Einstein–Podolsky–Rosen pair or a half-card, has its own "being-thus." This is characteristic of correlations that can be simulated with local classical resources. As I'll show in the following section, this is not the case for Popescu–Rohrlich bananas: the Popescu–Rohrlich correlations exclude a "being-thus" in this sense.

In effect, the Einstein–Podolsky–Rosen incompleteness argument rests on showing that the shared randomness required for the simulation—the common cause of the correlations—is missing from the quantum description. Einstein's "being-thus," represented by the value of a shared random variable, is the common cause of the correlations, the "instruction set" that tells the system how to respond to a measurement. What's shared is a random distribution of values of a common cause variable, like the values of the banana variable $\lambda$ or the random sequence of 0's and 1's from the *S*-list and the *T*-list in the simulation game. Each banana in a pair picked from the same bunch on an Einstein–Podolsky–Rosen tree shares a value of $\lambda$, and the values of $\lambda$ are distributed randomly over different bunches on the tree. Similarly, Alice and Bob each have copies of the *S*-list and the *T*-list, generated by some suitable random process. Picking a pair of bananas that share a value of $\lambda$ is like Alice and Bob each picking the same pair of corresponding bits from the *S*-list and the *T*-list for a round in the simulation game. Einstein's argument is simply that, if you have perfect correlations of this sort, and you can exclude a direct causal influence between the two systems (because the correlations persist if the systems are separated by any distance), then the *only* explanation for the correlations is the existence of a common cause. Since the quantum description of the correlated systems lacks a representation of the common cause, quantum mechanics must be incomplete. What's missing in at least some quantum state descriptions (like $|\phi^+\rangle$) is something corresponding to the banana variable $\lambda$ or the simulation variable with values defined by a pair of corresponding bits from the *S*-list and the *T*-list.

## The bottom line

- The Einstein–Podolsky–Rosen correlation can be simulated with local resources. A local simulation shows that it is possible for two systems to exhibit an Einstein–Podolsky–Rosen correlation if they both have their own "being-thus," in Einstein's sense, and if the separate "being-thuses" originate from a common source as the common cause of the correlation.
- In the case of Einstein–Podolsky–Rosen bananas, this will be the case if they originate on a banana tree as a particular pair, each with a shared value of some banana variable $\lambda$, the "being-thus" of the bananas.
- The tastes for all possible pairs of peelings are determined by the shared value of $\lambda$, and the distribution of shared $\lambda$ values for many Einstein–Podolsky–Rosen pairs of bananas provides an explanation of the correlation. In a simulation of the correlation, the variable $\lambda$ corresponds to the simulation variable.
- The Einstein–Podolsky–Rosen argument is that:
  (i) if you have a perfect correlation of this sort, and you can exclude a direct causal influence between the two systems (because the correlations persist if the systems are separated by any distance), then the *only* explanation for the correlation is the existence of a common cause or shared randomness;

(*continued*)

 **The bottom line** *(continued)*

(ii) since the description of the correlated systems in quantum mechanics lacks a representation of the common cause, quantum mechanics must be incomplete. What's missing in a quantum state description like $|\phi^+\rangle$ is something corresponding to the banana variable $\lambda$, or the simulation variable in a local simulation.

## 3.2 POPESCU–ROHRLICH BANANAS AND BELL'S THEOREM

The Einstein–Podolsky–Rosen argument lingered for about thirty years until John Bell showed that any common cause explanation of a probabilistic correlation between measurement outcomes on two separated systems would have to satisfy an inequality, now called Bell's inequality. He also showed that the inequality is violated by measurements of certain two-valued observables of a pair of quantum systems in an entangled state. So Einstein's intuition about the correlations of entangled quantum states, and the Einstein–Podolsky–Rosen argument that depends on this intuition, turn out to be wrong. There are correlations between quantum systems, where a causal influence from one system to the other can be excluded, that have no common cause explanation. The result is known as Bell's theorem.

It took a while before the implications of Bell's theorem penetrated mainstream physics, probably around the time that Alain Aspect and colleagues confirmed the violation of Bell's inequality in a series of experiments on entangled photons in the early 1980s. It took several more years for the result to trigger a revolution in quantum information in the 1990s, when it became respectable again to discuss foundational questions in quantum mechanics. Some of the initial discussion took place in a sort of samizdat publication that was hand-typed, mimeographed, and mailed to a select list of readers by a Swiss foundation, the Association F. Gonseth, as a "Written Symposium" on "Hidden Variables and Quantum Uncertainty" that went through 36 issues, from November, 1973 to October, 1984. The back of each issue carried the statement (in English, French, and German):

> "Epistemological Letters" are [sic] not a scientific journal in the ordinary sense. They want to create a basis for an open and informal discussion allowing confrontation and ripening of ideas before publishing in some adequate journal.

Einstein, Podolsky, and Rosen argued for the incompleteness of quantum mechanics from the perfect correlation between the outcomes of the same observable measured on two separated systems in an entangled state. Bell drew a different conclusion from the probabilistic correlation between the outcomes of measurements of different observables on the two systems. Instead of following Bell's argument, I'll derive a version of Bell's theorem as a limitation on the ability of Alice and Bob to simulate a correlation proposed by Sandu Popescu and Daniel Rohrlich if they are restricted to local resources.

Popescu and Rohrlich imagine a hypothetical device, now known as a PR box, with an input and output port for Alice and an input and output port for Bob.[13] There are two possible inputs to these ports, labeled 0 and 1, and two possible outputs, also labeled 0 and 1, that are randomly related to the inputs: the outputs 0 and 1 for Alice and Bob separately occur with equal probability for any input. In particular, Alice's output occurs randomly and independently of Bob's input, and similarly Bob's output occurs randomly and independently of Alice's input. A PR box can be used only once: after an input by Alice, no further Alice-input is possible, and similarly for Bob.

Now, Popescu and Rohrlich suppose that the Alice part of the box and the Bob part of the box can be separated, without any physical connection between them, and that even when separated by an arbitrary distance, the two parts of the box produce the correlation (see Figure 3.3):

- the outputs for Alice and Bob are the same, except when both inputs are 1, in which case the outputs are different;
- the marginal probabilities of the outputs 0 or 1 for any input by Alice or Bob separately are 1/2 (so the no-signaling condition is satisfied).

**Figure 3.3** PR box correlation.

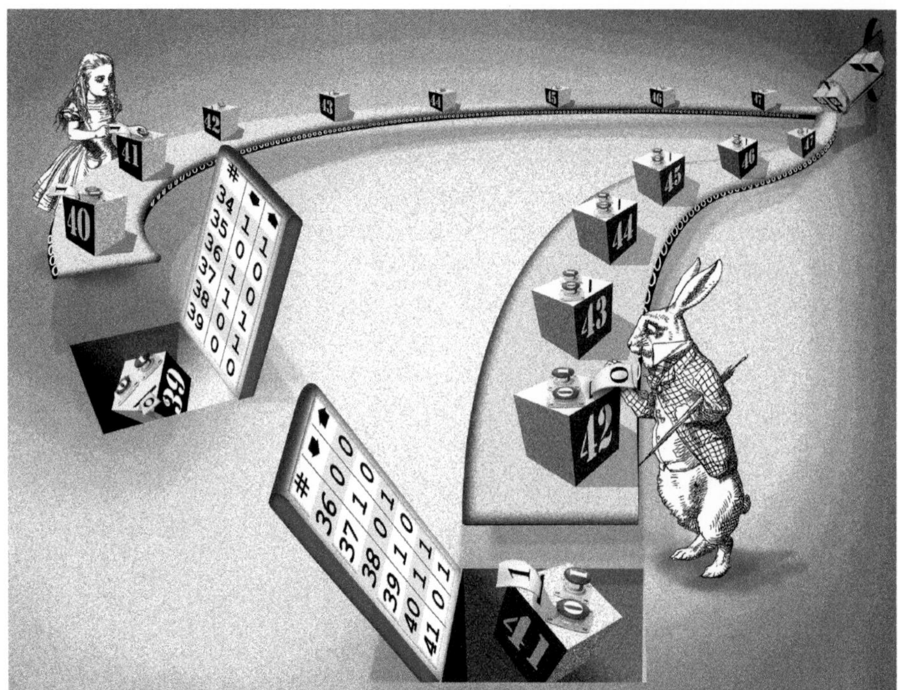

**Figure 3.4** Alice and Bob testing PR boxes on a production line. Similarly numbered half-boxes correspond to the same PR box. Inputs (represented by the down-arrows at the top of the displays) don't have to be coordinated and can occur at any time, or even not at all for one half-box of a pair. Inputs and outputs (represented by the up-arrows) in each display are separately completely random, but inputs and outputs are correlated according to the Popescu–Rohrlich correlation in Figure 3.3. After testing, a PR box is no longer available for further inputs, so the samples tested are discarded.

A PR box functions in such a way that if Alice inputs 0 or 1, her output is 0 or 1 with probability 1/2, irrespective of Bob's input, and irrespective of whether Bob inputs anything at all. Similarly for Bob. The requirement is simply that whenever there is in fact an input by Alice and an input by Bob, in whatever temporal order, the inputs and outputs are correlated in the required way. So for Alice and Bob separately, the output appears to be random for any input, but if Alice and Bob get together and compare inputs and outputs, they find that there are three cases when their outputs are the same, both 0 or both 1 (when they both input 0, or when Alice inputs 0 and Bob inputs 1, or when Alice inputs 1 and Bob inputs 0), and one case where their outputs are different, one output is 0 and the other output is 1 (when Alice and Bob both input 1). See Figure 3.4.

In Bananaworld, Popescu–Rohrlich pairs of bananas grow on Popescu–Rohrlich trees which, like Einstein–Podolsky–Rosen trees, have two-banana bunches. Tastes and peelings are correlated like the correlation of a PR box and persist if the bananas are separated by any distance (see Figure 3.5):

**Figure 3.5** Popescu–Rohrlich banana correlation.

- if the peelings are $SS$, $ST$, $TS$, the tastes are the same, 00 or 11;
- if the peelings are $TT$, the tastes are different, 01 or 10;
- the marginal probabilities for the tastes 0 or 1 if a banana is peeled $S$ or $T$ are $1/2$, irrespective of whether or not the paired banana is peeled (so the no-signaling principle is satisfied).

What could Alice and Bob do to simulate the correlation of Popescu–Rohrlich pairs of bananas? Here's a strategy that would work in three out of four rounds of the simulation game, on average, if the prompts are random. Alice and Bob generate a sufficiently long list of random 0's and 1's before the simulation game starts (in this case, just one random list), and they each take a copy of the list with them to consult during the game. So they share a list of random bits. If they each respond to a prompt with the bit on the shared list, in order, for each successive round, they will satisfy the condition on the marginal probabilities (because they each respond 0 or 1 with probability $1/2$, since the lists are random) and they will win all the rounds of the simulation game for which the prompts are $SS$, $ST$, or $TS$ (because they both ignore the prompt and respond with the same bit, either 0 or 1), and they will lose all the rounds for which the prompts are $TT$ (because they always respond the same, and the requirement for this combination of prompts is to respond differently). If the prompts are random, the probability of winning the game with this strategy is $3/4$.

It turns out that this is the *optimal* strategy for winning the PR simulation game with local resources, as I'll show below. This is a version of Bell's theorem. You might be puzzled about how a result about hypothetical bananas could be relevant to a theorem about quantum mechanics. First, this is not so much a result about hypothetical bananas as a result about a game with a moderator and two players. The optimal probability of winning the game with local resources is 3/4. This limitation on winning the game—on simulating the PR correlation with local resources—is a fact about the world, a limitation on correlations with a common cause. In a simulation with local resources, the correlation arises from a shared random variable, where each shared value of the random variable labels a local instruction set or local response strategy for Alice and Bob separately. Common causes are nature's local resources, so if there's no way in principle for Alice and Bob to simulate a correlation with local resources, then nature can't do it either. Secondly, Bell's theorem is about common cause correlations, not directly about quantum mechanics. Bell proved that a correlation between measurement outcomes on two separated systems satisfies a certain inequality if the correlation has a common cause. The relevance for quantum mechanics is that the inequality can be violated by the correlations of entangled quantum states, as Bell showed (and as I'll show in Section 3.3, "Simulating Popescu–Rohrlich Bananas"). In terms of the PR simulation game, it's also a fact about the world that Alice and Bob can do a better job of winning the game if they are allowed to use entangled quantum states instead of two copies of the same random list. It follows that there are nonlocal quantum correlations between separated systems that can't arise from a common cause (and can't arise from a direct causal influence between the systems either, if the systems are sufficiently far apart and the measurements are sufficiently close together in time, because any signal carrying information between the systems would have to move faster than the speed of light).

### The Popescu–Rohrlich correlation

- A PR box (after Popescu and Rohrlich) has an input and output port for Alice and an input and output port for Bob. There are two possible inputs to these ports, labeled 0 and 1, and two possible outputs, also labeled 0 and 1. The Alice part of the box and the Bob part of the box can be separated, without any physical connection between them, and even when separated the two parts of the box produce the correlation:
  (i) if the inputs are 00, 01, 10, the outputs are the same, but if the inputs are 11 the outputs are different;
  (ii) the marginal probabilities of the outputs 0 or 1 for any input by Alice or Bob separately are 1/2 (so the no-signaling condition is satisfied).

*(continued)*

 **The Popescu–Rohrlich correlation** *(continued)*

- In Bananaworld, Popescu–Rohrlich pairs of bananas occur on trees in bunches of just two bananas. Tastes and peelings are correlated like the correlation of a PR box and persist if the bananas are separated by any distance:
  (i) if the peelings are $SS$, $ST$, $TS$, the tastes are the same, but if the peelings are $TT$, the tastes are different;
  (ii) the marginal probabilities for the tastes 0 or 1 if a banana is peeled $S$ or $T$ are $1/2$ (so the no-signaling condition is satisfied).

There are lots of proofs of Bell's theorem in the literature on quantum mechanics, many of them informative only to the authors. The simplest proof I know is from a short but illuminating book for non-specialists, *Quantum Chance* by Nicolas Gisin, which, as the subtitle indicates, is about "nonlocality, teleportation and other quantum marvels."[14] Gisin's proof is the sort of immediately obvious, back-of-the-envelope proof you will be able to reproduce for your friends over a beer once you get the basic idea. Gisin was awarded the first Bell prize in 2009 "for his theoretical and experimental work on foundations and applications of quantum physics, in particular: quantum non-locality, quantum cryptography and quantum teleportation." With the group he leads, Gisin has performed some spectacular demonstrations of long-distance entanglement of photons using fiber optic cables under Lake Geneva.

The idea of the proof, applied to the simulation game for Popescu–Rohrlich bananas, is that there are really only four possible local strategies available to Alice at each round of the game, since she has no access to Bob's prompt or his response to the prompt, and similarly for Bob. Alice can respond 0 or 1 without reference to her prompt, or she can respond 0 or 1 depending on her prompt. Relabeling the prompt $S$ as 0 and the prompt $T$ as 1 (as in the original Popescu–Rohrlich labeling of the inputs to a PR box), this amounts to saying that there are only two ways to respond, depending on the prompt: Alice's response can either be the same as the prompt (0 for $S$ and 1 for $T$), or different from the prompt (1 for $S$ and 0 for $T$). Since there are two possible prompts, there are four possible local strategies for Alice (two for each prompt). Similarly, there are four possible local strategies for Bob, which yields 16 possible combined strategies. All you need to do now is count how many times each strategy is successful in simulating the Popescu–Rohrlich correlation for the four possible pairs of prompts.

Call the two possible strategies that are independent of the prompt 0 and 1, according to the response, and the two possible strategies that depend on the prompt "same" and "different," according to whether the strategy is to match the numerical value of the prompt (to respond 0 for $S$ and 1 for $T$), or differ from it (to respond 1 for $S$ and 0 for $T$). Table 3.1 shows that no combination of Alice's strategy and Bob's strategy is successful for more than three out of four possible pairs of prompts.

**Table 3.1** All possible response strategies for simulating the Popescu–Rohrlich correlation. The correct responses are indicated in bold.

| **Alice's strategy** | **Bob's strategy** | **response for input *SS*** | **response for input *ST*** | **response for input *TS*** | **response for input *TT*** | **Score** |
|---|---|---|---|---|---|---|
| 0 | 0 | **00** | **00** | **00** | 00 | 3 |
| 0 | 1 | 01 | 01 | 01 | **01** | 1 |
| 0 | same | **00** | 01 | **00** | **01** | 3 |
| 0 | different | 01 | **00** | 01 | 00 | 1 |
| 1 | 0 | 10 | 10 | 10 | **10** | 1 |
| 1 | 1 | **11** | **11** | **11** | 11 | 3 |
| 1 | same | 10 | **11** | 10 | 11 | 1 |
| 1 | different | **11** | 10 | **11** | **10** | 3 |
| same | 0 | **00** | **00** | 10 | **10** | 3 |
| same | 1 | 01 | 01 | **11** | 11 | 1 |
| same | same | **00** | 01 | 10 | 11 | 1 |
| same | different | 01 | **00** | **11** | **10** | 3 |
| different | 0 | 10 | 10 | **00** | 00 | 1 |
| different | 1 | **11** | **11** | 01 | **01** | 3 |
| different | same | 10 | **11** | **00** | **01** | 3 |
| different | different | **11** | 10 | 01 | 00 | 1 |

If the prompts are random, each of the four possible pairs of prompts $SS, ST, TS, TT$ occurs with probability $1/4$, so

$$p(\text{successful simulation}) = \frac{1}{4}(p(\text{responses same} \,|\, SS) + p(\text{responses same} \,|\, ST) + p(\text{responses same} \,|\, TS) + p(\text{responses different} \,|\, TT)).$$

The notation $p(\ldots)$ stands for the probability of the expression in parenthesis, and the vertical bar $|$ indicates a conditional probability, which can be read as "given that." So $p(\text{responses same} \,|\, SS)$ is the probability of "responses same," given that Alice peels $S$ and Bob peels $S$, and so on. Since the most any strategy is successful is for three out of four pairs of prompts (and the least any strategy is successful is for one out of four pairs of prompts), Table 3.1 can be summed up in the statement:

$$1/4 \leq p_L(\text{successful simulation}) \leq 3/4.$$

This is a version of Bell's inequality. It says that the probability of successfully simulating the correlation of Popescu–Rohrlich bananas with local resources available to Alice alone or to Bob alone ("$L$" for local), lies between $1/4$ and $3/4$. So the *optimal* probability is $3/4$. If Alice and Bob each separately respond randomly to the prompts, the success probability is $1/2$. In this case they will give the same response in half the rounds, and different responses in half the rounds (see below). In a simulation restricted to local resources, the correlation arises from a shared random variable, where the shared values of the random variable range over different common causes of the correlation (in this case, the 16 possible combined strategies). So the impossibility of perfectly simulating the Popescu–Rohrlich correlation if Alice and Bob are restricted to local resources means that the correlation can't arise from a common cause. If Alice and Bob are allowed to use entangled quantum states as a shared resource, they can break the 3/4 barrier (see the next section). So it must be impossible to simulate the correlations of entangled states with local resources, which means that, incredibly, we live in a nonlocal quantum world in which there are correlations without a common cause.

Where does a common cause come into the argument? Each row of Table 3.1 represents a deterministic Alice strategy and a deterministic Bob strategy for responding to the possible prompts. The combined strategy for Alice and Bob is equivalent to a deterministic common cause of the correlation in the responses, a local instruction set for Alice and a local instruction set for Bob on the basis of which they produce the correlated responses. The table lists the 16 possible combined deterministic strategies or deterministic common causes and the corresponding score, the number of correct responses, for the four possible input pairs. Alice and Bob will have to use the values of a shared random variable with 16 possible values, corresponding to the 16 combined strategies, to choose a combined strategy randomly for each round of the game. Computer scientists prefer the term "shared randomness" to "common cause," and refer to Bell's result as showing that the correlations of entangled quantum states can't arise from shared randomness in the correlated quantum systems.

If the values of the random variable occur with probabilities between 0 and 1 that sum to 1, then $1/4 \leq p_L(\text{successful simulation}) \leq 3/4$ for *any* probabilities, provided the values of the random variable are independent of the choice of prompts. For example, Alice and Bob could toss four fair coins many times before the start of the simulation game. The outcome of each quadruple toss is a quadruple of heads and tails (or a quadruple of bits, taking heads as 0 and tails as 1) and there are $2 \times 2 \times 2 \times 2 = 16$ possible quadruples, which occur with equal probability of 1/16 if the coins are fair. Alice and Bob each record the list of random quadruples and consult the shared list in order during successive rounds of the simulation game. In this case, the shared values of the random variable are the 16 quadruples, which occur with equal probability. Eight out of 16 strategies correspond to a score of 1 out of 4, and these are equally weighted with a probability of 1/16, and eight equally weighted strategies correspond to a score of 3 out of 4. So

$$p_L(\text{successful simulation}) = \frac{8}{16} \cdot \frac{1}{4} + \frac{8}{16} \cdot \frac{3}{4} = \frac{1}{2}.$$

If Alice and Bob use biased coins, the probabilities will be different, but even if the strategy represented by the top row has unit probability and all the other 15 strategies have zero probability, $p_L(\text{successful simulation}) = 3/4$. So Alice and Bob can achieve a successful simulation with probability 3/4, but there's no way the probability can exceed 3/4 for any probability distribution of strategies.

## Bell's theorem

- The optimal probability of successfully simulating the correlation of Popescu–Rohrlich bananas with local resources is 3/4.
- In a simulation with local resources, the correlation arises from a shared random variable, where the shared values of the random variable range over different common causes of the correlations.
- The impossibility of perfectly simulating the Popescu–Rohrlich correlation if Alice and Bob are restricted to local resources means that the correlation can't have a common cause explanation.
- If Alice and Bob are allowed to use entangled quantum states as a shared resource, they can break the 3/4 barrier and achieve an optimal probability of about .85.
- So, incredibly, we live in a nonlocal quantum world in which there are correlations without a common cause.

There are various Bell inequalities for different sets of observables besides the specific inequality in Bell's original 1964 paper. John Clauser, Michael Horne, Abner Shimony, and Richard Holt derived an inequality in 1969 that plays an important role in some of the subsequent chapters.[15] It's also easier to test the Clauser–Horne–Shimony–Holt inequality experimentally than Bell's original inequality. Here's a proof of the Clauser–Horne–Shimony–Holt version of Bell's theorem.

Consider again the problem of simulating the correlation of a pair of Popescu–Rohrlich bananas. It's convenient to relabel the responses $\pm 1$ instead of 0, 1. "Responses are the same" and "responses are different" mean the same thing whatever the units. So the probabilities—$p(\text{responses same} \,|\, A, B)$, the conditional probability that Alice and Bob produce the same response, given the prompts $A$ and $B$, and $p(\text{responses different} \,|\, A, B)$, the conditional probability that Alice and Bob produce different responses, given the prompts $A$ and $B$—take the same values whatever the units. Here, $A$ represents Alice's prompt, $S$ or $T$, and $B$ represents Bob's prompt, and I've separated the prompts by commas to avoid confusion with products in the following argument.

The Clauser–Horne–Shimony–Holt inequality involves the expectation value or average value of the *product of the responses* for a particular pair of prompts $A$ and $B$, denoted by the symbol $\langle AB \rangle$. The expectation value $\langle AB \rangle$ is the weighted sum of the products of the four possible pairs of responses, with the probabilities for the paired responses, $p(-1,-1 \,|\, A, B)$, $p(-1, 1 \,|\, A, B)$, $p(1,-1 \,|\, A, B)$, $p(1, 1 \,|\, A, B)$ as the weights. (So the expectation value, unlike the probabilities, does depend on the units for the responses, because the products of the responses are different for different units.) This sounds complicated, but there's a simple way of expressing the expectation value. The products of the responses are 1 if the responses are the same ($-1,-1$ or $1, 1$), and $-1$ if the responses are different

$(-1, 1 \text{ or } 1, -1)$, so the expectation value is just a sum of probabilities multiplied by $\pm 1$ appropriately:

$$\begin{aligned}\langle AB\rangle &= p(-1,-1\,|\,A,B) - p(-1,1\,|\,A,B) - p(1,-1\,|\,A,B) + p(1,1\,|\,A,B)\\ &= p(\text{responses same}\,|\,A,B) - p(\text{responses different}\,|\,A,B).\end{aligned}$$

Since $p(\text{responses same}\,|\,A,B) = 1 - p(\text{responses different}\,|\,A,B)$ and $p(\text{responses different}\,|\,A,B) = 1 - p(\text{responses same}\,|\,A,B)$, the expectation value $\langle AB\rangle$ can equally well be expressed as

$$\langle AB\rangle = 2p(\text{responses same}\,|\,A,B) - 1 = 1 - 2p(\text{responses different}\,|\,A,B),$$

so

$$\begin{aligned}p(\text{responses same}\,|\,A,B) &= \frac{1+\langle AB\rangle}{2},\\ p(\text{responses different}\,|\,A,B) &= \frac{1-\langle AB\rangle}{2}.\end{aligned}$$

To keep track of whether $S$ or $T$ refers to Alice or Bob in the following proof, I'll need to add subscripts: $S_A$ or $S_B$, $T_A$ or $T_B$. With the common cause or shared random variable $\lambda$ to consider as well, the notation becomes clumsy. So I'll simply write $S_A$ as $A$ and $T_A$ as $A'$, and $S_B$ as $B$ and $T_B$ as $B'$, indicating the prompts corresponding to Alice's two ways of peeling her banana as $A$ and $A'$ for peeling by the stem end ($S$) or the top end ($T$), and similarly I'll indicate the prompts corresponding to Bob's two ways of peeling his banana as $B$ and $B'$ for $S$ and $T$.

The relationship between the expectation value $\langle AB\rangle$ and the probabilities of the responses also holds for the expectation values $\langle AB'\rangle$, $\langle A'B\rangle$, and $\langle A'B'\rangle$. So

$$\begin{aligned}p(\text{successful simulation}) &= \frac{1}{4}(p(\text{responses same}\,|\,A,B) + p(\text{responses same}\,|\,A,B')\\ &\quad + p(\text{responses same}\,|\,A',B) + p(\text{responses different}\,|\,A',B'))\\ &= \frac{1}{4}\left(\frac{1+\langle AB\rangle}{2} + \frac{1+\langle AB'\rangle}{2} + \frac{1+\langle A'B\rangle}{2} + \frac{1-\langle A'B'\rangle}{2}\right)\\ &= \frac{1}{2}\left(1+\frac{K}{4}\right),\end{aligned}$$

where $K = \langle AB\rangle + \langle AB'\rangle + \langle A'B\rangle - \langle A'B'\rangle$.

Suppose Alice and Bob have a strategy for simulating the correlation of a pair of Popescu–Rohrlich bananas using a shared random variable $\lambda$. (The point of the theorem, of course, is to prove that there can be no such strategy.) The possible values of $\lambda$ needn't be the 16 values corresponding to the 16 combined strategies in Gisin's proof—they could be any values, with any probabilities, as long as the $\lambda$ probabilities are independent of the probabilities of the prompts, which are assumed to be freely chosen by a moderator or randomly chosen by a randomizing device. The $\lambda$ values could even be a continuous set with a probability distribution $\rho(\lambda)$ (a probability function defined on the continuous set of $\lambda$ values that integrates to 1 over the set, just as the set of probabilities for a finite set of $\lambda$ values sums to 1). The responses could be deterministic functions of $\lambda$, or it could be that, given prompts $A$ and $B$, Alice responds $a$ with a probability $p_\lambda(a\,|\,A)$ that depends on $\lambda$ and Bob responds $b$ with a probability $p_\lambda(b\,|\,B)$ that depends on $\lambda$. (I represent $\lambda$ as a subscript here because it's easier to read an expression like $p_\lambda(a\,|\,A)$ rather than $p(a\,|\,A,\lambda)$ in the proof below.)

The tastes and peelings of Popescu–Rohrlich bananas are correlated (statistically dependent) and not randomly related. So if $a, b$ represent tastes and $A, B$ peelings, the probability of a pair of tastes given a pair of peelings won't be equal to the product of the probability of the taste of Alice's banana given Alice's peeling and the probability of the taste of Bob's banana given Bob's peeling:

$$p(a, b\,|\,A, B) \neq p(a\,|\,A)p(b\,|\,B).$$

But in a simulation of these probabilities, the probability of a joint response to a pair of prompts, given a particular value of the shared random variable $\lambda$, *is* equal to the product of the probability of Alice's response to her prompt given $\lambda$ and the probability of Bob's response to his prompt given $\lambda$:

$$p_\lambda(a, b\,|\,A, B) = p_\lambda(a\,|\,A)p_\lambda(b\,|\,B).$$

That's because Alice's response to a prompt for a given value of $\lambda$ can't depend on Bob's prompt or on Bob's response to his prompt, and similarly for Bob's response with respect to Alice. So the responses, *conditional on the shared random variable* $\lambda$, are statistically independent, and the joint probability factorizes to a product of marginal probabilities for Alice and Bob separately.

This is where the classical resource of shared randomness plays a role in the argument. Putting it differently, this is where the assumption that the correlations are due to a common cause, or a local hidden variable, plays a role in limiting the extent to which Alice and Bob can simulate the correlation. The first equation says that the tastes and peelings are correlated or statistically *dependent*. The second equation says that the prompts and responses in a simulation with local resources are *conditionally* statistically *independent*, given the common cause represented by the shared random variable $\lambda$. (For more on conditional statistical independence, see the subsection "Correlations" in the "More" section at the end of the chapter.)

The expectation value or average value of the responses for a pair of prompts $A$ and $B$, and a particular value of $\lambda$, is $\langle AB\rangle_\lambda$. Assuming conditional statistical independence, the joint probabilities in the expression for $\langle AB\rangle_\lambda$ can each be expressed as a product of probabilities for Alice and Bob separately: $p_\lambda(-1,-1\,|\,A,B) = p_\lambda(-1\,|\,A)\cdot p_\lambda(-1\,|\,B)$, and so on, so:

$$\begin{aligned}
\langle AB\rangle_\lambda &= p_\lambda(-1,-1\,|\,A,B) - p_\lambda(-1,1\,|\,A,B) - p_\lambda(1,-1\,|\,A,B) + p_\lambda(1,1\,|\,A,B)\\
&= p_\lambda(-1\,|\,A)\cdot p_\lambda(-1\,|\,B) - p_\lambda(-1\,|\,A)\cdot p_\lambda(1\,|\,B) - p_\lambda(1\,|\,A)\cdot p_\lambda(-1\,|\,B) + p_\lambda(1\,|\,A)\cdot p_\lambda(1\,|\,B)\\
&= (p_\lambda(1\,|\,A) - p_\lambda(-1\,|\,A))(p_\lambda(1\,|\,B) - p_\lambda(-1\,|\,B))\\
&= \langle A\rangle_\lambda\langle B\rangle_\lambda,
\end{aligned}$$

where $\langle A\rangle_\lambda$ and $\langle B\rangle_\lambda$ represent averages over the two possible values $\pm1$ of $A$ and $B$, with weights equal to the $\lambda$-probabilities for 1 and –1.

### The Clauser–Horne–Shimony–Holt version of Bell's theorem

- The tastes and peelings of Popescu–Rohrlich bananas are correlated (statistically dependent) and not randomly related. So if $a, b$ represent tastes and $A, B$ peelings, the probability of a pair of tastes given a pair of peelings won't factorize to the product of the marginal probabilities for Alice and Bob separately: $p(a, b\,|\,A, B) \neq p(a\,|\,A)p(b\,|\,B)$.
- In a local simulation, the joint probability of a pair of responses to a pair of prompts, conditional on a shared random variable $\lambda$ (representing a common cause of the correlation), does factorize to a product of the marginal probabilities, conditional on $\lambda$, for Alice and Bob separately: $p_\lambda(a, b\,|\,A, B) = p_\lambda(a\,|\,A)p_\lambda(b\,|\,B)$.
- The condition that the joint probability factorizes is Bell's locality condition. It says that the prompts and responses are uncorrelated (or conditionally statistically independent), given the common cause represented by the shared random variable $\lambda$. That's because Alice's response to a prompt for a given value of $\lambda$ can't depend on Bob's prompt or on Bob's response to his prompt, and similarly for Bob's response with respect to Alice.
- Clauser, Horne, Shimony, and Holt consider the quantity $K = \langle AB\rangle + \langle AB'\rangle + \langle A'B\rangle - \langle A'B'\rangle$, where $\langle AB\rangle$ denotes the expectation value of the product of the responses for a pair of prompts $A, B$.
- The expectation value of the product of the responses for a pair of prompts factorizes for a particular value of $\lambda$. So $K_\lambda = \langle A\rangle_\lambda\langle B\rangle_\lambda + \langle A\rangle_\lambda\langle B'\rangle_\lambda + \langle A'\rangle_\lambda\langle B\rangle_\lambda - \langle A'\rangle_\lambda\langle B'\rangle_\lambda$.
- This can be expressed as $K_\lambda = \langle A\rangle_\lambda\left[\langle B\rangle_\lambda + \langle B'\rangle_\lambda\right] + \langle A'\rangle_\lambda\left[\langle B\rangle_\lambda - \langle B'\rangle_\lambda\right]$. Since each of the conditional expectation values is between –1 and +1, and so one of the bracketed expressions with the $B$ and $B'$ expectation values is 2 or –2 (in which case the other bracketed expression is 0), it follows that $-2 \leq K_\lambda \leq 2$.
- In a simulation of the Popescu–Rohrlich correlation with local resources, $K$ is $K_\lambda$ averaged over $\lambda$, with the probabilities of the $\lambda$ values as the weights. Since averaging over $\lambda$ won't change the inequality, $-2 \leq K_L \leq 2$ ("L" for local).

The quantity $K = \langle AB\rangle + \langle AB'\rangle + \langle A'B\rangle - \langle A'B'\rangle$ for a particular value of $\lambda$ can therefore be expressed as

$$\begin{aligned} K_\lambda &= \langle A\rangle_\lambda\langle B\rangle_\lambda + \langle A\rangle_\lambda\langle B'\rangle_\lambda + \langle A'\rangle_\lambda\langle B\rangle_\lambda - \langle A'\rangle_\lambda\langle B'\rangle_\lambda \\ &= \langle A\rangle_\lambda\left[\langle B\rangle_\lambda + \langle B'\rangle_\lambda\right] + \langle A'\rangle_\lambda\left[\langle B\rangle_\lambda - \langle B'\rangle_\lambda\right]. \end{aligned}$$

This says that $K_\lambda$ is equal to the product of $\langle A\rangle_\lambda$ and the sum $\left[\langle B\rangle_\lambda + \langle B'\rangle_\lambda\right]$, plus the product of $\langle A'\rangle_\lambda$ and the difference $\left[\langle B\rangle_\lambda - \langle B'\rangle_\lambda\right]$. Since $\langle B\rangle_\lambda, \langle B'\rangle_\lambda$ are both between –1 and 1, the sum in brackets $\left[\langle B\rangle_\lambda + \langle B'\rangle_\lambda\right]$ can take a maximum value of 2 (when the terms in the sum are both 1) or a minimum value of –2 (when the terms in the sum are both –1), and in both cases $\left[\langle B\rangle_\lambda - \langle B'\rangle_\lambda\right] = 0$. Similarly, $\left[\langle B\rangle_\lambda - \langle B'\rangle_\lambda\right]$ can take a maximum value of 2 (when $\langle B\rangle_\lambda = 1$ and $\langle B'\rangle_\lambda = -1$) or a minimum value of –2 (when $\langle B\rangle_\lambda = -1$ and $\langle B'\rangle_\lambda) = 1$), and in both these cases $\left[\langle B\rangle_\lambda + \langle B'\rangle_\lambda\right] = 0$. So the maximum and minimum values of one of the bracketed expressions with the $B$ and $B'$ expectation values is 2 or –2, in which case the other bracketed expression is 0. Since the smallest value of $\langle A\rangle_\lambda$ or $\langle A'\rangle_\lambda$ is –1 and the largest value is 1, it follows that $K_\lambda$ lies between –2 and 2.

The quantity $K$ is just $K_\lambda$ averaged over $\lambda$. Averaging over $\lambda$ won't change the inequality, because an average is just a weighted sum of the quantities $K_\lambda$ with the probabilities of the $\lambda$ values as the weights, and these are all positive numbers between 0 and 1. So in a simulation of the Popescu–Rohrlich correlation with local resources, $K_L$ ("L" for local) is similarly bounded:

$$-2 \leq K_L \leq 2.$$

This is the Clauser–Horne–Shimony–Holt version of Bell's theorem.

Since $p_L(\text{successful simulation}) = \frac{1}{2}(1+\frac{K_L}{4})$ as I showed above, and the maximum value of $K_L$ is 2, it follows, as before, that the optimal probability of simulating the Popescu–Rohrlich correlation with local resources is

$$p_L(\text{successful simulation}) \leq \frac{1}{2}\left(1 + \frac{2}{4}\right) = 3/4.$$

I'll show in the next section that if Alice and Bob are allowed to base their strategy on shared entangled quantum states prepared before they separate, then they can achieve a value of $K_Q = 2\sqrt{2}$. So they can do better than classical players or players restricted to local resources and win the simulation game with probability $\frac{1}{2}(1 + \frac{2\sqrt{2}}{4}) \approx .85$. The value $K_Q = 2\sqrt{2}$ is called the Tsirelson bound, after Boris Tsirelson (sometimes spelled Cirel'son) who first proved that this is in fact the optimal quantum value.[16] The demonstration in the next section doesn't show this. Rather, I show only the possibility of simulating the correlations of Popescu–Rohrlich bananas with probability $\frac{1}{2}(1 + \frac{2\sqrt{2}}{4})$ using entangled quantum states, and so of achieving a value $K = 2\sqrt{2}$ with quantum resources—not that this is the *optimal* quantum value, which is harder to prove. The optimal quantum value can be achieved with appropriate measurements on quantum states that are said to be "maximally entangled." The state $|\phi^+\rangle$ is maximally entangled, and appropriate measurements on $|\phi^+\rangle$ produce the maximal violation of the Clauser–Horne–Shimony–Holt inequality $-2 \leq K \leq 2$.

It's convenient to write $E = K/4$. Then the optimal value of $E$ for classical common cause correlations, or correlations that can be simulated with local resources, is $E_L = 1/2$, and the optimal quantum value is $E_Q = 1/\sqrt{2}$. The difference is the square root in the quantum case! After a similar analysis showing why a classical computer can't simulate arbitrary quantum correlations efficiently, Feynman commented:[17]

> I've entertained myself always by squeezing the difficulty of quantum mechanics into a smaller and smaller place, so as to get more and more worried about this particular item. It seems to be almost ridiculous that you can squeeze it to a numerical question that one thing is bigger than another. But there you are—it is bigger than any logical argument can produce, if you have this kind of [classical] logic.

The Popescu–Rohrlich correlation is designed to achieve a value of $E_{PR} = 1$, which is the maximum value of $E$ for no-signaling correlations (corresponding to $K = 4$, when the first three terms take the value 1 and the term with the – sign takes the value –1).

In Bananaworld there are bananas that exhibit classical correlations, with a value of $E$ that is less than or equal to $1/2$, as well as quantum correlations, where this correlation can take values up to $1/\sqrt{2}$, and superquantum no-signaling correlations, which can take values between $1/\sqrt{2}$ and 1. The question raised by Popescu and Rohrlich was why we live in a world in which correlations are limited by the Tsirelson bound $1/\sqrt{2}$, rather than the no-signaling bound 1. Is there some principle about our world that limits the Clauser–Horne–Shimony–Holt correlation to values that are less than or equal to $1/\sqrt{2}$? I'll take up this question in Chapter 9, "Why the Quantum?"

**The bottom line**

- The probability of successfully simulating the Popescu–Rohrlich correlation is $\frac{1}{2}(1+\frac{K}{4})$. In a simulation with local resources, $K_L = 2$, so the optimal probability of a successful simulation with local resources is $3/4$.
- If Alice and Bob are allowed to base their simulation strategy on shared entangled quantum states, they can achieve a value of $K_Q = 2\sqrt{2}$ with maximally entangled states like $|\phi^+\rangle$. So with quantum resources, the optimal probability of a successful simulation increases to $\frac{1}{2}(1 + \frac{2\sqrt{2}}{4}) \approx .85$.
- The quantum value $K_Q = 2\sqrt{2}$ is called the Tsirelson bound. It's convenient to write $E = K/4$. Then the optimal value of $E$ for classical common cause correlations, or equivalently correlations that can be simulated with local resources, is $E_L = 1/2$, the optimal quantum value is $E_Q = 1/\sqrt{2}$, and the optimal value for a PR box or Popescu–Rohrlich bananas is $E_{PR} = 1$, which is also the maximum value for no-signaling correlations.

## 3.3 SIMULATING POPESCU–ROHRLICH BANANAS

In the simulation game, a moderator gives Alice and Bob separate prompts, $S$ or $T$, at each round of the game, and Alice and Bob are each supposed to respond with a 0 or a 1.

They win the round if their responses agree with the correlations for Popescu–Rohrlich bananas:

- if the peelings are $SS, ST, TS$, the tastes are the same, 00 or 11;
- if the peelings are $TT$, the tastes are different 01 or 10;
- the marginal probabilities for the tastes 0 or 1 if a banana is peeled $S$ or $T$ are $1/2$ (so the no-signaling constraint is satisfied).

If Alice and Bob are allowed access to local resources, like shared random lists of 0's and 1's prepared before the start of the simulation, the probability of winning a round of the game is at most .75. With entangled quantum states as a resource, Alice and Bob can achieve a success probability of approximately .85.

Here's the way they do it. Before the simulation starts, Alice and Bob prepare many pairs of photons in the maximally entangled state $|\phi^+\rangle = |0\rangle|0\rangle + |1\rangle|1\rangle$. They each store one photon from each pair for later measurements during the simulation, and they keep track of the different pairs of entangled photons, so that the photons they measure at each round belong to the same entangled state. The strategy is for Alice to measure the polarization of her photons in directions 0 and $\pi/4$, represented by the polarization observables $A$ and $A'$, when she gets the prompt $S$ or $T$, respectively, and for Bob to measure the polarization of his photons in directions $\pi/8$ and $-\pi/8$, represented by the polarization observables $B$ and $B'$, when he gets the prompt $S$ or $T$, respectively. (Up to now, I've indicated the angles of polarization directions in degrees. It's more convenient here and in subsequent sections to use radians, where $90° = \pi/2$, $45° = \pi/4$, and $22.5° = \pi/8$.) See Figure 3.6.

For measurements of a pair of polarization observables $A, B$ or $A, B'$ or $A', B$ on two entangled photons, the angle between the polarization directions is $\pi/8$, and so the probability that they get the same outcome, both horizontally polarized or both vertically polarized in the directions in which the polarizations are measured, is $\cos^2(\frac{\pi}{8}) \approx .85$, and the probability that they get different outcomes is $\sin^2(\frac{\pi}{8}) \approx .15$. For a measurement of

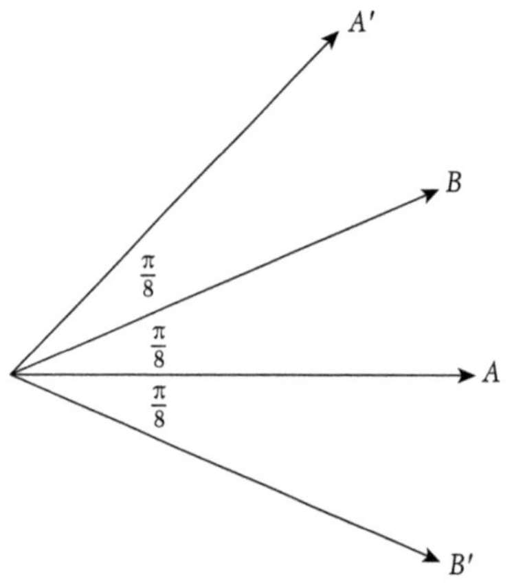


**Figure 3.6** The polarization directions $A, A'$ and $B, B'$ for the optimal quantum simulation of Popescu–Rohrlich bananas.

the polarization observables $A'$, $B'$, though, the angle between the polarization directions is $3\pi/8$, and so the probability that they get the same outcome is $\cos^2(\frac{3\pi}{8}) = \sin^2(\frac{\pi}{8})$ and the probability that they get different outcomes is $\sin^2(\frac{3\pi}{8}) = \cos^2(\frac{\pi}{8})$. (If your trigonometry is a little rusty, see the summary of useful trigonometric relations in the subsection "Some Useful Trigonometry" in the "More" section at the end of the chapter.)

If the possible values of the polarization observables $A$ and $A'$ for Alice's photons are represented as 1 for horizontal polarization in the directions of the polarization measurements, and –1 for vertical polarization in the directions orthogonal to the polarization measurements, and similarly for the polarization observables $B$ and $B'$ for Bob's photons, then the expectation value of $AB$ in the state $|\phi^+\rangle$ is:

$$\begin{aligned}\langle AB\rangle_{|\phi^+\rangle} &= p(\text{outcomes same}\,|\,A,B) - p(\text{outcomes different}\,|\,A,B)\\ &= \cos^2\left(\frac{\pi}{8}\right) - \sin^2\left(\frac{\pi}{8}\right)\\ &= \cos\left(2\cdot\frac{\pi}{8}\right)\\ &= \cos\frac{\pi}{4} = \frac{1}{\sqrt{2}},\end{aligned}$$

and similarly for $\langle AB'\rangle_{|\phi^+\rangle}$ and $\langle A'B\rangle_{|\phi^+\rangle}$. But $\langle A'B'\rangle_{|\phi^+\rangle} = \sin^2(\frac{\pi}{8}) - \cos^2(\frac{\pi}{8}) = -\cos\frac{\pi}{4} = -\frac{1}{\sqrt{2}}$.

The Clauser–Horne–Shimony–Holt quantity $K$ for these polarization measurements is therefore:

$$\begin{aligned}K &= \langle AB\rangle_{|\phi^+\rangle} + \langle AB'\rangle_{|\phi^+\rangle} + \langle A'B\rangle_{|\phi^+\rangle} - \langle A'B'\rangle_{|\phi^+\rangle}\\ &= 4\cdot\frac{1}{\sqrt{2}} = 2\sqrt{2}.\end{aligned}$$

So with this strategy, exploiting polarization measurements on many pairs of photons in the maximally entangled state $|\phi^+\rangle$, Alice and Bob can successfully simulate the Popescu–Rohrlich correlation with probability $\frac{1}{2}(1+\frac{K}{4}) = \frac{1}{2}(1+\frac{2\sqrt{2}}{4}) \approx .85$. This turns out to be the optimal quantum probability, and $2\sqrt{2}$ is the maximum quantum value of $K$, the Tsirelson bound.

## 3.4 LOOPHOLES

There are two possible loopholes to worry about in an experiment designed to test Bell's inequality: a locality loophole and a detection loophole. If the time interval between measurements on the two entangled photons isn't short enough to prevent a signal from transmitting information between the photons, there's a locality loophole. The first attempt to close the locality loophole was an experiment by Aspect, Dalibard, and Roger in 1982.[18] They used an ingenious setup where the directions in which the polarizations of the two photons are measured changes rapidly while the photons are moving towards the analyzers, so information about the polarization observable measured on one photon, and the outcome of the measurement, would have to be transmitted faster than light to reach the remote system by the time that system's polarization is measured.

To get a feel for the detection loophole, suppose that in simulating the correlation of Popescu–Rohrlich bananas Alice is allowed to reject half her prompts and ask for a new prompt. Then Alice and Bob could successfully simulate the correlation with local resources. They begin by sharing two random variables, $\lambda$ and $\mu$, which can each be 0 or 1. For half the rounds, on average, $\mu = 0$. The strategy is for Alice to accept the prompt $T$ for these rounds, but to reject the prompt $S$, in which case the round is aborted. Then for all the accepted rounds of the simulation when $\mu = 0$, the prompts are either $TS$ or $TT$. Alice responds with the value $\lambda$. Bob responds with $\lambda$ if his prompt is $S$, but if his prompt is $T$, he flips the value of $\lambda$ and responds with 1 if $\lambda = 0$ and 0 if $\lambda = 1$. So Alice and Bob agree for the prompts $TS$ and disagree for the prompts $TT$. When $\mu = 1$, the strategy is for Alice to reject the prompt $T$. Then for all accepted rounds of the simulation with $\mu = 1$, the prompts are $SS$ or $ST$. For these rounds, Alice and Bob both respond with the value of $\lambda$. So they agree for the prompts $SS$ and $ST$. Taking the two cases together, Alice and Bob agree for the prompts $SS, ST, TS$, and disagree for the prompt $TT$, as required.

The point here is that if a photon counter or other detection device simply fails to register an outcome in half the cases in an experimental test of Bell's inequality—corresponding to Alice rejecting her prompt and aborting the round—the correlation of entangled photons could be derived from a theory in which the correlated photons share hidden variables. In other words, a common cause wouldn't be ruled out. Of course, a 50% efficiency is pretty bad, but it raises the question about how efficient measuring instruments need to be to detect a violation of Bell's inequality. Nicolas Gisin and Bernard Gisin have shown that if Alice and Bob are each allowed to reject a prompt in $1/3$ of the rounds, corresponding to detectors with an efficiency of as much as $2/3$, they could simulate the Popescu–Rohrlich correlation perfectly.[19] Good photon detectors are better than this and can achieve an efficiency of more than 90%.

There's considerable literature on loophole-free tests of Bell's inequality, involving the question of detector efficiency as well as the problem of ensuring that the entangled quantum systems are spacelike separated at the times of their respective measurements, so that any signal carrying information between them would have to violate the relativistic constraint on traveling faster than the speed of light. There are experiments that close the locality loophole and experiments that close the detection loophole. An ingenious loophole-free 2015 experiment performed by Hensen *et al*[20] with entangled electrons separated by more than a kilometer resolves the technical difficulties of closing both loopholes in the same experiment.

# 3.5 MORE

### 3.5.1 Some Useful Trigonometry

For those who have forgotten their trigonometry, here are some trigonometric relations that are used at various places in the book. I'll represent angles as radians rather than degrees: $\pi = 180°, \pi/2 = 90°, \pi/4 = 45°, \pi/8 = 22.5°$.

To begin:

$$\cos(x \pm y) = \cos x \cos y \mp \sin x \sin y.$$

So

$$\cos 2\theta = \cos^2\theta - \sin^2\theta = 1 - 2\sin^2\theta = 2\cos^2\theta - 1,$$

since $\cos^2\theta + \sin^2\theta = 1$.

Here's another way of putting these expressions, in terms of the relation between the angle and the half angle, rather than the relation between twice the angle and the angle:

$$1 - \cos\theta = 2\sin^2\frac{\theta}{2},$$
$$1 + \cos\theta = 2\cos^2\frac{\theta}{2}.$$

The expression for $\cos(x + y)$ as $\cos x \cos y - \sin x \sin y$ is useful for deriving relations like $\cos(\theta + \frac{\pi}{2}) = \cos\theta\cos\frac{\pi}{2} - \sin\theta\sin\frac{\pi}{2} = -\sin\theta$, because $\cos\frac{\pi}{2} = 0$ and $\sin\frac{\pi}{2} = 1$. From the expression for $\cos(x - y)$ as $\cos x \cos y + \sin x \sin y$ you can derive

$$\cos\frac{3\pi}{8} = \cos\left(\frac{\pi}{2} - \frac{\pi}{8}\right) = \cos\frac{\pi}{2}\cos\frac{\pi}{8} + \sin\frac{\pi}{2}\sin\frac{\pi}{8} = \sin\frac{\pi}{8}.$$

Also,

$$\sin(x \pm y) = \sin x \cos y \pm \cos x \sin y$$

so $\sin 2\theta = 2\sin\theta\cos\theta$ and

$$\sin\left(\frac{3\pi}{8}\right) = \sin\left(\frac{\pi}{2} - \frac{\pi}{8}\right) = \cos\frac{\pi}{8}$$

Finally, $\cos\frac{\pi}{4} = \sin\frac{\pi}{4} = \frac{1}{\sqrt{2}}$ and

$$\cos^2\frac{\pi}{8} = \frac{2 + \sqrt{2}}{4},$$
$$\sin^2\frac{\pi}{8} = \frac{2 - \sqrt{2}}{4}.$$

### 3.5.2 Correlations

A remarkable feature of quantum mechanics is that there are nonlocal quantum correlations associated with entangled quantum states that are inconsistent with any explanation by a direct causal connection between the events or by a common cause.

Correlations that have a common cause are *conditionally statistically independent* with respect to the common cause. I'll show below that conditional statistical independence is equivalent to two conditions, sometimes called "outcome independence" and "parameter independence."

First a brief review of some elementary background on conditional probability to clarify terminology and notation. Consider again the playing card example. A card is randomly selected from a deck of cards and cut in half. The two half-cards are placed in envelopes and mailed to different addresses. Opening the envelopes after their arrival will reveal perfectly correlated colors, but nothing travels between the two addresses as the cause of the correlation. The common cause in this case is a variable associated with the card from which the two half cards were cut, with values, black or red, that fix the color of the half-cards. In the general case, a common cause could be probabilistic and fix only the probabilities of the states of the correlated systems, with probabilities between 0 and 1.

Call the color observables of the half-cards mailed to different addresses $A$ and $B$. I'll represent particular colors, values of $A$ and $B$, by $a$ and $b$. The probability that a half-card at the first address has a particular color, $p(a)$, where $a$ is "red" or "black," is $1/2$. But the conditional probability, $p(a \,|\, b)$, that a half-card at the first address has a particular color $a$, given that the half-card at the second address has a particular color $b$, is 1 if $a = b$ and 0 if $a \neq b$, because the two half-cards always have the same color. The vertical bar "$|$" indicates a conditional probability: in this case, the probability of the event $a$ to the left of the slash, given the condition indicated by the event $b$ to the right of the slash.

The joint probability $p(a, b)$ that $A$ has the value $a$ and $B$ has the value $b$, and the conditional probability, are related in the following way:

$$p(a, b) = p(a \,|\, b)p(b) = p(b \,|\, a)p(a),$$

where $p(a \,|\, b)$ is the conditional probability that $A$ has the value $a$, given that $B$ has the value $b$, and $p(b \,|\, a)$ is the conditional probability that $B$ has the value $b$, given that $A$ has the value $a$.

The colors are *uncorrelated* or *statistically independent* if the color of a half-card is irrelevant to the color of the half card at the remote address—if $p(a \,|\, b) = p(a)$ or $p(b \,|\, a) = p(b)$. These conditions are equivalent: one follows from the other because of their relation to the joint probability. Equivalently, the colors are uncorrelated if $p(a, b) = p(a)p(b)$. This follows if you replace $p(a \,|\, b)$ by $p(a)$ or $p(b \,|\, a)$ by $p(b)$ in the expression for the joint probability in terms of the conditional probability.

The colors are *correlated* or *statistically dependent* if $p(a, b) \neq p(a)p(b)$—if the joint probability of a pair of colors, $a$ and $b$, is not equal to the product of the probabilities of the colors at the two locations.

In the scenario considered, where a card is cut in half and the two half-cards are mailed to different addresses, the colors of the half-cards are correlated according to this definition. Labeling the two addresses as 1 and 2, $p(\text{red}, \text{red}) = 1/2$ and $p_1(\text{red})p_2(\text{red}) = 1/4$, because $p_1(\text{red}) = p_2(\text{red}) = 1/2$, so $p(\text{red}, \text{red}) \neq p_1(\text{red})p_2(\text{red})$, and similarly, $p(\text{black}, \text{black}) \neq p_1(\text{black})p_2(\text{black})$. Also, $p(\text{red}, \text{black}) \neq p_1(\text{red})p_2(\text{black})$ because $p(\text{red}, \text{black}) = 0$ and $p_1(\text{red})p_2(\text{black}) = 1/4$. So $p(a, b) \neq p(a)p(b)$, since $p(a, b)$ is either $1/2$ or 0, and $p(a)p(b) = 1/4$.

Now for conditional statistical independence. To leave open the possibility that observables might be indefinite before they are measured, $p(a, b)$ could written explicitly as $p(a, b \mid A, B)$, the conditional probability that the values of the two observables are $a$ and $b$, given that the observables $A$ and $B$ are measured at the separate locations. This notation is redundant for the playing card example, since cards have a definite color even before opening the envelope at either address. In the case of Bananaworld correlations the notation isn't redundant, because a banana only has a definite taste after it's peeled a certain way, which corresponds to measuring an observable and recording a definite outcome. Conditional statistical independence is expressed formally as

$$p(a, b \mid A, B, \lambda) = p(a \mid A, \lambda)p(b \mid B, \lambda).$$

Given the common cause $\lambda$, peelings and tastes are uncorrelated: the joint probability of a pair of tastes for the pair of bananas, conditional on the common cause $\lambda$ and a pair of peelings, $p(a, b \mid A, B, \lambda)$, is equal to the product of the conditional probabilities for the separate bananas. Here $A$ and $B$ denote Alice's peeling and Bob's peeling, respectively, each $S$ or $T$, and $a$ and $b$ denote the respective tastes of Alice's banana and Bob's banana, ordinary (0) or intense (1). The joint conditional probability is said to be *factorizable*. This is Bell's locality condition.

Averaging over $\lambda$—adding the terms weighted by the $\lambda$-probabilities—produces the correlations for the averaged probabilities. So the joint probability—the observed or operational or "surface" joint probability (as opposed to the "hidden" conditional $\lambda$ probability)—is not equal to the product of the marginal probabilities for Alice and Bob separately:

$$p(a, b \mid A, B) \neq p(a \mid A)p(b \mid B).$$

Conditional statistical independence, the mark of a common cause, is equivalent to two conditions on the probabilities:[21]

- *outcome independence*: the probability that a banana tastes ordinary or intense, given $\lambda$ and a particular peeling ($S$ or $T$), is independent of the taste of the paired banana after it is peeled;

- *parameter independence*: the probability that a banana tastes ordinary or intense, given $\lambda$ and a particular peeling ($S$ or $T$) is independent of how you peel the paired banana.

The point of the distinction is that the taste of a peeled banana (the outcome of a particular peeling) is not under Alice's or Bob's control, but we suppose that how Alice or Bob peel their bananas is something they can choose freely (the "parameter" choice can be $S$ or $T$). See the next chapter, "Really Random," for more on this "free choice" assumption.

What's the difference between parameter independence and the no-signaling principle? Parameter independence is the condition "no signaling, given $\lambda$." What's excluded is the possibility of signaling by exploiting access to $\lambda$. Parameter independence refers explicitly to $\lambda$, the common cause. The no-signaling principle is an observational or operational condition characterizing the surface phenomenon—it doesn't refer to $\lambda$. If there is a common cause $\lambda$, the no-signaling principle is "no signaling, conditional on $\lambda$, averaged over $\lambda$."

I'll write $p_\lambda^{AB}(a,b)$ for the joint probability $p(a,b\,|\,A,B,\lambda)$, and $p_\lambda^{AB}(a\,|\,b)$ for the conditional probability $p(a\,|\,b,A,B,\lambda)$. Since $A$, $B$, and $\lambda$ appear throughout, it's easier to see what's going on if these symbols are represented as subscripts and superscripts in this way, rather then putting them on the right-hand side of the conditional symbol "$|$." Then outcome independence is the condition

$$p_\lambda^{AB}(a\,|\,b) = p_\lambda^{AB}(a)$$
$$p_\lambda^{AB}(b\,|\,a) = p^{AB}(b),$$

and parameter independence is the condition

$$p_\lambda^{AB}(a) = p_\lambda^{A}(a)$$
$$p_\lambda^{AB}(b) = p_\lambda^{B}(b).$$

The "parameter" here is the remote observable.

To derive conditional statistical independence from parameter independence and outcome independence, begin by expressing the joint probability $p_\lambda^{AB}(a,b)$ in terms of the conditional probability $p_\lambda^{AB}(a\,|\,b)$ as

$$p_\lambda^{AB}(a,b) = p_\lambda^{AB}(a\,|\,b)p_\lambda^{AB}(b).$$

(Don't be thrown off by the superscripts and subscripts here. Think of the superscripts and subscripts in $p_\lambda^{AB}(a\,|\,b)$ and $p_\lambda^{AB}(b)$ as defining new probabilities $p'(a\,|\,b)$ and $p'(b)$. Then the equation reads: $p'(a,b) = p'(a\,|\,b)p'(b)$, which is just the definition of conditional probability.) By outcome independence, $p_\lambda^{AB}(a,b) = p_\lambda^{AB}(a)p_\lambda^{AB}(b)$, and by parameter independence, $p_\lambda^{AB}(a,b) = p_\lambda^{A}(a)p_\lambda^{B}(b)$, which is conditional statistical independence (expressed above as $p(a,b\,|\,A,B,\lambda) = p(a\,|\,A,\lambda)p(b\,|\,B,\lambda)$).

To go the other way and derive parameter independence from conditional statistical independence, begin with conditional statistical independence:

$$p_\lambda^{AB}(a, b) = p_\lambda^{A}(a)p_\lambda^{A}(b).$$

In Bananaworld, the two values of $b$ are 0 and 1, so $p_\lambda^{AB}(a, 0) = p_\lambda^{A}(a)p_\lambda^{B}(0)$ and $p_\lambda^{AB}(a, 1) = p_\lambda^{A}(a)p_\lambda^{B}(1)$. Adding $p_\lambda^{AB}(a, 0)$ and $p_\lambda^{AB}(a, 1)$ gives the marginal probability $p_\lambda^{AB}(a)$, the sum of the probabilities of $a$ for all possible values of $b$. So $p_\lambda^{AB}(a) = p_\lambda^{A}(a)(p_\lambda^{B}(0) + p_\lambda^{B}(1))$. Since $p_\lambda^{B}(0) + p_\lambda^{B}(1) = 1$ (because 0 and 1 are the only two possibilities), it follows that

$$p_\lambda^{AB}(a) = p_\lambda^{A}(a),$$

which is parameter independence for the outcome $a$. Similarly, $p_\lambda^{AB}(b) = p_\lambda^{B}(b)$ follows by adding the two expressions $p_\lambda^{AB}(0, b) = p_\lambda^{A}(0)p_\lambda^{B}(b)$ and $p_\lambda^{AB}(1, b) = p_\lambda^{A}(1)p_\lambda^{B}(b)$ to get the marginal probability $p_\lambda^{AB}(b)$.

To derive outcome independence from conditional statistical independence, begin again with conditional statistical independence, $p_\lambda^{AB}(a, b) = p_\lambda^{A}(a)p_\lambda^{A}(b)$, and use the equalities just derived for parameter independence, $p_\lambda^{AB}(a) = p_\lambda^{A}(a)$ and $p_\lambda^{AB}(b) = p_\lambda^{B}(b)$, to write conditional statistical independence as

$$p_\lambda^{AB}(a, b) = p_\lambda^{AB}(a)p_\lambda^{AB}(b).$$

From the relation between joint probability and conditional probability, $p_\lambda^{AB}(a, b)$ can also be expressed as $p_\lambda^{AB}(a, b) = p_\lambda^{AB}(a)p_\lambda^{AB}(b \mid a)$. Taking these two expressions together gives

$$p_\lambda^{AB}(b \mid a) = p_\lambda^{AB}(b),$$

which is outcome independence for the outcome $b$ with respect to the outcome $a$. Similarly, $p_\lambda^{AB}(a, b)$ can be expressed as $p_\lambda^{AB}(a, b) = p_\lambda^{AB}(a \mid b)p_\lambda^{AB}(b)$ from the relation between joint probability and conditional probability, which gives $p_\lambda^{AB}(a \mid b) = p_\lambda^{AB}(a)$, outcome independence for the outcome $a$ with respect to the outcome $b$.

## The bottom line

- For correlations with a common cause $\lambda$, the joint probabilities are conditionally statistically independent with respect to $\lambda$, which is to say that the joint probabilities conditional on $\lambda$ factorize to a product of the marginal probabilities conditional on $\lambda$: $p(a, b \mid A, B, \lambda) = p(a \mid A, \lambda)p(b \mid B, \lambda)$.

*(continued)*

 **The bottom line** *(continued)*

- Conditional statistical independence is equivalent to two conditions on the probabilities. In Bananaworld, the two conditions are:

  *Outcome independence*: the probability that a banana tastes ordinary or intense, given $\lambda$ and a particular peeling ($S$ or $T$), is independent of the taste of the paired banana after it is peeled;

  *Parameter independence*: the probability that a banana tastes ordinary or intense, given $\lambda$ and a particular peeling ($S$ or $T$), is independent of how you peel the paired banana.
- The point of the distinction is that the taste of a peeled banana (the outcome of a particular peeling) is not under Alice's or Bob's control, but we suppose that how Alice and Bob peel their bananas is something they can can choose freely (the "parameter" choice can be $S$ or $T$).
- Parameter independence is the condition "no signaling, given $\lambda$." The no-signaling principle is an observational or operational condition characterizing the surface phenomenon—it doesn't refer to $\lambda$. If there is a common cause $\lambda$, the no-signaling principle is "no signaling, conditional on $\lambda$, averaged over $\lambda$."

### 3.5.3 Boolean Algebras

In everyday language and classical theories, we talk about objects as having properties that fit together in a certain way when we use words like "and," "or," and "not," and we assume that events fit together in a corresponding way. There's an implicit structure here that we take for granted without thinking about it.

Consider the playing card example again, and suppose the state of a card is defined by whether it's a club, spade, diamond, or heart, so the state space of a card is a set of four states: {club, spade, diamond, heart}. The symbol {...} denotes the set of elements represented by the entries between the braces. The properties "black" or "red" divide the state space into two subsets: black = {club, spade} and red = {diamond, heart}.

Suppose Alice and Bob play a game of cards where they each play a card at each round of the game. Since a round is associated with a pair of cards, the state space of the game is the Cartesian product of the four-element state space of Alice and the four-element state space of Bob—the set of ordered pairs, $S$, where the first member of each pair is a state in Alice's state space and the second member is a state in Bob's state space. There are $4 \times 4 = 16$ states in the state space $S$: {(club, club), (club, spade), (club, diamond), (club, heart), (spade, club), (spade, spade), ..., (heart, heart)}. Every event—that Alice plays a heart, or Bob plays a black card—is represented by a subset of the state space. For example, the event "Alice plays a heart" is represented by the subset {(heart, club), (heart, spade), (heart, diamond), (heart, heart)}.

A classical disjunction ("or," in the inclusive sense: "this or that, or both"), say "Alice played a heart or a red card," is represented by the set-theoretical *union* of the subsets associated with "Alice played a heart" and "Alice played a red card." The union of two subsets $S_1$ and $S_2$ is the subset, denoted by $S_1 \cup S_2$, consisting of all the elements (ordered pairs of states) that are in either $S_1$ or $S_2$. In this case the union is just the eight-element subset associated with "Alice played a red card," since the "Alice played a heart" subset is included in the "Alice played a red card" subset: {(diamond, club), (diamond, spade), (diamond, diamond), (diamond, heart), (heart, club), (heart, spade), (heart, diamond), (heart, heart)}. This eight-element subset of the state space is the set-theoretical *complement* of the subset associated with "Alice played a black card." The complement of a subset $S$ is the subset $S'$ of elements not in $S$, and every card that is not a red card is a black card.

A classical conjunction ("and"), say "Alice played a club and Bob played a red card," is represented by the set-theoretical *intersection* of the subsets associated with "Alice played a club" and "Bob played a red card." The intersection of two subsets $S_1$ and $S_2$ is the subset, denoted by $S_1 \cap S_2$, consisting of all the elements that are in both $S_1$ and $S_2$. In this case the intersection is the two-element subset {(club, diamond), (club, heart)}.

If the proposition "Alice played a club" is true, then it follows that the proposition "Alice played a club or a red card" is true. This logical relation is represented by the inclusion of the "Alice played a club" subset $T$ in the "Alice played a club or a red card" subset $U$, denoted by $T \subseteq U$.

These set-theoretical structural relations were formalized by George Boole in the mid-19th century as an algebraic calculus, now called a Boolean algebra.[22] Here's the idea. The union of two subsets, $S_1$ and $S_2$, is the smallest subset containing both $S_1$ and $S_2$. So in the collection of subsets of the state space $S$, it's the least upper bound of $S_1$ and $S_2$ with respect to the ordering defined by set inclusion $\subseteq$. It's an upper bound of $S_1$ and $S_2$ because $S_1 \subseteq (S_1 \cup S_2)$ and $S_2 \subseteq (S_1 \cup S_2)$. It's the *least* upper bound because $S_1 \cup S_2$ is the smallest subset that includes both $S_1$ and $S_2$. Similarly, since the intersection of $S_1$ and $S_2$ is the largest subset contained in both $S_1$ and $S_2$, it's the greatest lower bound of $S_1$ and $S_2$ with respect to the ordering defined by set inclusion. It's a lower bound of $S_1$ and $S_2$ because $(S_1 \cap S_2) \subseteq S_1$ and $(S_1 \cap S_2) \subseteq S_2$. Since $S_1 \cap S_2$ is the largest subset that is included in both $S_1$ and $S_2$, it's the *greatest* lower bound.

A Boolean algebra $B$, defined abstractly, is a set of elements $a, b, \ldots$ with a partial ordering relation $\leq$ that holds for some elements of the algebra, $x \leq y$, just as the relation $\subseteq$ holds for some subsets of the set $S$. A least upper bound, denoted by $x \vee y$, and a greatest lower bound, denoted by $x \wedge y$, can be defined for any two elements in the algebra with respect to this partial ordering, just as a least upper bound and a greatest lower bound was defined for subsets of the set $S$ with respect to the partial ordering defined by set inclusion $\subseteq$. So now you have a structure with two binary operations $\vee$ and $\wedge$. To complete the algebra you need a complement, $x'$, for every element $x$ in the algebra (corresponding to the complement of a subset in the state space $S$), a maximum element with respect to the ordering, represented by $\mathbf{1}$ (corresponding to the entire state space $S$), and a minimum element with respect to the ordering, represented by **0** (corresponding to the empty set): $x \leq \mathbf{1}$ and $\mathbf{0} \leq x$ for every element in the algebra.

The Boolean algebra corresponding to the four-element state space {club, spade, diamond, heart} of a playing card is a 16-element Boolean algebra. The minimal non-**0** elements in a Boolean algebra are referred to as Boolean atoms. These are the one-element subsets {club}, {spade}, {diamond}, {heart}. They are minimal elements in the algebra, because there are no elements between the one-element subsets and the empty set or **0** element. Apart from the four atoms, there are the two minimum and maximum elements **0** and **1**, the four complements of the atoms (called co-atoms) corresponding to the subsets representing not-club ({spade, diamond, heart}), not-spade ({club, diamond, heart}), not-diamond ({club, spade, heart}), not-heart ({club, spade, diamond}), and the six elements corresponding to the subsets {club, spade}, {club, diamond}, {club, heart}, {spade, diamond}, {spade, heart}, {diamond, heart}.

The operations $\wedge$ and $\vee$ are associative: $(a \wedge b) \wedge c = a \wedge (b \wedge c)$, and similarly for $\vee$, so you can drop the parentheses without ambiguity. If the atoms corresponding to club, spade, diamond, heart are denoted by $c, s, d, h$, the co-atoms are represented by $c' = s \vee d \vee h, s' = c \vee d \vee h, d' = c \vee s \vee h, h' = c \vee s \vee d$. If there are $n$ atoms in a Boolean algebra, the number of elements in the algebra is $2^n$, so the Boolean algebra corresponding to the four-atom state space of a card has $2^4 = 16$ elements, and the 16-atom state space of the card game has $2^{16} = 65,536$ elements.

A Boolean algebra defined in this way as a partially ordered structure is referred to as a Boolean lattice. There's a standard way of representing a Boolean lattice called a "Hasse diagram," where points represent the Boolean elements and lines represent the partial ordering. If two points are connected by a line, the Boolean element represented by the top point is above the element represented by the lower point in the ordering. To illustrate, the Boolean algebra with four atoms is represented as the Hasse diagram in Figure 3.7.

Classical logic has the structure of a Boolean algebra, and every Boolean algebra is isomorphic to an algebra of subsets of some set. Formally, this means that the semantics of classical logic is set-theoretic: we interpret classical propositions as subsets of a set. A classical proposition that says that a system has a certain property is true if and only if the state of the system is in the set of states associated with the property. The dynamics of a theory describes how the state evolves in time and so describes how a system's properties change in time. Classical probability theory assumes that the underlying property structure over which probabilities are defined is a Boolean algebra.

This might all seem rather abstract, but here's the connection with physics: if the observables of a theory all commute, the properties, represented by "yes–no" observables, form a Boolean algebra. The observables in classical Newtonian mechanics all commute, so the properties of classical systems form a Boolean algebra. The state of a classical particle is specified by the position and momentum of the particle, represented by a point in the six-dimensional state space or "phase space" of the particle, with three coordinates for position in the $x$, $y$, and $z$ directions, and three corresponding momentum coordinates.

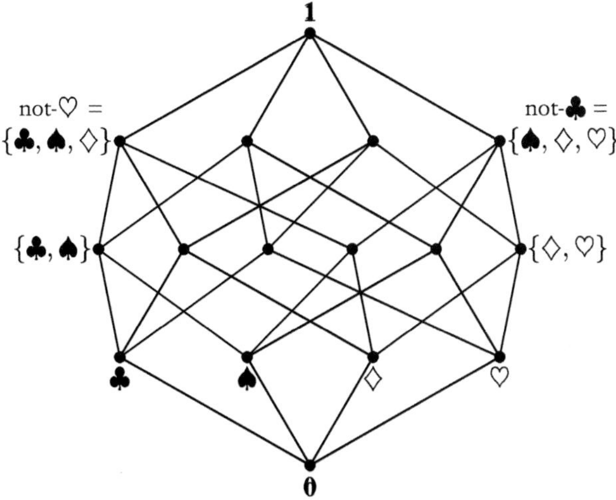


**Figure 3.7** Hasse diagram of a Boolean lattice with four atoms, ♣, ♠, ◇, ♡. The two elements on the top line without labels are (in order, from left to right): not-◇ = {♣, ♠, ♡} and not-♠ = {♣, ◇, ♡}. The four elements without labels on the line below that are (in order, from left to right): {♣, ◇}, {♣, ♡}, {♠, ◇}, {♠, ♡}.

The dynamics of classical mechanics describes the trajectory of the particle in phase space, and so how the position and momentum change over time. Other quantities, like energy, angular momentum, and so on, are functions of position and momentum. So specifying that the energy lies in a certain range amounts to specifying the subset of the state space containing all the states in which the energy, as a function of the state, lies in this range, just as specifying the color of a card as red amounts to specifying the subset of card states, diamonds and hearts, that have the property red. Position and momentum take a continuous range of values, but you could consider just two possible positions for a particle, whether it is in a certain spatial region or not, and two possible momenta, whether the momentum is in a certain range or not. Then the state space is a four-element space, and the structure of properties is the same as the Boolean algebra represented in Figure 3.7. Taking finer subdivisions for position and momentum, or many particles, increases the number of elements in the state space and the number of elements in the associated Boolean algebra, but it doesn't change any structural feature. Going from a finite or denumerable set of values to a continuum of values introduces technical issues involving infinities, but again this doesn't change the Boolean character of the underlying structure.

For a playing card as an entity in a game, the most precise specification of the state of a card is the suit and the value of the card: a two of hearts, or an ace of clubs. Any other features are irrelevant. You could consider an even more precise specification, given by the positions and momenta of all the particles in a card, but this would be massively redundant if the only distinctions are those relevant to the game. The notion of a "system" and its associated state space depends in part on the set of properties you want to include in a theory. What the playing card example has in common with classical mechanics is that a state corresponds to a complete list of all the properties of a system. Since each property is associated with a subset of the state space of the system, whether the system is a playing card or a classical particle, specifying all the properties of a system at a particular time amounts to specifying all the subsets containing the state, which is equivalent to specifying the state.

A measurement in a classical theory is then simply a procedure for finding out whether or not a system has a certain property, and since the state is associated with a list of all the properties of the system, nothing prevents you from finding out as many properties of a system as you like, in principle, and so fixing the classical state to an arbitrary degree of accuracy. Even if a measurement procedure involves a dynamical interaction that correlates a property of a system with the pointer value of a measuring instrument and the interaction disturbs the state of the system, you could reverse the measurement interaction and restore the original state, since the dynamics of classical mechanics is reversible, and so build up the list by further measurements.

The point here is that quantum observables don't all commute, and quantum properties don't have the structure of a Boolean algebra. The way in which quantum probabilities and quantum correlations differ from classical probabilities and classical correlations arises because of the specific non-Boolean character of the underlying quantum property structure. For example, a classical disjunction is true if and only if one or both disjuncts is true (one or both of the propositions on either side of the "or"), but this is not the case for a quantum disjunction. A quantum disjunction can be true without either of the disjuncts having a well-defined truth value, and this difference between the classical and quantum case shows up in the enhanced computational power of quantum computers relative to classical computers. (See Section 8.1, "Quantum Computation," in Chapter 8, "Quantum Feats.") The structure of quantum properties is not even embeddable into a Boolean algebra—there is no one-to-one structure-preserving map from the quantum property structure into a Boolean algebra, which is another way of saying that the probabilities of quantum mechanics do not arise because something has been left out of a Boolean or classical story. Instead of a Boolean algebra, what you have is a collection of Boolean algebras, each corresponding to a mutually commuting set of "yes–no" observables. Observables in different Boolean algebras generally don't commute, but the Boolean algebras are "entwined" in a particular way, so that an observable can belong to more than one Boolean algebra. What's meant by a "measurement" is then not the same thing as a measurement in a classical theory, because there is no state in the sense corresponding to a classical list of properties in a non-Boolean theory.

The non-embeddability result is known as the Kochen–Specker theorem. Simon Kochen and Ernst Specker,[23] and Bell in an independent argument,[24] proved that no "noncontextual" hidden variable theory, in which the observables all have definite values prior to measurement fixed by the hidden variables, can reproduce the quantum probabilities of a qutrit or any quantum system more complex than a qubit. (A context is defined by a Boolean algebra of mutually commuting "yes–no" observables. In a *contextual* theory, the value of an observable can be different for each of the Boolean contexts to which it belongs.) This is equivalent to the non-embeddability of the quantum property structure into a Boolean algebra. See Sections 6.2 and 6.3, "The Aravind–Mermin Magic Pentagram" and "The Kochen–Specker Theorem and Klyachko Bananas," in the chapter "Quantum Magic" for more on contextuality in quantum mechanics.

## The bottom line

- If the observables of a theory all commute, as in classical Newtonian physics, the properties, represented by "yes–no" observables, form a Boolean algebra. Classical logic has the structure of a Boolean algebra, and classical probability theory assumes that the underlying property structure over which probabilities are defined is a Boolean algebra.
- A measurement in a classical theory is a procedure for finding out whether or not a system has a certain property, and since the state is associated with a list of all the properties of the system, nothing in principle prevents you from finding out as many properties of a system as you like, and so fixing the classical state to an arbitrary degree of accuracy.
- Quantum properties don't have the structure of a Boolean algebra, and the way in which quantum probabilities and quantum correlations differ from classical probabilities and classical correlations arises because of the specific non-Boolean character of the underlying quantum property structure.
- Instead of a Boolean algebra, what you have is a collection of Boolean algebras, each corresponding to a mutually commuting set of "yes–no" observables. Observables in different Boolean algebras generally don't commute, but the Boolean algebras are "entwined" in a particular way, so that an observable can belong to more than one Boolean algebra.
- Simon Kochen and Ernst Specker, and Bell in an independent argument, proved that no "noncontextual" hidden variable theory, in which the observables all have definite values prior to measurement fixed by the hidden variables, can reproduce the quantum probabilities of a qutrit or any quantum system more complex than a qubit. (A context is defined by a Boolean algebra of mutually commuting "yes–no" observables. In a *contextual* theory, the value of an observable can be different for each of the Boolean contexts to which it belongs.) This result, known as the Kochen–Specker theorem, is equivalent to the impossibility of embedding the quantum property structure into a Boolean algebra.

## Notes

1. The Einstein–Podolsky–Rosen paper: Albert Einstein, Boris Podolsky, and Nathan Rosen, "Can quantum-mechanical description of reality be considered complete?" *Physical Review* 47, 777–780 (1935).
2. Niels Bohr's reply: "Can quantum-mechanical description of physical reality be considered complete?" *Physical Review* 48, 696–702 (1935).
3. Richard Feynman's quote "It has not yet become obvious to me that there's no real problem ..." is from his paper "Simulating physics with computers," *International Journal of Theoretical Physics* 21, 467 (1982). The quote is on p. 471.
4. David Bohm's hidden variable theory was published as a two-part paper: "A suggested interpretation quantum theory in terms of 'hidden' variables," I and II, *Physical Review* 85, 166–179, 180–193 (1952). See Sheldon Goldstein's review article "Bohmian mechanics" in *The Stanford Encyclopedia of Philosophy* (Spring 2013 edition), Edward N. Zalta (ed.). Bohm's book on quantum theory is *Quantum Theory* (Prentice Hall, Englewood Cliffs, NJ, 1951).
5. The Bohm–Bub hidden variable theory: David Bohm and Jeffrey Bub, "A proposed solution of the measurement problem in quantum mechanics by a hidden variable theory," *Reviews of Modern Physics* 38, 453–469 (1966). Abner Shimony comments on the Bohm–Bub hidden variable theory in an article "The status of hidden-variable theories" in P. Suppes, L. Henkin, A. Joja, and G. C. Moisil (eds.), *Logic, Methodology and Philosophy of Science IV*, Proceedings of the Fourth International Congress for Logic, Methodology and Philosophy of Science, Bucharest, 1971 (North-Holland Publishing Company, Amsterdam, 1973), p. 599:

> One elegant family of nonlocal hidden variable theories has been investigated in some detail by Bohm and Bub (1966) (their interest in this family having probably arisen more from its amenability to mathematical analysis than from physical heuristics). This family has the property of agreeing statistically with quantum mechanics after the hidden variables become "randomized." However, immediately after a measurement is performed the hidden variables are in a kind of non-equilibrium distribution, and hence new measurements performed before a certain relaxation time has elapsed can be expected to disagree statistically with quantum mechanics. Bohm and Bub conjecture that this relaxation time is of the order of $h/kT$, where $T$ is the temperature of the apparatus. At room temperature the relaxation time would be about $10^{-13}$ seconds. Papaliolios ["Experimental test of a hidden-variable quantum theory," *Physical Review Letters* 18, 622–625 (1967)] ingeniously tested their conjecture by measuring the intensity of light passing through a stack of three extremely thin sheets of polaroid with varying orientations of their axes of polarization. The results were entirely in accordance with quantum mechanics, even though the transit time of light through each sheet of polaroid was about $7.5 \times 10^{-14}$ seconds. He remarks that "it is also possible to perform a more definitive test of Bohm and Bub's choice of $h/kT$ as the relaxation time, by repeating the experiment at lower temperatures. The lack of a theoretical understanding of this choice of $\tau$, however, does not at this time justify cooling the apparatus to liquid air (or lower) temperatures."

6. The original version of the Ghirardi–Rimini–Weber theory was published in GianCarlo C. Ghirardi, Alberto Rimini, and T. Weber, "Unified dynamics for microscopic and macroscopic

systems," *Physical Review* D34, 470–491 (1986). For later variants and extensions of the theory, see Ghirardi's review article "Collapse theories," in Edward N. Zalta (ed.), *The Stanford Encyclopedia of Philosophy* (Winter 2011 edition).

7. Einstein's comment about Bohm's theory being "too cheap for me" is in a letter to Max Born dated May 12, 1952. The letter is printed in Max Born, *The Born–Einstein Letters* (Walker and Co., London, 1971), p. 192.
8. Bell's critical review of "no go" results about the impossibility of hidden variable extensions of quantum mechanics satisfying various constraints: "On the problem of hidden variables in quantum mechanics," *Reviews of Modern Physics* 38, 447–452 (1966). Bell's comment, referring to Bohm's hidden variable theory, that the theory evades these no go results because the equations of motion are nonlocal, so that "an explicit causal mechanism exists whereby the disposition of one piece of apparatus affects the results obtained with a distant piece," is on p. 452. His further comment that "the Einstein–Podolsky–Rosen paradox is resolved in the way in which Einstein would have liked least," and his query whether "*any* hidden variable account of the quantum mechanics *must* have this extraordinary character" are both on p. 452. Bell's analysis concludes with the proposal:

   > It would therefore be interesting, perhaps, to pursue some further "impossibility proofs," replacing the arbitrary axioms objected to above by some condition of locality, or of separability of distant systems.

   His 1964 proof—Bell's theorem—was his response to this proposal. It was published two years prior to the long-delayed publication of the paper that raised the issue of locality as a constraint on any hidden variable extension of quantum mechanics.
9. Bell's theorem: J. S. Bell, "On the Einstein–Podolsky–Rosen paradox," *Physics* 1, 195–200 (1964). Bell's conclusion that a hidden variable theory reproducing the quantum statistics must introduce "a mechanism whereby the setting of one measuring device can influence the reading of another instrument, however remote" is on p. 199.
10. Einstein's article "Quantenmechanik und Wirklichkeit," appeared in *Dialectica* 2, 320–324 (1948). There is an English translation entitled "Quantum mechanics and reality" by Max Born in M. Born, *The Born–Einstein Letters* (Walker and Co., London, 1971), pp. 168–173. Einstein's remark, "Unless one makes this kind of assumption about the independence of the existence (the 'being-thus') of objects which are far apart from one another in space—which stems from everyday thinking—physical thinking in the familiar sense would not be possible," is on p. 170.
11. Einstein's comment in a letter to Max Born, "That which really exists in *B* should therefore not depend on what kind of measurement is carried out in part of space *A* ..." is on p. 164 of *The Born–Einstein Letters*. Born reproduces the comment from Einstein's marginal "caustic comments" (including "Ugh!" and "Blush, Born, Blush") to the last chapter, "Metaphysical Conclusions," of Born's book *The Natural Philosophy of Cause and Chance*, The Waynflete Lectures (Clarendon Press, Oxford, 1949).
12. Nicolas Gisin's characterization of "no signaling" as "no communication without transmission" is from his book *Quantum Chance: Nonlocality, Teleportation and Other Quantum Marvels* (Copernicus, Göttingen, 2014).
13. Sandu Popescu and Daniel Rohrlich introduced the nonlocal box now referred to as a PR box in their paper "Quantum nonlocality as an axiom," *Foundations of Physics* 24, 379 (1994). They pointed out that relativistic causality by itself does not rule out simulating a PR-box with a probability greater than 3 / 4. This was actually shown earlier by L. A. Khalfi and B. S. Tsirelson,

"Quantum and quasi-classical analogs of Bell inequalities," in P. Lahti and P. Mittelstaedt (eds.), *Symposium on the Foundations of Modern Physics* (World Scientific, Singapore, 1985), pp. 441–460, and independently by Peter Rastall, "Locality, Bell's theorem, and quantum mechanics," *Foundations of Physics* 15, 963–972 (1985). Other nonlocal boxes with interesting correlations include the millionaire box introduced by Andrew C. Yao in "Protocols for secure computations," *SFCS '82, Proceedings of the 23rd Annual Symposium on Foundations of Computer Science*, pp. 160–164 (1982). The millionaire box is a generalization of a PR box, where the inputs can take any value in the interval between 0 and 1, as well as 0 and 1 (which are the only two possible inputs in the case of a PR box). If Alice's input is less than or equal to Bob's input, the outputs of the box are the same. If Alice's input is greater than Bob's input, the outputs are different. So two millionaires, Alice and Bob, who are not equally rich, could use such a box to decide who is richer without revealing how much money they each have. They both input the amount of money they own (converted to decimals between 0 and 1). If the outputs are the same, Bob is richer. If the outputs are different, Alice is richer. Needless to say, although the correlation is logically possible, the box is an imaginary device.

14. Nicolas Gisin's proof of Bell's theorem is from *Quantum Chance: Nonlocality, Teleportation and Other Quantum Marvels* (Copernicus, Göttingen, 2014).
15. The Clauser–Horne–Shimony–Holt correlation is from John F. Clauser, Michael A. Horne, Abner Shimony, and Richard A. Holt, "Proposed experiment to test local hidden-variable theories," *Physical Review Letters* 23, 880–884 (1969).
16. Derivation of the Tsirelson bound: B. S. Cirel'son, "Quantum generalizations of Bell's inequality," *Letters in Mathematical Physics* 4, 93–100 (1980).
17. Feynman's comment about "squeezing the difficulty of quantum mechanics into a smaller and smaller place" is from Feynman's article "Simulating physics with computers," *International Journal of Theoretical Physics* 21, 467 (1982); p. 485.
18. The experiment in which the measurement settings are changed during the time of flight of the photons: Alain Aspect, Jean Dalibard, and Gérard Roger, "Experimental test of Bell's inequalities using time varying analyzers," *Physical Review Letters* 49, 1804–1807 (1982). There were earlier results by Alain Aspect, Philipe Grangier, and Gérard Roger: "Experimental tests of realistic local theories via Bell's theorem," *Physical Review Letters* 47, 460–467 (1982); "Experimental realization of EPR Gedankenexperiment: a new violation of Bell's inequalities," *Physical Review Letters* 49, 91–94 (1982).
19. The detection loophole: Nicolas Gisin and Bernard Gisin, "A local hidden variable model of quantum correlation exploiting the detection loophole," *Physics Letters A* 260, 323–327 (1999).
20. Bas Hensen *et al*, 'Experimental loophole-free violation of a Bell inequality using entangled electron spins separated by 1.3 km,' arXiv quant/ph1508.05949.
21. The terms "parameter independence" and "outcome independence" are due to Abner Shimony, from "Contextual hidden variables and Bell's inequalities," *British Journal for the Philosophy of Science* 3, 24–45 (1984).
22. George Boole: *An Investigation of the Laws of Thought: On which are Founded the Mathematical Theories of Logic and Probability* (Cambridge University Press, New York, 2009), first published in 1854.
23. The Kochen–Specker theorem: Simon Kochen and Ernst P. Specker, "On the problem of hidden variables in quantum mechanics," *Journal of Mathematics and Mechanics* 17, 59–87 (1967).
24. Bell's version of the Kochen–Specker theorem: John Stuart Bell, "On the problem of hidden variables in quantum mechanics," *Review of Modern Physics* 38, 447–452 (1966). Reprinted in John Stuart Bell, *Speakable and Unspeakable in Quantum Mechanics* (Cambridge University Press, Cambridge, 1989).

# 4

# Really Random

In this chapter, I show that the tastes of Popescu–Rohrlich bananas peeled from the stem end or the top end, and measurement outcomes on entangled qubits, are *intrinsically random* events. That turns out to be a feature of nonclassical correlations. To see this, it's illuminating to display correlations as arrays of joint probabilities. So the first section of this chapter is a bit of a technical digression on correlation arrays, which are really quite straightforward once you get the picture, and very useful for what follows. In the second section, "Really Random Bananas," I'll show that the taste of a Popescu–Rohrlich banana for a particular peeling is an intrinsically random event. Astonishingly, the *correlated* tastes of a pair of Popescu–Rohrlich bananas, no matter how far apart, arise from intrinsically random events. I'll explain what I mean by "intrinsically random" or "really random" in this section. The final section of the chapter, "Really Random Qubits," is about the intrinsic randomness of qubits and other quantum systems.

## 4.1 CORRELATION ARRAYS

For two possible peelings for Alice and two possible peelings for Bob, and two possible tastes for Alice and two possible tastes for Bob, there are 16 joint conditional probabilities in total. These can be represented as a correlation array, as in Table 4.1.

A word about notation here: instead of writing $p(0_A, 1_B \mid S_A, T_B)$ for the probability that Alice gets the outcome 0 and Bob gets the outcome 1 when Alice peels $S$ and Bob peels $T$, I leave out the subscripts and commas, unless they are necessary to avoid ambiguity, and write this expression as $P(01 \mid ST)$. The order—Alice first, Bob second—indicates

**Table 4.1** A correlation array of joint conditional probabilities.

| Bob \ Alice | | $S$ | | $T$ | |
|---|---|---|---|---|---|
| | | 0 | 1 | 0 | 1 |
| $S$ | 0 | $p(00 \mid SS)$ | $p(10 \mid SS)$ | $p(00 \mid TS)$ | $p(10 \mid TS)$ |
| | 1 | $p(01 \mid SS)$ | $p(11 \mid SS)$ | $p(01 \mid TS)$ | $p(11 \mid TS)$ |
| $T$ | 0 | $p(00 \mid ST)$ | $p(10 \mid ST)$ | $p(00 \mid TT)$ | $p(10 \mid TT)$ |
| | 1 | $p(01 \mid ST)$ | $p(11 \mid ST)$ | $p(01 \mid TT)$ | $p(11 \mid TT)$ |

whether the reference is to Alice or to Bob. So the first two slots in $p(--\,|\,--)$ before the conditionalization sign $|$ represent the two possible tastes for Alice's banana and Bob's banana, respectively, and the second two slots after the conditionalization sign represent the two possible peelings that Alice and Bob choose, respectively.

Take a look at the top left cell in Table 4.1. These four entries are the probabilities of the four possible combinations of tastes, given that Alice peels $S$ and Bob peels $S$, so they must sum to 1. The same applies to the other four cells, which display the probabilities for the four possible combinations of tastes for the peelings $ST$, $TS$, and $TT$. The $S$'s and $T$'s and 0's and 1's at the tops of the columns and to the left of the rows are there as guides to the coordinates of the entry in the appropriate position (the two peelings, in the order Alice first, Bob second, and the two tastes, in the same order). So, for the top left cell, the entry in the top left position of the cell is the $00\,|\,SS$ entry (corresponding to the two 0 tastes for the two $S$ peelings), the entry in the top right position is the $10\,|\,SS$ entry (corresponding to the taste 1 for Alice and 0 for Bob for the two $S$ peelings), and so on. The order is always: Alice first, Bob second, for the peelings and the tastes.

The marginal probability for Alice's banana tasting 0 when she and Bob both peel $S$, written $p(0_A\,|\,SS)$ (the subscript $A$ is added here to avoid ambiguity), is just the sum of the probabilities that Alice's banana tastes 0 for the two possible tastes of Bob's banana: $p(00\,|\,SS) + p(01\,|\,SS)$. This is the sum of the two entries in the left column of the $SS$ cell (the top left cell). The no-signaling principle says that it makes no difference to Alice's marginal probability whether Bob peels $S$ or $T$: $p(00\,|\,SS) + p(01\,|\,SS) = p(00\,|\,ST) + p(01\,|\,ST)$. I'll write this marginal probability as $p(0_A\,|\,S_A)$, dropping the reference to Bob's peeling, so

$$p(0_A\,|\,S_A) = p(0_A\,|\,SS) = p(0_A\,|\,ST),$$

where, by definition

$$p(0_A\,|\,SS) = p(00\,|\,SS) + p(01\,|\,SS)$$
$$p(0_A\,|\,ST) = p(00\,|\,ST) + p(01\,|\,ST).$$

This says that the sum of the two entries in the left column of the $SS$ cell should be the same as the sum of the two entries in the left column of the $ST$ cell (the bottom left cell).

Going through all four cells similarly for the remaining three marginal probabilities for Alice, $p(1_A\,|\,S_A)$, $p(0_A\,|\,T_A)$, and $p(1_A\,|\,T_A)$, and Bob's four marginal probabilities, no signaling requires that:

- the sum of the two entries in each of the *top four columns* (the four columns in the $SS$ and $TS$ cell, the two top cells) should be equal to the sum of the two entries in each of the corresponding *bottom four columns*, respectively (the four columns in the $ST$ and $TT$ cells, the bottom two cells);
- the sum of the two entries in each of the *left four rows* (the four rows in the $SS$ and $ST$ cells, the two left cells) should be equal to the sum of the two entries in each of the corresponding *right four rows*, respectively (the four rows in the $TS$ and $TT$ cells, the two right cells).

Don't worry if you have trouble keeping this straight. What's important is how to read a correlation array. The detailed descriptions are here just so you can check that you are reading a correlation array correctly.

There's a neat way of expressing the PR box correlation, or the correlation of a pair of Popescu–Rohrlich bananas, using addition modulo 2, or mod 2 for short (Boolean addition, or bit-wise addition), symbolized by "$\oplus$," and multiplication in the usual sense, symbolized by "$\cdot$." The addition operation "$\oplus$" applies to bits, 0 or 1, and yields 0 when you add 0 to 0, and 1 when you add 1 to 0 or 0 to 1, as you might expect, but it yields 0 when you add 1 to 1 (a multiple of 2 takes you back to 0, hence "addition mod 2"). The multiplication operation "$\cdot$" yields 0 when you multiply 0 by 0, or 0 by 1, or 1 by 0, and 1 when you multiply 1 by 1.

If Alice's and Bob's inputs to a PR box are $A$ and $B$, with possible values 0 and 1, and their outputs are $a$ and $b$, also with possible values 0 and 1, the PR box correlation is just:

$$a \oplus b = A \cdot B.$$

This is also the correlation between tastes $a, b$ and peelings $A, B$ for a pair of Popescu–Rohrlich bananas if the two possible peelings, from the stem end or the top end, are represented by 0 and 1.

To see this, notice that the equality holds when both sides of the equation are 0, or when both sides are 1. So consider the cases separately. Because the addition is mod 2, the left-hand side is 0 when the outputs, $a$ and $b$, are the same, either both 0 or both 1. The right-hand side is 0 when one of the inputs, $A$ or $B$, is 0, or when they are both 0. This case expresses the condition: the outputs are the same when one of the inputs is 0, or when both inputs are 0. The left-hand side is 1 when the outputs are different: one output is 0 and the other output is 1. The right-hand side is 1 when both inputs are 1. This case expresses the condition: the outputs are different when both inputs are 1. The two cases together express the PR box correlation: the outputs are same when the inputs are 00, 01, or 10, and the outputs are different when the inputs are 11.

Table 4.2 displays the Popescu–Rohrlich correlation defined by $a \oplus b = A \cdot B$ as a correlation array. If it's stipulated that the marginal probabilities are all 1/2, the probabilities satisfy the no-signaling principle: the probability that Alice's banana tastes 0 (or 1) when she peels $S$ is 1/2, regardless of whether Bob peels $S$ or $T$, and similarly when she peels $T$, and the same goes for Bob's banana. In Section 4.2, "Really Random Bananas," I'll show

**Table 4.2** The standard Popescu–Rohrlich correlation array.

| Alice / Bob | | $S$: 0 | $S$: 1 | $T$: 0 | $T$: 1 |
|---|---|---|---|---|---|
| $S$ | 0 | $p(00\,\vert\,SS) = 1/2$ | $p(10\,\vert\,SS) = 0$ | $p(00\,\vert\,TS) = 1/2$ | $p(10\,\vert\,TS) = 0$ |
| | 1 | $p(01\,\vert\,SS) = 0$ | $p(11\,\vert\,SS) = 1/2$ | $p(01\,\vert\,TS) = 0$ | $p(11\,\vert\,TS) = 1/2$ |
| $T$ | 0 | $p(00\,\vert\,ST) = 1/2$ | $p(10\,\vert\,ST) = 0$ | $p(00\,\vert\,TT) = 0$ | $p(10\,\vert\,TT) = 1/2$ |
| | 1 | $p(01\,\vert\,ST) = 0$ | $p(11\,\vert\,ST) = 1/2$ | $p(01\,\vert\,TT) = 1/2$ | $p(11\,\vert\,TT) = 0$ |

**Table 4.3** The standard Popescu–Rohrlich correlation array without the probability labels.

| Bob | Alice | S 0 | S 1 | T 0 | T 1 |
|---|---|---|---|---|---|
| $S$ | 0 | 1/2 | 0 | 1/2 | 0 |
| | 1 | 0 | 1/2 | 0 | 1/2 |
| $T$ | 0 | 1/2 | 0 | 0 | 1/2 |
| | 1 | 0 | 1/2 | 1/2 | 0 |

that if the correlations are required to satisfy the no-signaling principle, it follows that the marginal probabilities are all 1/2.

It's easier to read the correlation as the array in Table 4.3 without the conditional probability labels, which are redundant anyway since the same information is implicit in the $S, T, 0, 1$ coordinates at the tops of the rows and to the left of the columns. Having explained how to read a correlation array, I'll leave this information out in future. If you want to refresh your memory, you can always refer back to Table 4.1.

In the standard Popescu–Rohrlich correlation array in Table 4.3, three of the cells, for $SS$, $ST$, and $TS$, have the 1/2 entries along the northwest to southeast diagonal, and one cell, for $TT$, has the 1/2 entries along the southwest to northeast diagonal.

You get a similar Popescu–Rohrlich correlation array by switching "same" and "different" in the definition of the standard Popescu–Rohrlich correlation:

- if the peelings are $SS, ST, TS$, the tastes are different, 01 or 10;
- if the peelings are $TT$, the tastes are the same, 00 or 11;
- the marginal probabilities for the tastes 0 or 1 if a banana is peeled $S$ or $T$ are 1/2, irrespective of whether or not the paired banana is peeled (so the no-signaling constraint is satisfied).

This is equivalent to switching the directions of the diagonals where the 1/2 entries appear, as you can see by comparing Table 4.4 with Table 4.3. Expressed as an equation, the correlation in Table 4.4 is $a \oplus b \oplus 1 = A \cdot B$, rather than $a \oplus b = A \cdot B$ for the standard Popescu–Rohrlich correlation in Table 4.3.

**Table 4.4** The Popescu–Rohrlich correlation array for the correlation $a \oplus b \oplus 1 = A \cdot B$.

| Bob | Alice | S 0 | S 1 | T 0 | T 1 |
|---|---|---|---|---|---|
| $S$ | 0 | 0 | 1/2 | 0 | 1/2 |
| | 1 | 1/2 | 0 | 1/2 | 0 |
| $T$ | 0 | 0 | 1/2 | 1/2 | 0 |
| | 1 | 1/2 | 0 | 0 | 1/2 |

You get three more Popescu–Rohrlich correlation arrays by switching the entries in the *TT* cell with the entries in one of the other three cells in the standard Popescu–Rohrlich correlation array in Table 4.3, and three more again by switching the entries in the *TT* cell with the entries in one of the other three cells in the Popescu–Rohrlich correlation array in Table 4.4. That's a total of eight Popescu–Rohrlich correlation arrays.

 **The bottom line**

- I show how to display correlations as arrays of joint probabilities, which is useful for what follows.
- I show how to express a PR box correlation, or the correlation of a pair of Popescu–Rohrlich bananas, using addition modulo 2 (mod 2 for short) symbolized by "$\oplus$," and multiplication in the usual sense, symbolized by "$\cdot$." The correlation "the tastes are the same for peelings *SS*, *ST*, *TS*, but different for the peelings *TT*," can be expressed as $a \oplus b = A \cdot B$, where $a$ and $b$ stand for the taste of Alice's banana and the taste of Bob's banana (each represented by 0 or 1), and $A$ and $B$ stand for Alice's and Bob's peelings, also represented by 0 (for *S*) or 1 (for *T*).

## 4.2 REALLY RANDOM BANANAS

The notion of intrinsic randomness[1] is tricky because the intuition is that an intrinsically random event should be independent of any information available before the event occurs, and if the event is one among several possible events that could occur—say, the outcome of a measurement with several possible outcomes—the probabilities should be uniformly distributed over the alternative possibilities so that they all have the same probability. In a relativistic universe, different inertial observers will disagree about who measured first, or who peeled first. I'll follow a proposal by Roger Colbeck and Renato Renner that an event is intrinsically random relative to a time order, and relative to a specific event in this time order, but I'll drop the uniform probability requirement.[2] The possibility of events that are intrinsically random or "free" (to use the Colbeck–Renner term), in the sense that they are uncorrelated with any prior events, is the significant difference between a quantum world and a classical world.

In special relativity, the time order is defined by the light cone structure. (See the discussion of Figure 1.11 in the subsection "Special Relativity" in the "More" section at the end of Chapter 1 for the light cone structure.) For a pair of entangled photons, the measurement outcomes can be shown to be intrinsically random with respect to the preparation of the entangled state, in the sense that they are uncorrelated with any event that is not in the future light cone of the preparation event, so independent of any event that the preparation couldn't have caused. I'll show below that the taste of a Popescu–Rohrlich banana for a particular peeling is intrinsically random with respect to the preparation of a two-banana bunch on a Popescu–Rohrlich tree at the moment the bananas are ripe enough to exhibit a Popescu–Rohrlich correlation. The tastes are separately intrinsically random, in the sense that they are independent of any events at all that occur, in any reference frame, before the bananas are formed as a Popescu–Rohrlich pair, even though the taste of Alice's banana for a particular peeling is predictable from the taste of Bob's banana for a particular peeling, and conversely. The tastes of the bananas for particular peelings are events that occur after the preparation of the pair in every reference frame.

It follows from the no-signaling principle or the principle of free choice that you can't clone or copy the state of a Popescu–Rohrlich banana so that the clone behaves exactly like the original for both possible peelings. Suppose you could. If Bob peels his banana from the stem end and a cloned banana from the top end, then he could infer Alice's choice of peeling from the tastes of his banana and the cloned banana. If the tastes are the same, Bob infers that Alice peeled $S$, because otherwise, to satisfy the Popescu–Rohrlich correlation, the taste of Alice's banana would have to be the same as the taste of Bob's banana (which Bob peeled $S$), but different from the taste of Bob's cloned banana (which he peeled $T$). If the tastes are different, Bob infers that Alice peeled $T$, because if Alice peeled $S$, the tastes of Bob's banana and the cloned banana would have to be the same as the taste of Alice's banana. So Bob can infer whether Alice peeled $S$ or $T$ by checking whether the tastes of his banana and the cloned banana are the same or different.

For an inertial observer who sees Alice peeling her banana before Bob peels his bananas, the no-signaling principle is violated: if Bob can infer Alice's choice of peeling by checking to see whether the tastes of his banana and the cloned banana are the same or different, then Alice can signal instantaneously to Bob. For an inertial observer who sees Bob peeling his banana before Alice peels her banana, there is a constraint on Alice's choice of peeling: Alice must peel $S$ if the tastes of Bob's banana and the cloned banana are the same, and she must peel $T$ if the tastes are different. So what appears to be a violation of no signaling for some inertial observers appears to be a violation of free choice for other inertial observers. Assuming free choice or no signaling, cloning the state of a Popescu–Rohrlich banana is impossible.

Here's a proof of the no-cloning theorem for Popescu–Rohrlich bananas using the correlation $a \oplus b = A \cdot B$. Suppose Bob is able to clone the state of his banana, and suppose he peels his banana from the stem end and the clone from the top end. Represent the peeling of Bob's banana by $B$ (which takes the numerical value 0 if Bob peels from the stem end) and the peeling of the cloned banana by $B'$ (which takes the numerical value 1 if Bob peels from the top end), and the two tastes correspondingly by $b$ and $b'$ (which take the values 0 or 1 for "ordinary" and "intense").

The Popescu–Rohrlich correlation requires that

$$a \oplus b = A \cdot B$$
$$a \oplus b' = A \cdot B'.$$

Adding the two left-hand sides of the equations mod 2 (so $1 \oplus 1 = 0$) and equating with the sum of the right-hand sides gives

$$(a \oplus b) \oplus (a \oplus b') = b \oplus b' = A \cdot (B \oplus B') = A,$$

because $a \oplus a = 0$ for any value of $a$, and $B \oplus B' = 1$ if $B = 0$ and $B' = 1$. So Bob can read off the value of $A$—Alice's peeling—from $b \oplus b'$, the sum of the two values of the tastes of his banana and the clone. As before, this violates free choice or no signaling, depending on whether Bob peels and tastes his banana and the cloned banana before or after Alice peels her banana.

The no-cloning proof can be adapted to show that the Popescu–Rohrlich correlation is "monogamous": a third banana can't share the correlation between two Popescu–Rohrlich bananas. In the proof, suppose $B$ represents Bob's peeling and $B'$ represents Clio's peeling of a third banana that shares the Popescu–Rohrlich correlation with Alice's banana. Then Bob and Clio together could figure out the value of $A$ from the tastes of their bananas, so instantaneous signaling between Alice and Bob–Clio would be possible, or Alice's free choice would be violated.

There's a similar no-cloning theorem for quantum states. A copying machine takes some input, say a page of print, and a blank page, and produces as output two pages of the same print. The quantum no-cloning theorem says that there is no quantum dynamical process that will take an arbitrary unknown input quantum state $|\psi\rangle$ and some quantum system in a standard state (the analogue of the blank page) and produce as output two quantum systems in the state $|\psi\rangle$. Of course, the no-cloning theorem doesn't say that you can't prepare as many copies as you like of a known quantum state $|\psi\rangle$, which is always possible.

Entangled quantum states, like PR boxes, are monogamous. Two qubits in the entangled state $|\phi^+\rangle$ can't share the correlation defined by the entanglement with a third qubit—which doesn't prevent three qubits, or many qubits, from being entangled in other ways.

The impossibility of cloning the state of a Popescu–Rohrlich banana means that it must be impossible to gain enough information about a Popescu–Rohrlich banana state by measuring some banana variables to produce a copy without irreversibly changing the state. So there must be a *necessary information loss on measurement*. The argument is similar to the no-cloning argument. Suppose there are some banana variables that Bob could measure that would enable him to know the tastes of both possible peelings of his banana. If the tastes are the same, Bob infers that Alice peels $S$, and if the tastes are different, Bob infers that Alice peels $T$. So Bob could infer whether Alice peels $S$ or $T$ if he could find out, before he peels his banana, whether the tastes of his banana would be the same or different for both possible peelings. Again, either there is a violation of free choice if Bob gains this information before Alice peels her banana, or the no-signaling principle is violated if Bob gains the information after Alice peels her banana. It follows that if Bob obtains some information, before peeling his banana, about what the taste of his banana would be if he peels $T$, there must be a *necessary information loss* about the taste of his banana if he peels $S$, and conversely; and similarly for Alice. That's an uncertainty principle in Bananaworld. You might say that any measurement that produces information about the taste of a banana for an $S$-peeling leads to an irreducible and uncontrollable disturbance of information in the banana about the taste for a $T$-peeling, which is the Bananaworld counterpart to what Niels Bohr calls the "irreducible and uncontrollable measurement disturbance" in quantum mechanics.

To see that the Popescu–Rohrlich correlation forces the taste of a banana for a particular peeling to be be intrinsically random, it's convenient to consider the correlation defined by $a \oplus b = (A \oplus 1) \cdot B$, which corresponds to the correlation array in Table 4.5. For this Popescu–Rohrlich correlation, the tastes are the same for the peelings $SS$, $TS$, and $TT$, but different for the peeling $ST$.

I'll use the no-signaling principle and a chain of equalities to show that the marginal probabilities are all $1/2$. This follows from the Popescu–Rohrlich correlation even if the probabilities are conditional on events that occur, in any reference frame, before the bananas are formed as a Popescu–Rohrlich pair. So the taste of a Popescu–Rohrlich banana for a particular peeling is intrinsically random with respect to the preparation of the pair, in the sense that the tastes are independent of any events at all that occur, in any reference frame, before the bananas are formed as a Popescu–Rohrlich pair.

From no signaling, the probability that Alice's banana tastes 0, given that she peels $S$, doesn't depend on how Bob peels his banana. So this probability is just the sum of the probabilities that her banana tastes 0 for the two possible tastes of Bob's banana, given that he peels $S$:

$$p(0_A \,|\, S_A) = p(0_A 0_B \,|\, SS) + p(0_A 1_B \,|\, SS).$$

Similarly, the probability that Bob's banana tastes 0 given that he peels $S$ is the sum of the probabilities for the two possible tastes of Alice's banana:

$$p(0_B \,|\, S_B) = p(0_A 0_B \,|\, SS) + p(1_A 0_B \,|\, SS).$$

The first terms in the sums on the right-hand sides of these two equations are the same and the second terms are zero (because the tastes are the same for the pair of peelings $SS$), so

$$p(0_A \,|\, S_A) - p(0_B \,|\, S_B) = 0.$$

**Table 4.5** The Popescu–Rohrlich correlation array for the correlation $a \oplus b = (A \oplus 1) \cdot B$.

| | Alice | $S$ | | $T$ | |
|---|---|---|---|---|---|
| Bob | | 0 | 1 | 0 | 1 |
| $S$ | 0 | 1/2 | 0 | 1/2 | 0 |
| | 1 | 0 | 1/2 | 0 | 1/2 |
| $T$ | 0 | 0 | 1/2 | 1/2 | 0 |
| | 1 | 1/2 | 0 | 0 | 1/2 |

Similarly,

$$p(0_B \,|\, S_B) - p(0_A \,|\, T_A) = 0,$$
$$p(0_A \,|\, T_A) - p(0_B \,|\, T_B) = 0,$$
$$p(0_B \,|\, T_B) - p(1_A \,|\, S_A) = 0.$$

These four equations are "chained" in the sense that the second term on the left-hand side of each equation, except the last, is the same as the first term on the left-hand side of the following equation, but with a minus sign. So these terms cancel if you add the left-hand sides of the four equations. The sum of the terms on the left-hand sides is equal to the sum of the terms on the right-hand sides, which is zero. So from this chain of four equations you get

$$p(0_A \,|\, S_A) - p(1_A \,|\, S_A) = 0.$$

Since $p(1_A \,|\, S_A) = 1 - p(0_A \,|\, S_A)$, it follows that $p(0_A \,|\, S_A) = 1/2$ and $p(0_A \,|\, S_A) = 1/2$. There's nothing special about the marginal probabilities when Alice peels $S$, and the same reasoning shows that all the marginal probabilities are $1/2$.

Now suppose $e$ represents any event that occurs, in any reference frame, before the bananas are formed as a Popescu–Rohrlich pair. If you conditionalize the probabilities in the previous argument on $e$, the equations all remain true—if Alice and Bob can't violate the no-signaling principle by exploiting $e$. For example, $p(0_A \,|\, S_A, e)$ doesn't depend on how Bob peels his banana, so

$$p(0_A \,|\, S_A, e) = p(0_A 0_B \,|\, SS, e) + p(0_A 1_B \,|\, SS, e),$$

and so on. So $p(0_A \,|\, S_A, e) = 1/2$, and all the other marginal probabilities, conditional on $e$, are equal to $1/2$, which means that the marginal probabilities are independent of $e$.

**The bottom line**

- The "free choice" principle in Bananaworld says that Alice is free to choose whether she peels the stem end or the top end of her banana, in the sense that, for every inertial observer, her choice is uncorrelated with events in the past of her choice.
- It follows from the no-signaling principle, or the principle of free choice, that you can't clone the state of a Popescu–Rohrlich banana, and that the correlation is monogamous: a third banana can't share the correlation between two Popescu–Rohrlich bananas. It also follows that if Bob obtains information, before peeling his banana, about the taste of his banana if he peels from the top end, there must be a necessary information loss about the taste if he peels from the stem end.

*(continued)*

 **The bottom line** *(continued)*

- I show that the marginal probabilities for a pair of Popescu–Rohrlich bananas are all 1 / 2, and this is so even if the probabilities are conditional on events that occur, in any reference frame, before the bananas are formed as a Popescu–Rohrlich pair. So the taste of a Popescu–Rohrlich banana for a particular peeling is intrinsically random with respect to the preparation of the pair, in the sense that the tastes are independent of any events at all that occur, in any reference frame, before the bananas are formed as a Popescu–Rohrlich pair.

## 4.3 REALLY RANDOM QUBITS

Suppose you have two systems, like entangled photons or Popescu–Rohrlich bananas, with some observed correlation between two possible inputs (measurements or peelings) $A, A'$ for Alice's system, two possible inputs $B, B'$ for Bob's system, and two possible outputs (measurement outcomes or tastes, represented by 0 or 1) for each input. Artur Ekert and Renato Renner consider to what extent a correlation can limit the amount of information an adversary or eavesdropper, Eve, could have about Alice's and Bob's outputs for given inputs.[3] Specifically, is it possible for Eve to build a device that outputs values that are the same as Alice's outputs for her input $A$, say? Ekert and Renner show that this is impossible for a particular correlation of two entangled qubits in the state $|\phi^+\rangle$: the probability of a device successfully predicting Alice's outputs, based on any information held by Eve about events before the preparation of the correlated systems, in any reference frame, can be no better than a random guess. The observed correlation itself imposes this constraint on Eve's information, irrespective of whether quantum mechanics or some post-quantum theory is true as an explanation of the correlation.

The correlation Ekert and Renner consider is as close as you can get with entangled qubits to the Popescu–Rohrlich correlation in Table 4.5. Here's what I mean. The probability of Alice and Bob successfully simulating the Popescu–Rohrlich correlation in Table 4.5 for random prompts $A, A'$ for Alice and $B, B'$ for Bob (so each pair of prompts comes up with probability 1 / 4) is

$$\begin{aligned} p(\text{successful simulation}) = \frac{1}{4}\big(&p(\text{responses same} \,|\, AB) + p(\text{responses same} \,|\, A'B) \\ &+ p(\text{responses same} \,|\, A'B') + p(\text{responses different} \,|\, AB')\big). \end{aligned}$$

As in the corresponding simulation in Section 3.2, "Popescu–Rohrlich Bananas and Bell's Theorem," in the "Bananaworld" chapter, the prompts corresponding to Alice's two ways of peeling her banana are indicated as $A$ and $A'$ for peeling by the stem end ($S$) or the top end ($T$), and similarly the prompts corresponding to Bob's two ways of peeling his banana are indicated as $B$ and $B'$ for $S$ and $T$.

For the optimal simulation of the correlation in Table 4.5 with quantum resources, Alice and Bob share many copies of pairs of photons in the entangled state $|\phi^+\rangle = |0\rangle|0\rangle + |1\rangle|1\rangle$, Alice measures the polarization of her photons in directions $A = 0$ or $A' = \pi/4$, and Bob measures the polarization of his photons in directions $B = \pi/8$ or $B' = 3\pi/8$. Figure 4.1 shows the polarization directions.

Table 4.6 shows the correlation array for this quantum correlation, with 0 and 1 representing measurement outcomes "horizontal" and "vertical." (Here, $A$ and $A'$ represent polarization measurements in specific directions for Alice (0 and $\pi/4$), and $B$ and $B'$ represent polarization measurements in specific directions for Bob ($\pi/8$ and $3\pi/8$). In the previous correlation arrays, $A$ and $A'$ stood for Alice's two peelings, from the stem end or the top end, and $B$ and $B'$ for Bob's two peelings.)

To compare with the Popescu–Rohrlich probabilities of 0 and 1/2 in Table 4.5, the approximate values of the probabilities for the maximally entangled state $|\phi^+\rangle$ for these

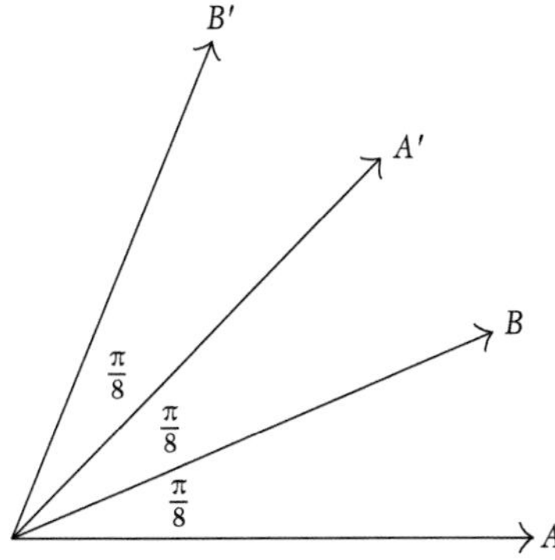


**Figure 4.1** The polarization directions $A, A'$ and $B, B'$ for the optimal quantum simulation of the correlation $a \oplus b = (A \oplus 1) \cdot B$ in Table 4.6.

**Table 4.6** The quantum correlation array for pairs of photons in the state $|\phi^+\rangle$ when Alice measures polarization in directions $A = 0$, $A' = \pi/4$, and Bob measures polarization in directions $B = \pi/8$, $B' = 3\pi/8$.

| Bob | Alice | $A$ 0 | $A$ 1 | $A'$ 0 | $A'$ 1 |
|---|---|---|---|---|---|
| $B$ | 0 | $\frac{1}{2}\cos^2\frac{\pi}{8}$ | $\frac{1}{2}\sin^2\frac{\pi}{8}$ | $\frac{1}{2}\cos^2\frac{\pi}{8}$ | $\frac{1}{2}\sin^2\frac{\pi}{8}$ |
| | 1 | $\frac{1}{2}\sin^2\frac{\pi}{8}$ | $\frac{1}{2}\cos^2\frac{\pi}{8}$ | $\frac{1}{2}\sin^2\frac{\pi}{8}$ | $\frac{1}{2}\cos^2\frac{\pi}{8}$ |
| $B'$ | 0 | $\frac{1}{2}\sin^2\frac{\pi}{8}$ | $\frac{1}{2}\cos^2\frac{\pi}{8}$ | $\frac{1}{2}\cos^2\frac{\pi}{8}$ | $\frac{1}{2}\sin^2\frac{\pi}{8}$ |
| | 1 | $\frac{1}{2}\cos^2\frac{\pi}{8}$ | $\frac{1}{2}\sin^2\frac{\pi}{8}$ | $\frac{1}{2}\sin^2\frac{\pi}{8}$ | $\frac{1}{2}\cos^2\frac{\pi}{8}$ |

**Table 4.7** Approximate values for the probabilities in Table 4.6.

| Bob | Alice | $A$ 0 | $A$ 1 | $A'$ 0 | $A'$ 1 |
|---|---|---|---|---|---|
| $B$ | 0 | .427 | .073 | .427 | .073 |
| | 1 | .073 | .427 | .073 | .427 |
| $B'$ | 0 | .073 | .427 | .427 | .073 |
| | 1 | .427 | .073 | .073 | .427 |

measurements are shown in Table 4.7. The entries $\frac{1}{2}\cos^2\frac{\pi}{8}$ and $\frac{1}{2}\sin^2\frac{\pi}{8}$ in the correlation array in Table 4.6 are approximately .427 and .073, respectively, rather than 1/2 and 0 in the correlation array in Table 4.5. So $p(\text{outcomes same}) = \cos^2\frac{\pi}{8} \approx .854$ for the three cases $AB, A'B, A'B'$ when the angle between the polarizations is $\pi/8$, and $p(\text{outcomes different}) = \sin^2\frac{3\pi}{8} = \cos^2\frac{\pi}{8} \approx .854$ for the case $AB'$ when the angle between the polarizations is $3\pi/8$. These correspond to the Popescu–Rohrlich probabilities $p(\text{tastes same}) = 1$ for the peelings $SS, TS, TT$ and $p(\text{tastes different}) = 1$ for peeling $ST$. This is as close as you can get with entangled photons to the Popescu–Rohrlich correlation in Table 4.5.

Here's the Ekert–Renner argument. Suppose Eve's device outputs values $z$, where $z$ can be either 0 or 1. I'll write $p(z\,|\,A)$ for the probability that Alice's $A$-output is $z$. By the no-signaling principle, $p(z\,|\,A)$ is equal to $p(z\,|\,AB)$ for any input $B$ by Bob, which is the sum of the probabilities that Alice's $A$-input has the output $z$ and Bob's $B$-input has either the output $z$ or the alternative possible output $1 - z$, which I'll write as $z'$:

$$p(z\,|\,A) = p(zz\,|\,AB) + p(zz'\,|\,AB).$$

To be clear, $p(zz\,|\,AB)$ is the probability that Alice's $A$-input has the output $z$ and Bob's $B$-input has the output $z$, and $p(zz'\,|\,AB)$ is the probability that Alice's $A$-input has the output $z$ and Bob's $B$-input has the output $z'$. The order is always Alice first, Bob second.

The probability that Bob's $B$-input has the output $z$ is the sum of the probabilities that Bob's $B$-input has the output $z$ and Alice's $A$-input has either the output $z$ or the alternative possible output $z'$:

$$p(z\,|\,B) = p(zz\,|\,AB) + p(z'z\,|\,AB).$$

The first term on the right-hand side of each of these two equations is the same, so subtracting the two equations gives

$$p(z\,|\,A) - p(z\,|\,B) = p(zz'\,|\,AB) - p(z'z\,|\,AB).$$

The right-hand side is less than or equal to $p(0_A 1_B\,|\,AB) + p(1_A 0_B\,|\,AB) = p(\text{outcomes different}\,|\,AB)$, because probabilities are positive numbers. Putting this together,

$$p(z\,|\,A) - p(z\,|\,B) \leq p(\text{outcomes different}\,|\,AB).$$

You can derive similar inequalities for the inputs $A'B$ and $A'B'$:

$$p(z\,|\,B) - p(z\,|\,A') \le p(\text{outcomes different}\,|\,A'B),$$
$$p(z\,|\,A') - p(z\,|\,B') \le p(\text{outcomes different}\,|\,A'B').$$

For the input $AB'$,

$$p(z\,|\,B') = p(zz\,|\,AB') + p(z'z\,|\,AB'),$$
$$p(z'\,|\,A) = p(z'z\,|\,AB') + p(z'z'\,|\,AB').$$

The second term on the right-hand side of the first equation is the same as the first term on the right-hand side of the second equation, so subtracting the two equations gives

$$p(z\,|\,B') - p(z'\,|\,A) = p(zz\,|\,AB') - p(z'z'\,|\,AB').$$

The right-hand side of this equation is less than or equal to $p(zz\,|\,AB') + p(z'z'\,|\,AB')$, which is $p(\text{outcomes same}\,|\,AB')$, and so

$$p(z\,|\,B') - p(z'\,|\,A) \le p(\text{outcomes same}\,|\,AB').$$

The four inequalities are linked in a chain:

$$p(z\,|\,A) - p(z\,|\,B) \le p(\text{outcomes different}\,|\,AB),$$
$$p(z\,|\,B) - p(z\,|\,A') \le p(\text{outcomes different}\,|\,A'B),$$
$$p(z\,|\,A') - p(z\,|\,B') \le p(\text{outcomes different}\,|\,A'B'),$$
$$p(z\,|\,B') - p(z'\,|\,A) \le p(\text{outcomes same}\,|\,AB').$$

The second term on the left-hand side of each inequality, except the last, is the same as the first term on the left-hand side of the following inequality, but with a minus sign, so these terms cancel each other when you add the terms on the left-hand sides of the inequalities. The sum of the terms on the left-hand sides is $p(z\,|\,A) - p(z'\,|\,A)$, and since $p(z'\,|\,A) = 1 - p(z\,|\,A)$, the sum is equal to $2p(z\,|\,A) - 1$. The sum of the terms on the left-hand sides is less than or equal to the sum of the terms on the right-hand sides, which I'll call $P_{2-}$:

$$P_{2-} = p(\text{outcomes different}\,|\,AB) + p(\text{outcomes different}\,|\,A'B) + p(\text{outcomes different}\,|\,A'B') + p(\text{outcomes same}\,|\,AB').$$

The subscript 2 indicates that Alice and Bob each receive two prompts. Later I'll consider $n$ prompts and the corresponding quantity $P_{n-}$, where the responses are the same for all pairs of prompts except one, when the responses are different.

So $2p(z\,|\,A) - 1 \le P_{2-}$, which gives

$$p(z\,|\,A) \le \frac{1}{2}(1 + P_{2-}).$$

This condition on $p(z|A)$ is a new Bell inequality, called a chained Bell inequality.[4] The derivation, as for Bell's original inequality, holds for any theory that satisfies the no-signaling principle.

What is $P_{2-}$? I'll write

$$P_{2+} = p(\text{outcomes same}\,|\,AB) + p(\text{outcomes same}\,|\,A'B) + p(\text{outcomes same}\,|\,A'B') + p(\text{outcomes different}\,|\,AB').$$

Since $p(\text{outcomes same}) + p(\text{outcomes different}) = 1$ for any pair of measurements, $P_{2+} + P_{2-} = 4$. So $P_{2-} = 4 - P_{2+}$.

The maximum value of $P_{2+}$ for a simulation of the Popescu–Rohrlich correlation in Table 4.5 with resources that don't violate the no-signaling principle is 4 (when $p(\text{successful simulation}) = 1$, so for a PR box, or Popescu–Rohrlich bananas). The optimal probability of a successful simulation with quantum resources is $\frac{1}{2}(1 + \frac{1}{\sqrt{2}})$, as I showed in Section 3.3, "Simulating Popescu–Rohrlich Bananas" in the "Bananaworld" chapter, so the maximum value of $P_{2+}$ for quantum resources is $\frac{4}{2}(1 + \frac{1}{\sqrt{2}}) = 2 + \sqrt{2} \approx 3.414$. The optimal probability of a successful simulation with classical or local resources is $3/4$, so the maximum value of $P_{2+}$ for classical or local resources is 3.

Since $P_{2-} = 4 - P_{2+}$, the *minimum* value of $P_{2-}$ is 0 (for a PR box, or Popescu–Rohrlich bananas), the *minimum* value of $P_{2-}$ for quantum resources is $2 - \sqrt{2} \approx .586$, and the *minimum* value of $P_{2-}$ for classical or local resources is 1. From Table 4.7 you can see that each of the four terms in $P_{2-}$ is equal to $\sin^2 \frac{\pi}{8} \approx 2 \times .073 = .146$, so $P_{2-} = 4 \sin^2 \frac{\pi}{8} \approx .586$ for the polarization measurements $AB, AB', A'B, A'B'$.

For the Popescu–Rohrlich correlation in Table 4.5, $P_{2-}$ is zero, so $p(z|A) \leq 1/2$. By the same argument, $p(z'|A) \leq 1/2$. But $p(z|A) + p(z'|A) = 1$, from which it follows that $p(z|A) = p(z'|A) = 1/2$. So Eve's device would be useless: she might as well guess randomly. What's notable here is that this nonclassical correlation limits the amount of information an eavesdropper Eve could have about the tastes of the bananas. That's not the case for any classical correlation, where $P_{2-} = 1$ and $1/2(1 + P_{2-}) = 1$, so all you could conclude is that $p(z|A) \leq 1$, which is no constraint at all on Eve's device.

For the photon correlation in Table 4.7, Alice and Bob can estimate $P_{2-}$ as approximately .586 from the statistics of measurements on many pairs of entangled photons and calculate $p(z|A)$ as $1/2(1 + .586) \approx .793$, which means that they can be confident that Eve's device will fail in just over 20% of cases. They can reduce the accuracy of Eve's device by making more measurements and extending the chain of inequalities. If Alice and Bob divide the interval between 0° and 90° into intervals $\theta = \pi/4n$ for some large number $n$, and Alice measures the polarizations of her photons randomly in directions $0, 2\theta, 4\theta, \ldots, (2n-2)\theta$, so for even multiples of $\theta$, and Bob measures the polarizations of his paired photons randomly in directions $\theta, 3\theta, 5\theta, \ldots, (2n-1)\theta$, so for odd multiples of $\theta$, they can extend the chain to $2n$ inequalities.

Here's the chain, where I've indicated the angles of Alice's and Bob's measurements by $A = 0$ and $B = \theta$, and so on:

$$\begin{aligned}
p(z\,|\,A = 0) - p(z\,|\,B = \theta) &\leq p(\text{outcomes different}\,|\,A = 0, B = \theta),\\
p(z\,|\,B = \theta) - p(z\,|\,A = 2\theta) &\leq p(\text{outcomes different}\,|\,A = 2\theta, B = \theta),\\
p(z\,|\,A = 2\theta) - p(z\,|\,B = 3\theta) &\leq p(\text{outcomes different}\,|\,A = 2\theta, B = 3\theta),\\
&\vdots\\
p(z\,|\,B = (2n-1)\theta) - p(z'\,|\,A = 0) &\leq p(\text{outcomes same}\,|\,A = 0, B = (2n-1)\theta).
\end{aligned}$$

As before, corresponding probabilities cancel out when you add the terms on the left-hand sides of the inequalities because they appear with plus and minus signs in consecutive inequalities. The sum of the terms on the left-hand sides is less than or equal to the sum of the terms on the right-hand sides. So, noting that $p(z'\,|\,A = 0) = 1 - p(z\,|\,A = 0)$, this gives $p(z\,|\,A = 0) \leq \frac{1}{2}(1 + P_{n-})$, where $P_{n-}$ is the sum of the terms on the right-hand sides and the subscript $n$ indicates that Alice and Bob each measure one of $n$ observables. I'll call $A$ the polarization observable for the angle zero (rather than writing $A = 0$).

Now $\theta = \pi/4n$, so each of the terms in $P_{n-}$ (from the right-hand sides) is $\sin^2 \frac{\pi}{4n}$, and there are $2n$ of these terms from the $2n$ inequalities. So $P_{n-} = 2n \sin^2 \frac{\pi}{4n} \leq 2/n$ because $\sin^2 \frac{\pi}{4n} \leq 1/n^2$. In the limit as $n$ tends to infinity and the measurements are made at angles $\theta$ that are closer and closer together, $P_{n-}$ tends to 0, and you get $p(z\,|\,A) \leq 1/2$, from which it follows, as before, that $p(z\,|\,A) = 1/2$. So, as for the Popescu–Rohrlich correlation, Eve's information about the outcome of Alice's polarization measurement is no better than a random guess.

For two photons in an entangled state like $|\phi^+\rangle = |0\rangle|0\rangle + |1\rangle|1\rangle$, Colbeck and Renner use a chained Bell inequality to show that there can't be a variable, $z$, associated with the history of the photons before the preparation of the entangled state in the reference frame of any inertial observer, that provides information about the outcomes of polarization measurements on the photons, so that Alice's and Bob's marginal probabilities conditional on $z$ are closer to 1 than the probabilities of the entangled quantum state $|\phi^+\rangle$.[5] In other words, the information in $|\phi^+\rangle$ about the probabilities of measurement outcomes is as complete as it could be. (The argument at the end of the last section shows the same thing for Popescu–Rohrlich bananas.) Next, they show that the same conclusion follows for any entangled state, not necessarily a maximally entangled state like $|\phi^+\rangle$ with equal coefficients for the product states in the sum. Finally, they extend the argument to any quantum state, for measurement outcome probabilities between 0 and 1, by exploiting the fact that a measurement interaction described by the dynamics of quantum mechanics leads to an entangled state of the measured system and the measuring instrument. If you're interested in the details, see the references to the Colbeck–Renner papers in the "Notes" at the end of the chapter.

 **The bottom line**

- Following an argument by Ekert and Renner, I use what's called a chained Bell inequality to prove that the ability of an eavesdropper, Eve, to predict the outcomes of polarization measurements on entangled pairs of photons, on the basis of information about events before their preparation, in any inertial reference frame, is no better than a random guess. The argument is independent of quantum mechanics and depends only on the observed correlation between measurement outcomes.
- For two photons in an entangled state like $|\phi^+\rangle = |0\rangle|0\rangle + |1\rangle|1\rangle$, Colbeck and Renner use a similar chained Bell inequality to show that there can't be a variable, $z$, associated with the history of the photons before the preparation of the entangled state in the reference frame of any inertial observer, that provides information about the outcomes of polarization measurements on the photons, so that Alice's and Bob's marginal probabilities conditional on $z$ are closer to 1 than the probabilities of the entangled quantum state $|\phi^+\rangle$. They extend the argument to any entangled state, and then to any quantum state.

# 4.4 MORE

## 4.4.1 Simulating Entanglement in Bananaworld

Here's another way of seeing that measurement outcomes on entangled photons are intrinsically random. In the "Bananaworld" chapter, I considered simulating a Popescu–Rohrlich correlation in Bananaworld with quantum resources. Here I'll show how to simulate a correlation between measurement outcomes on a pair of entangled photons in a quantum world with Bananaworld resources. You can do it with a combination of approximately 30% local deterministic resources and 70% Popescu–Rohrlich bananas, a nonlocal resource available in Bananaworld. Since the tastes for particular peelings of Popescu–Rohrlich bananas are intrinsically random, you could say that, in Bananaworld, about 70% of the measurement outcomes for the quantum correlation are intrinsically random. In our quantum world there are no Popescu–Rohrlich correlations to mix with deterministic correlations to get the photon correlation, so measurement outcomes on the photons are intrinsically random. I'll show this for the correlations in Table 4.6, which I repeat here for convenience as Table 4.8.[6]

A deterministic correlation array is a correlation array in which all the probabilities are 0 or 1: for every pair of inputs (peelings or measurements) there is a single pair of outputs (tastes or measurement outcomes). The correlation array in Table 4.8 can be expressed

**Table 4.8** The quantum correlation array for pairs of photons in the state $|\phi^+\rangle$ when Alice measures polarization in directions $A = 0$, $A' = \pi/4$, and Bob measures polarization in directions $B = \pi/8$, $B' = 3\pi/8$.

| Alice | | A | | A′ | |
|---|---|---|---|---|---|
| Bob | | 0 | 1 | 0 | 1 |
| B | 0 | $\frac{1}{2}\cos^2\frac{\pi}{8}$ | $\frac{1}{2}\sin^2\frac{\pi}{8}$ | $\frac{1}{2}\cos^2\frac{\pi}{8}$ | $\frac{1}{2}\sin^2\frac{\pi}{8}$ |
| | 1 | $\frac{1}{2}\sin^2\frac{\pi}{8}$ | $\frac{1}{2}\cos^2\frac{\pi}{8}$ | $\frac{1}{2}\sin^2\frac{\pi}{8}$ | $\frac{1}{2}\cos^2\frac{\pi}{8}$ |
| B′ | 0 | $\frac{1}{2}\sin^2\frac{\pi}{8}$ | $\frac{1}{2}\cos^2\frac{\pi}{8}$ | $\frac{1}{2}\cos^2\frac{\pi}{8}$ | $\frac{1}{2}\sin^2\frac{\pi}{8}$ |
| | 1 | $\frac{1}{2}\cos^2\frac{\pi}{8}$ | $\frac{1}{2}\sin^2\frac{\pi}{8}$ | $\frac{1}{2}\sin^2\frac{\pi}{8}$ | $\frac{1}{2}\cos^2\frac{\pi}{8}$ |

as a mixture or probability distribution of the deterministic correlation arrays $D_1$ and $D_2$ in Table 4.10 and $D_3$ and $D_4$ in Table 4.11, each weighted with a probability $1/2\sin^2\pi/8$, and the Popescu–Rohrlich correlation array in Table 4.5, repeated below as Table 4.9, weighted with probability $1 - 2\sin^2\pi/8$. Symbolically,

$$\frac{1}{2}\sin^2\frac{\pi}{8}(D_1 + D_2 + D_3 + D_4) + \left(1 - 2\sin^2\frac{\pi}{8}\right) PR,$$

where $PR$ here stands for the correlation array in Table 4.9.

**Table 4.9** The Popescu–Rohrlich correlation array for the correlation $a \oplus b = (A \oplus 1) \cdot B$.

| Alice | | S | | T | |
|---|---|---|---|---|---|
| Bob | | 0 | 1 | 0 | 1 |
| S | 0 | 1/2 | 0 | 1/2 | 0 |
| | 1 | 0 | 1/2 | 0 | 1/2 |
| T | 0 | 0 | 1/2 | 1/2 | 0 |
| | 1 | 1/2 | 0 | 0 | 1/2 |

**Table 4.10** Deterministic correlation arrays $D_1$ and $D_2$.

| Alice | | A | | A′ | |
|---|---|---|---|---|---|
| Bob | | 0 | 1 | 0 | 1 |
| B | 0 | 1 | 0 | 1 | 0 |
| | 1 | 0 | 0 | 0 | 0 |
| B′ | 0 | 1 | 0 | 1 | 0 |
| | 1 | 0 | 0 | 0 | 0 |

| Alice | | A | | A′ | |
|---|---|---|---|---|---|
| Bob | | 0 | 1 | 0 | 1 |
| B | 0 | 0 | 0 | 0 | 0 |
| | 1 | 0 | 1 | 0 | 1 |
| B′ | 0 | 0 | 0 | 0 | 0 |
| | 1 | 0 | 1 | 0 | 1 |

**Table 4.11** Deterministic correlation arrays $D_3$ and $D_4$.

| Bob | Alice | $A$ 0 | $A$ 1 | $A'$ 0 | $A'$ 1 |
|---|---|---|---|---|---|
| $B$ | 0 | 0 | 0 | 0 | 0 |
| | 1 | 1 | 0 | 1 | 0 |
| $B'$ | 0 | 0 | 0 | 0 | 0 |
| | 1 | 1 | 0 | 1 | 0 |

| Bob | Alice | $A$ 0 | $A$ 1 | $A'$ 0 | $A'$ 1 |
|---|---|---|---|---|---|
| $B$ | 0 | 0 | 1 | 0 | 1 |
| | 1 | 0 | 0 | 0 | 0 |
| $B'$ | 0 | 0 | 1 | 0 | 1 |
| | 1 | 0 | 0 | 0 | 0 |

To see this, for each of the 16 entries in Table 4.8, add the values of the entries in the corresponding positions in the four deterministic correlation arrays and the Popescu–Rohrlich correlation array in Table 4.9, weighted in the appropriate way. So, for example, adding the appropriately weighted values for the outcome 00 for the measurement pair $AB$ (the entry in the top left position in the top left cell) gives $1/2\sin^2\pi/8 \times 1 + 1/2\sin^2\pi/8 \times 0 + 1/2\sin^2\pi/8 \times 0 + 1/2\sin^2\pi/8 \times 0 + (1 - 2\sin^2\pi/8) \times 1/2 = 1/2(1 - \sin^2\pi/8) = 1/2\cos^2\pi/8$, and similarly for the other entries.

The deterministic correlations defined by the arrays $D_1, D_2, D_3, D_4$ are local correlations. Alice and Bob could simulate the correlation $D_1$ by both responding 0 for all prompts, and they could simulate the correlation $D_2$ by both responding 1 for all prompts. To simulate the correlation $D_3$, Alice responds 0 for all prompts and Bob responds 1 for all prompts, and to simulate the correlation $D_4$ Alice responds 1 for all prompts and Bob responds 0 for all prompts. In Bananaworld, if Alice and Bob had access to Popescu–Rohrlich bananas, they could simulate the correlations of the entangled quantum state $|\phi^+\rangle$ with a combination of local resources corresponding to the deterministic correlations $D_1, D_2, D_3, D_4$ and a nonlocal intrinsically random resource corresponding to Popescu–Rohrlich bananas correlated as in Table 4.9. They could do this by sharing a list of random numbers $1, \ldots, 5$, where the numbers $1, \ldots, 4$ occur with equal probability $\frac{1}{2}\sin^2\frac{\pi}{8}$ and 5 occurs with probability $1 - 2\sin^2\frac{\pi}{8}$. For the numbers $1, \ldots, 4$, they ignore the prompts and respond according to the deterministic array $D_1, D_2, D_3$, or $D_4$, and for the number 5 they respond with the tastes of Popescu–Rohrlich bananas correlated as in Table 4.9 when they each peel their banana according to the prompt for that round ($A = S$ or $A' = T$ for Alice, and $B = S$ or $B' = T$ for Bob).

As Bell showed, the correlation in Table 4.8 can't be simulated with local resources. (See Section 3.2, "Popescu–Rohrlich Bananas and Bell's Theorem," in the "Bananaworld" chapter.) In Bananaworld, the correlation can be simulated with a mixture of local resources and $1 - 2\sin^2\frac{\pi}{8} \approx 70\%$ intrinsically random Popescu–Rohrlich bananas, a nonlocal resource.

## Notes

1. My discussion of intrinsic randomness for Popescu–Rohrlich bananas owes a lot to Nicolas Gisin's book, *Quantum Chance: Nonlocality, Teleportation, and Other Quantum Marvels* (Copernicus, Göttingen, 2014). See especially Chapter 3, "Non-Locality and True Randomness."

2. The notions of "free choice" and "intrinsic randomness" are adapted (with some changes) from the discussion in Roger Colbeck and Renato Renner, "Free randomness can be amplified," *Nature Physics* 8, 450–454 (2012); "A short note on the concept of free choice," arXiv quant/ph1302.4446v1. See also Daniela Frauchiger, Renato Renner, and Matthias Troyer, "True randomness from realistic quantum devices," arXiv quant/ph1311.4547v1. Frauchiger, Renner, and Troyer say: "We call a process ***truly random*** if the outcome is uniformly distributed ***and*** independent of all information available in advance." My notion of "intrinsic randomness" corresponds to this notion of "true randomness," except that I drop the uniform distribution requirement. As I see it, what's significantly new about quantum mechanics is the possibility of events that are intrinsically random, in the sense that they are independent of all information available in advance. Intrinsically random events in this sense are "free" in the same sense that choices are "free" for Colbeck and Renner.
3. The Ekert–Renner argument is from Artur Ekert and Renato Renner, "The ultimate physical limits of privacy," in *Nature Physics* 507, 443–447 (2014); see "Eavesdropping Quantified" on p. 445.
4. The idea of a chained Bell inequality was introduced in a paper by Samuel L. Braunstein and Carlton M. Caves, "Wringing out better Bell inequalities," *Annals of Physics* 202, 22–56 (1990). Jonathan Barrett, Lucien Hardy, and Adrian Kent used a chained Bell inequality to show that two parties can share a secret key securely, where privacy is guaranteed by the no-signaling principle alone, even against eavesdroppers using post-quantum theories, in "No signaling and quantum key distribution," *Physical Review Letters* 95, 010503 (2005).
5. The Colbeck–Renner argument: Roger Colbeck and Renato Renner, "No extension of quantum theory can have improved predictive power," *Nature Communications* 2, 411 (2011). See also "Is a system's wave function in one-to-one correspondence with its elements of reality?" *Physical Review Letters* 108, 150402 (2012), and "The completeness of quantum theory for predicting measurement outcomes," arXiv quant-ph/1208.4123. A related result: John H. Conway and Simon Kochen, "The free will theorem," *Foundations of Physics* 36, 1441–1473 (2006) and "The strong free will theorem," *Notices of the American Mathematical Society* 56, 226–232 (2009).
6. The simulation of the correlations of the Bell state $|\phi^+\rangle$ by deterministic correlation arrays and Popescu–Rohrlich bananas is based on an exercise in a series of lectures, "Advanced Topics in Quantum Information Theory," by Matthias Christandl and Renato Renner at the Swiss Federal Institute of Technology in Zürich in the spring semester, 2013.

# 5

# The Big Picture

This chapter is about how to see the relation between classical, quantum, and superquantum correlations geometrically. The key is the diagram in Figure 5.2, which is a schematic representation of the different sorts of correlations in Bananaworld. Architects refer to the basic organizing scheme of an architectural design, the big picture concept presented in a sketch, as a "parti." Think of Figure 5.2 as the parti for the narrative in the book.[1]

The next section is about how different information-theoretic structures that satisfy the no-signaling principle are related. The point is to show that the features of the theory that led Feynman to say that "nobody understands quantum mechanics" aren't peculiar to quantum mechanics but arise for a large class of nonclassical theories, including theories that allow superquantum correlations like Popescu–Rohrlich correlations. From this broader perspective, divorced from the particular mathematical formalism of quantum mechanics, it's easier to see what the problems are and what would count as solutions. The final section of the chapter is about classical correlations, and the sense in which they are rather special.

## 5.1 NO-SIGNALING CORRELATIONS

First, some terminology. A *polygon* is a plane figure with finite straight line segments (sides or edges) joining vertices in a closed loop, like an equilateral triangle (three edges joining three vertices), a square (four edges joining four vertices), or a pentagram (five edges joining five vertices). The equilateral triangle and the square are examples of *regular* polygons, which is to say that the edges are all the same length, and they are *convex*: you can draw a straight line between any two points in the polygon without going outside the polygon. A circle is a convex set, but not a circle with a hole in the interior. A pentagon is a convex set, but not a pentagram (a five-pointed star): you can't draw a straight line between two adjacent vertices without going outside the pentagram.

A *polyhedron* is the three-dimensional analogue of a polygon, a solid in three dimensions, with vertices, edges, and flat faces, like a tetrahedron (four faces), a cube (six faces), or an octahedron (eight faces), as illustrated in Figure 5.1. A *polytope* is the multi-dimensional analogue of a polyhedron in more than three dimensions, but the term is also applied quite generally to include similar geometric objects in any number of

**Figure 5.1** Regular polyhedra: a tetrahedron, a cube, and an octahedron.

dimensions—so a square is a two-dimensional polytope. A *facet* of a polytope is the multi-dimensional analogue of a face of a polyhedron, one dimension less than the dimension of the polytope.

Here's the gist of it, to be elaborated below. Correlations that can be simulated with classical or local resources are represented by the points in the innermost square in Figure 5.2, labeled $\mathcal{L}$. The square is a schematic representation of a polytope, the local correlation polytope $\mathcal{L}$. For correlations between tastes and peelings for two bananas peeled by Alice and Bob in Bananaworld, the local correlation polytope is a four-dimensional hyperoctahedron (the four-dimensional analogue of a three-dimensional octahedron), which

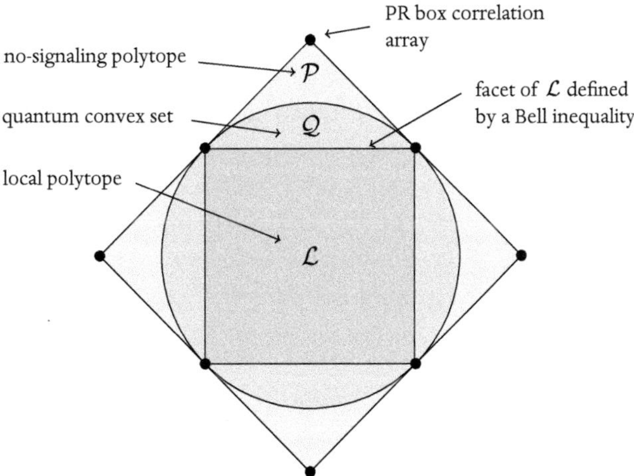


**Figure 5.2** A schematic representation of the different sorts of correlations in Bananaworld.

has eight vertices. These eight vertices represent all the local deterministic correlations (with 0, 1 probabilities for the outcomes of $S$ or $T$ peelings) in a particular representation proposed by Itamar Pitowsky that I'll explain below.[2] In Pitowsky's representation, a correlation array is reduced to a correlation vector. Adding another eight vertices representing the eight Popescu–Rohrlich correlations defined at the end of Section 4.1, "Correlation Arrays," in the previous chapter extends the four-dimensional hyperoctahedron $\mathcal{L}$, represented by the innermost square, to the no-signaling polytope $\mathcal{P}$, represented by the outermost square. The no-signaling polytope $\mathcal{P}$ is a four-dimensional hypercube, with 16 vertices. Points inside the hypercube $\mathcal{P}$ represent all the no-signaling probabilistic correlations between tastes and peelings for a pair of bananas. Quantum correlations are represented by the points inside a convex set $\mathcal{Q}$ with a continuous boundary between the octahedron $\mathcal{L}$ represented by the innermost square and the hypercube $\mathcal{P}$ represented by the outermost square. Points between the quantum convex set $\mathcal{Q}$ and the boundary of the hypercube $\mathcal{P}$ represent superquantum no-signaling correlations.

So classical, quantum, and superquantum correlations in Bananaworld between tastes and peelings for a pair of bananas peeled by Alice and Bob can be represented by three nested sets in a four-dimensional space: an eight-vertex hyperoctahedron $\mathcal{L}$ inside a convex set $\mathcal{Q}$ inside a 16-vertex hypercube $\mathcal{P}$. The representation in Figure 5.2 is only schematic—so, in particular, the boundary of the quantum set $\mathcal{Q}$ is a complicated three-dimensional region, not uniformly spherical, as Figure 5.2 suggests.

The diagram illustrates the flawed logic underlying the Einstein–Podolsky–Rosen argument for the incompleteness of quantum mechanics. The local polytope represents correlations for which something has been left out of the story: the common causes or "hidden variables" that explain how the correlations arise. The nonlocal probabilistic correlations in the region between the boundary of the local polytope $\mathcal{L}$ and the boundary of the no-signaling polytope $\mathcal{P}$ are, as von Neumann put it, "perfectly new and *sui generis* aspects of physical reality" and don't represent ignorance about hidden variables that have been left out of the story.[3] What's wrong with the Einstein–Podolsky–Rosen argument, as Bell saw, is that it imposes separability and locality requirements on quantum correlations that are only appropriate for the correlations in the local correlation polytope $\mathcal{L}$. "Nobody understands quantum mechanics" because the sorts of correlations represented by points between the boundary of the local correlation polytope $\mathcal{L}$ and the boundary of the no-signaling polytope $\mathcal{P}$ can't be explained by common causes or by direct causes, and these are the only sorts of causal explanations we have for how correlations arise.

If you're not used to thinking geometrically, the brief explanation of how to read Figure 5.2 is probably bewildering. In the rest of this section, I'll unpack the summary in the previous paragraphs.

The correlation between two Bananaworld bananas can be represented as a correlation array. The *deterministic* correlation arrays are arrays with probabilities 0 or 1 only. Each deterministic correlation array provides a complete instruction set for a pair of bananas to respond with a particular pair of tastes for a particular pair of peelings.

Ignoring no-signaling constraints for the moment, there are four possible arrangements of 0's and 1's that sum to 1 in each cell of the correlation array (four possible arrangements of one 1 and three 0's), and four cells, so there are $4^4 = 256$ deterministic arrays.

Of these 256 deterministic correlation arrays, 240 arrays violate the no-signaling principle and so don't represent correlations in Bananaworld. They represent nonlocal deterministic correlations: the joint probabilities can't be expressed as products of marginal or local probabilities for Alice and Bob separately. For example, Table 5.1 represents a nonlocal deterministic correlation array in which the taste of Alice's banana depends on Bob's choice of peeling. If Bob peels $S$, the probability is 1 that both bananas taste 0 if Alice peels $S$, and the probability is 1 that Alice's banana tastes 0 and Bob's banana tastes 1 if Alice peels $T$. So if Bob peels $S$, Alice's banana tastes 0, regardless of whether she peels $S$ or $T$. If Bob peels $T$, the probability is 1 that Alice's banana tastes 1 and Bob's banana tastes 0 if Alice peels S, and the probability is 1 that both bananas taste 1 if Alice peels $T$. So if Bob peels $T$, Alice's banana tastes 1, regardless of whether she peels $S$ or $T$. Similarly, the taste of Bob's banana corresponds to Alice's choice of peeling. So a choice of peeling by Alice or Bob is instantaneously revealed in the taste of a remote banana.

The remaining 16 deterministic correlation arrays satisfy the no-signaling principle and represent *local* deterministic correlations: cases where the joint 0, 1 probabilities for Alice and Bob can be generated from two correlation arrays for Alice and Bob separately. For example, the no-signaling deterministic correlation array in which the tastes are both ordinary (0) for all possible peeling combinations, represented by Table 5.2, can be generated from the correlation arrays for Alice and for Bob in Table 5.3 that each assign probability 1 to the taste 0 for either peeling, $S$ or $T$.

**Table 5.1** Nonlocal (signaling) deterministic correlation array.

| | Alice | $S$ | | $T$ | |
|---|---|---|---|---|---|
| Bob | | 0 | 1 | 0 | 1 |
| $S$ | 0 | 1 | 0 | 0 | 0 |
| | 1 | 0 | 0 | 1 | 0 |
| $T$ | 0 | 0 | 1 | 0 | 0 |
| | 1 | 0 | 0 | 0 | 1 |

**Table 5.2** Local (no-signaling) deterministic correlation array.

| | Alice | $S$ | | $T$ | |
|---|---|---|---|---|---|
| Bob | | 0 | 1 | 0 | 1 |
| $S$ | 0 | 1 | 0 | 1 | 0 |
| | 1 | 0 | 0 | 0 | 0 |
| $T$ | 0 | 1 | 0 | 1 | 0 |
| | 1 | 0 | 0 | 0 | 0 |

**Table 5.3** Deterministic correlation arrays for Alice and for Bob.

| Alice | $S$ | $T$ |
|---|---|---|
| 0 | 1 | 1 |
| 1 | 0 | 0 |

| Bob | $S$ | $T$ |
|---|---|---|
| 0 | 1 | 1 |
| 1 | 0 | 0 |

The 16 local deterministic arrays can be represented as 16 points in an eight-dimensional space—each array corresponds to a point. Why eight dimensions? There are 16 0's or 1's in each array. If there were no constraints, these 16 0's or 1's could be represented as the coordinates of a point with respect to a coordinate system of 16 orthogonal axes. But there are four probability constraints (the sum of the 0's and 1's in each cell of the array must be 1), and four independent no-signaling constraints. (At the beginning of Section 4.1, "Correlation Arrays," in the previous chapter, I listed eight no-signaling constraints in terms of sums of column entries and sums of row entries. Specifying four of them fixes the remaining four.) So there are only eight independent variables, which can be associated with the eight directions of an orthogonal coordinate system in an eight-dimensional space.

The 16 points representing the local deterministic correlation arrays are the vertices of a regular eight-dimensional polytope with 16 vertices. The edges all have length 1 because the vertex coordinates are 0's or 1's. The local correlation polytope $\mathcal{L}$ is the smallest closed convex set containing the vertices representing the local deterministic correlation arrays: the closed convex "hull" of the 16 vertices in technical jargon. The interior points represent mixtures or probability distributions of deterministic arrays. The vertices are extremal points of the polytope and can't be represented as mixtures.

For example, the Einstein–Podolsky–Rosen correlation.

- if the peelings are the same (*SS* or *TT*), the tastes are the same, with equal probability for 00 and 11;
- if the peelings are different (*ST* or *TS*), the tastes are uncorrelated, with equal probability for 00, 01, 10, 11;
- the marginal probabilities for the tastes 0 or 1 if a banana is peeled $S$ or $T$ are $1/2$, irrespective of whether or not the paired banana is peeled (so the no-signaling constraint is satisfied).

corresponds to the correlation array in Table 5.4, which can be represented as a mixture of the four deterministic correlation arrays $D_1, D_2, D_3, D_4$ in Tables 5.5 and Table 5.6 with equal weights of $1/4$. Each deterministic array represents a common cause of the correlation, or an instruction set telling Alice and Bob how to respond to prompts in a simulation of the correlation. From the positions of the four probabilities 1 in the array $D_1$ you can see that in this case Alice and Bob each responds 0 to $S$ and 0 to $T$. Similarly, for the array $D_2$, each responds 1 to $S$ and 1 to $T$. For the array $D_3$, each responds 0 to $S$ and 1 to $T$, and for the array $D_4$, each responds 1 to $S$ and 0 to $T$. Mixing these randomly with

**Table 5.4** Einstein–Podolsky–Rosen correlation array.

| Bob \ Alice | | S 0 | S 1 | T 0 | T 1 |
|---|---|---|---|---|---|
| S | 0 | 1/2 | 0 | 1/4 | 1/4 |
| | 1 | 0 | 1/2 | 1/4 | 1/4 |
| T | 0 | 1/4 | 1/4 | 1/2 | 0 |
| | 1 | 1/4 | 1/4 | 0 | 1/2 |

**Table 5.5** Deterministic correlation arrays $D_1$ and $D_2$.

| Bob \ Alice | | S 0 | S 1 | T 0 | T 1 |
|---|---|---|---|---|---|
| S | 0 | 1 | 0 | 1 | 0 |
| | 1 | 0 | 0 | 0 | 0 |
| T | 0 | 1 | 0 | 1 | 0 |
| | 1 | 0 | 0 | 0 | 0 |

| Bob \ Alice | | S 0 | S 1 | T 0 | T 1 |
|---|---|---|---|---|---|
| S | 0 | 0 | 0 | 0 | 0 |
| | 1 | 0 | 1 | 0 | 1 |
| T | 0 | 0 | 0 | 0 | 0 |
| | 1 | 0 | 1 | 0 | 1 |

**Table 5.6** Deterministic correlation arrays $D_3$ and $D_4$.

| Bob \ Alice | | S 0 | S 1 | T 0 | T 1 |
|---|---|---|---|---|---|
| S | 0 | 1 | 0 | 0 | 1 |
| | 1 | 0 | 0 | 0 | 0 |
| T | 0 | 0 | 0 | 0 | 0 |
| | 1 | 1 | 0 | 0 | 1 |

| Bob \ Alice | | S 0 | S 1 | T 0 | T 1 |
|---|---|---|---|---|---|
| S | 0 | 0 | 0 | 0 | 0 |
| | 1 | 0 | 1 | 1 | 0 |
| T | 0 | 0 | 1 | 1 | 0 |
| | 1 | 0 | 0 | 0 | 0 |

equal probability amounts to Alice and Bob each having a copy of an $S$-list and a $T$-list of random 0's and 1's, and responding from the appropriate list for the prompt, in order, for each round of a simulation game, which is the strategy for simulating the Einstein–Podolsky–Rosen correlation in Section 3.1, "Einstein–Podolsky–Rosen Bananas," in the "Bananaworld" chapter.

To see the relation between the local correlation polytope $\mathcal{L}$ and the no-signaling polytope $\mathcal{P}$ in Figure 5.2, it's convenient to replace the correlation arrays with correlation vectors, as proposed by Itamar Pitowsky. This reduces the number of dimensions from eight to four—just one more dimension than three, which is easier to visualize.

Take the correlation arrays $D_1$ and $D_2$. For the peeling combinations $SS, ST, TS, TT$, the tastes are all the same (both 0 in the case of $D_1$, and both 1 in case of $D_2$). So these two correlation arrays can be characterized by the correlation vector (same, same, same, same) for $SS, ST, TS, TT$, in that order. For the correlation arrays $D_3$ and $D_4$, the tastes are the same for $SS$ and $TT$, and different for $ST$ and $TS$, so the correlation vector characterizing $D_2$ and $D_3$ is (same, different, different, same). Writing 1 for "same" and –1 for "different," these are the correlation vectors $(1, 1, 1, 1)$ and $(1, -1, -1, 1)$. The remaining 12 deterministic correlation arrays also come in pairs and have the six correlation vectors

$$(1, 1, -1, -1), (1, -1, 1, -1), (-1, -1, -1, -1), (-1, -1, 1, 1), (-1, 1, -1, 1), (-1, 1, 1, -1).$$

The eight correlation vectors of the 16 local deterministic correlation arrays (eight pairs) define the vertices of an octahedron in four dimensions: a four-dmensional hyperoctahedron. The hyperoctahedron is the closed convex hull of the eight vertices: the smallest closed convex set containing the vertices.

Here's how to picture it. An octahedron in three dimensions has six vertices, as illustrated in Figure 5.1. To get the octahedron from a square in a two-dimensional plane, imagine a line drawn through the center of the square in the third dimension orthogonal to the plane. Add two vertices on the line symmetrically on either side of the square, the same distance from adjacent vertices as the length of an edge of the square. The octahedron is the closed convex hull of the six vertices. To get the eight-vertex four-dimensional hyperoctahedron, imagine a line drawn through the center of the octahedron along a fourth dimension. Add two vertices on the line symmetrically on either side of the octahedron, the same distance from adjacent vertices as the length of an edge in the octahedron. The four-dimensional hyperoctahedron is the closed convex hull of the six vertices of the three-dimensional octahedron, together with the additional two vertices.

A note for polytope enthusiasts: the standard "canonical" representation of a four-dimensional hyperoctahedron is the closed convex hull of the vertices

$$(1, 0, 0, 0), (0, 1, 0, 0), (0, 0, 1, 0), (0, 0, 0, 1)$$

$$(-1, 0, 0, 0), (0, -1, 0, 0), (0, 0, -1, 0), (0, 0, 0, -1).$$

The hyperoctahedron representing the local correlation polytope is a rotated version of the canonical hyperoctahedron.

The eight correlation vectors of the four-dimensional hyperoctahedron have an even number of 1's (four, two, or zero 1's):

$$(1, 1, 1, 1), (1, 1, -1, -1), (1, -1, 1, -1), (1, -1, -1, 1)$$

$$(-1, -1, -1, -1), (-1, -1, 1, 1), (-1, 1, -1, 1), (-1, 1, 1, -1).$$

The remaining eight possible correlation vectors have an odd number of 1's:

$$(1, 1, 1, -1), (1, 1, -1, 1), (1, -1, 1, 1), (-1, 1, 1, 1)$$

$$(-1, -1, -1, 1), (-1, -1, 1, -1), (-1, 1, -1, -1), (1, -1, -1, -1).$$

These turn out to be the correlation vectors of PR boxes defined in Section 4.1, "Correlation Arrays," in the previous chapter, and are outside the local polytope. For example, the standard PR box defined by the correlation array in Table 4.3 is represented by the correlation vector (same, same, same, different) or $(1, 1, 1, -1)$. Moving "different" to one of the other three slots produces three more PR boxes with correlation vectors $(-1,1,1,1)$, $(1,-1,1,1)$, $(1,1,-1,1)$, and switching "same" for "different" produces the four PR box correlations defined by the correlation vectors $(1,-1,-1,-1)$, $(-1,1,-1,-1)$, $(-1,-1,1,-1)$, $(-1,-1,-1,1)$.

The eight correlation vectors defining the vertices of $\mathcal{L}$, together with the eight correlation vectors representing all possible PR box correlations, define the 16 vertices of a four-dimensional hypercube $\mathcal{P}$ that contains $\mathcal{L}$. (A hypercube in four dimensions is related to a three-dimensional cube in the same way that a cube is related to a square in two dimensions.) The PR box correlations are extremal no-signaling correlations, so the points in the hypercube represent all possible no-signaling correlation arrays—outside the hypercube $\mathcal{P}$ the correlation arrays are all signaling arrays.

The correlations between measurement outcomes on a pair of qubits, where one of two possible observables is measured on each qubit, are limited by the Tsirelson bound and are represented by a convex set $\mathcal{Q}$ that is not a polytope, with a continuous boundary of extremal points or quantum pure states (indicated by the circular curved line in Figure 5.2) between the hyperoctahedron $\mathcal{L}$ and the hypercube $\mathcal{P}$. As I mentioned before, the quantum boundary is not perfectly spherical, as Figure 5.2 suggests—remember, it's just a schematic representation of the different sorts of correlations you can get by peeling and tasting two bananas in Bananaworld—but a rather complicated three-dimensional region of points, where the furthest points from the center are at the Tsirelson bound.

You might wonder why the vertices of the local correlation polytope $\mathcal{L}$ are included in the quantum convex set $\mathcal{Q}$. That's because the correlations represented by these vertices can be simulated by Alice and Bob with shared quantum states. For example, the Einstein–Podolsky–Rosen correlation, which is classical, can be simulated with classical resources, but also by quantum measurements on pairs of entangled states.

## The bottom line

- Figure 5.2, the key diagram for the narrative in the book, is a schematic representation of the different sorts of probabilistic correlations in Bananaworld.
- Classical, quantum, and superquantum correlations in Bananaworld between tastes and peelings for a pair of bananas peeled by Alice and Bob can be represented by three nested sets: a polytope $\mathcal{L}$ representing classical correlations, inside a convex set $\mathcal{Q}$ representing quantum correlations, inside a polytope $\mathcal{P}$ representing all no-signaling correlations. Superquantum no-signaling correlations are represented by the region between the boundary of the quantum convex set $\mathcal{Q}$ and the boundary of no-signaling polytope $\mathcal{P}$. The representation in Figure 5.2 is only schematic. So, in particular, the boundary of the quantum set $\mathcal{Q}$ is a complicated three-dimensional manifold, not spherical, as Figure 5.2 suggests.

*(continued)*

 **The bottom line** *(continued)*

- The local polytope represents probabilistic correlations for which something has been left out of the story: the common causes or "hidden variables" that explain how the correlations arise. The nonlocal probabilistic correlations in the region between the boundary of the local polytope $\mathcal{L}$ and the boundary of the no-signaling polytope $\mathcal{P}$ are, as von Neumann put it, "perfectly new and *sui generis* aspects of physical reality" and don't represent ignorance about hidden variables that have been left out of the story.
- What's wrong with the Einstein–Podolsky–Rosen argument, as Bell saw, is that it imposes separability and locality requirements on quantum correlations that are only appropriate for the correlations in the local correlation polytope $\mathcal{L}$. The sorts of correlations represented by points between the boundary of the local correlation polytope $\mathcal{L}$ and the boundary of the no-signaling polytope $\mathcal{P}$ can't arise this way.

## 5.2 CLASSICAL CORRELATIONS

Correlations represented by points (correlation arrays) inside the local polytope $\mathcal{L}$ can be simulated by Alice and Bob with classical or local resources. Correlations outside $\mathcal{L}$ can't be simulated in this way. Here's the core insight expressed geometrically: a classical simulation generates classical correlations represented by the points in a *simplex*, where the vertices of the simplex represent "local hidden variable" deterministic states that are the common causes of the correlations. Probability distributions over these vertices (the extremal states of the classical simulation) are mixed states represented by points in the interior of the simplex.

A simplex is a regular convex polytope: a triangle as opposed to a square, or a tetrahedron as opposed to a cube. A one-dimensional simplex or one-simplex has two vertices joined by a line segment or edge. To construct a two-dimensional simplex or two-simplex, take a new vertex as a point outside the line in a two-dimensional plane containing the line, the same distance from each of the original vertices as the length of the edge joining them, and join this new vertex to the original two vertices with two new edges. You get an equilateral triangle. To construct a three-simplex, continue in this way by taking a point in a third dimension outside the plane of the triangle, the same distance from each of the triangle vertices as the length of the edges joining them, and join this new vertex to the three vertices with three new edges. You get a tetrahedron, and so on as you take more points in more dimensions.

A simplex is the geometric counterpart of a Boolean algebra (see Subsection 3.5.3, "Boolean Algebras," in the "More" section of the "Bananaworld" chapter). In fact, there's a one-to-one correspondence between the vertices, edges, and faces of a simplex and the elements of a Boolean algebra. The vertices correspond to Boolean atoms, and the edges and faces correspond to compound Boolean elements.

The crucial fact about simplices relevant here is that any interior point of a simplex can be represented *uniquely* as what's called a "convex combination" of the vertices: a linear sum of the vertices, with non-negative coefficients between 0 and 1 that sum to 1. *No other polytope or convex set has this feature.*

For example, the simplest polytope is a line segment, with vertices or end points $x_1$ and $x_2$ along a single coordinate dimension labeled $x$. Any point on a line segment between the vertices $x_1$ and $x_2$ can be represented as $p_1x_1 + p_2x_2$, where $p_1$ and $p_2$ are non-negative numbers that sum to 1. If $p_1 = 1$ and $p_2 = 0$, the point is the vertex $x_1$. If $p_1 = 0$ and $p_2 = 1$, the point is the vertex $x_2$. For values of $p_1, p_2$ between 0 and 1 that sum to 1, the point is an interior point between $x_1$ and $x_2$. If $p_1 = p_2 = 1/2$, the point is halfway between $x_1$ and $x_2$. The convex combination here is unique—just one pair of $p_1, p_2$ values is associated with each point on the line. A line segment is an instance of a simplex, and this uniqueness is characteristic of polytopes that are simplices.

In the case of a square or a cube, neither of which are simplices, you can represent the interior points in terms of the vertices in a similar way, but the representation isn't necessarily unique. The center of a square can be represented as an equal convex combination ($p_1 = p_2 = 1/2$) of the two vertices of either diagonal, or as an equal convex combination ($p_1 = p_2 = p_3 = p_4 = 1/4$) of all four vertices. Similarly, the center of a cube and the centers of the square faces of a cube can be represented in several ways as convex combinations of the vertices.

So far, this is all just some technical stuff about simplices. Here's how it applies to classical or local correlations, like the Einstein–Podolsky–Rosen correlation represented in Table 5.4. The Einstein–Podolsky–Rosen correlation can be simulated with classical resources—there is a common cause explanation of the correlation, represented by a shared random variable. In the classical simulation in Section 3.1 of the "Bananaworld" chapter, the common cause is represented by a pair of values selected from two lists of random bits, the *S*-list and the *T*-list, shared by Alice and Bob and consulted during the simulation. So the random variable representing the common cause ranges over four quadruples of bits: $\{00; 00\}$, $\{01; 01\}$, $\{10; 10\}$, $\{11; 11\}$, where the first two bits in each quadruple represent a possible pair of tastes for *both* of Alice's peelings, and the second two bits in each quadruple represent a possible pair of tastes for *both* of Bob's peelings. The first two bits are the same as the second two bits because in this simulation Alice and Bob share the same lists of random bits.

By contrast, the correlation array describing the Einstein–Podolsky–Rosen *phenomena* assigns tastes to *alternative* peelings by Alice and by Bob. It's a summary of what you find for all possible alternative pairs of peelings, a particular peeling for Alice and a particular peeling for Bob. Even though, for each pair of bananas, Alice and Bob can only peel their own banana in one particular way, in a simulation with local or classical resources Alice and Bob need to share an instruction set that specifies a response (a taste) for *both* prompts (both peelings), because the moderator of the simulation game is free to choose a prompt corresponding to either of the two possible ways of peeling a banana. The instructions in the instruction set represent the common causes of the observed correlation in the Einstein–Podolsky–Rosen correlation array. The Einstein–Podolsky–Rosen correlation is generated from the instruction set in a simulation by selecting a pair of tastes in each quadruple, one for Alice and one for Bob, corresponding to the prompt for Alice and the prompt for Bob.

Now here's the connection with simplices. A common cause explanation of the Einstein–Podolsky–Rosen correlation is an explanation of how the correlation could arise in peeling and tasting pairs of bananas that have definite tastes for *both* possible peelings, prior to being peeled $S$ or $T$. The common causes are represented by quadruples of 0's and 1's that specify a pair of tastes for both of Alice's peelings and a pair of tastes for both of Bob's peelings. There are four common causes represented by four quadruples that label the four vertices of a simplex, specifically a three-simplex or tetrahedron. The correlation is generated as an equal weight probability distribution or mixture over the common causes represented by the vertices of the tetrahedron (a convex combination of the vertices).

The distinction to hold onto here is between an *observed correlation* in the phenomena, described by a correlation array or a correlation vector represented by a point in the local correlation polytope $\mathcal{L}$, and a *common cause explanation* of the observed correlation, represented by a simplex. Classical or local correlations are rather special, because these are just the correlations that can be explained by a simplex theory. The fact that a correlation represented by a point in the interior of the simplex can be expressed *uniquely* as a mixture of the common causes represented by the vertices of a simplex means that there is no ambiguity in a simplex explanation. The explanation is complete—nothing is left unexplained.

The local correlation polytope $\mathcal{L}$ associated with the observed correlation phenomena is not a simplex, so there needn't be a unique representation of a point in the interior of $\mathcal{L}$ as a mixture or probability distribution of the correlations represented by the vertices of $\mathcal{L}$. For example, the local correlation array represented by Table 5.7 can be represented as an equal weight mixture of the correlation arrays $D_1, D_2$ with correlation vector $(1, 1, 1, 1)$ in Table 5.5 and the correlation arrays $D_5, D_6$ with correlation vector $(-1, -1, -1, -1)$ in

**Table 5.7** Correlation array associated with the two equivalent mixtures $m_1$ and $m_2$.

| Bob | Alice | $S$ 0 | $S$ 1 | $T$ 0 | $T$ 1 |
|---|---|---|---|---|---|
| $S$ | 0 | 1/4 | 1/4 | 1/4 | 1/4 |
| | 1 | 1/4 | 1/4 | 1/4 | 1/4 |
| $T$ | 0 | 1/4 | 1/4 | 1/4 | 1/4 |
| | 1 | 1/4 | 1/4 | 1/4 | 1/4 |

**Table 5.8** Deterministic correlation arrays $D_5$ and $D_6$ with correlation vector $(-1, -1, -1, -1)$.

| Bob \ Alice | | S 0 | S 1 | T 0 | T 1 |
|---|---|---|---|---|---|
| S | 0 | 0 | 0 | 0 | 0 |
| | 1 | 1 | 0 | 1 | 0 |
| T | 0 | 0 | 0 | 0 | 0 |
| | 1 | 1 | 0 | 1 | 0 |

| Bob \ Alice | | S 0 | S 1 | T 0 | T 1 |
|---|---|---|---|---|---|
| S | 0 | 0 | 1 | 0 | 1 |
| | 1 | 0 | 0 | 0 | 0 |
| T | 0 | 0 | 1 | 0 | 1 |
| | 1 | 0 | 0 | 0 | 0 |

**Table 5.9** Deterministic correlation arrays $D_7$ and $D_8$ with correlation vector $(-1, 1, 1, -1)$.

| Bob \ Alice | | S 0 | S 1 | T 0 | T 1 |
|---|---|---|---|---|---|
| S | 0 | 0 | 0 | 0 | 0 |
| | 1 | 1 | 0 | 0 | 1 |
| T | 0 | 1 | 0 | 0 | 1 |
| | 1 | 0 | 0 | 0 | 0 |

| Bob \ Alice | | S 0 | S 1 | T 0 | T 1 |
|---|---|---|---|---|---|
| S | 0 | 0 | 1 | 1 | 0 |
| | 1 | 0 | 0 | 0 | 0 |
| T | 0 | 0 | 0 | 0 | 0 |
| | 1 | 0 | 1 | 1 | 0 |

Table 5.8, or alternatively as an equal weight mixture of the correlation arrays $D_3, D_4$ with correlation vector $(1, -1, -1, 1)$ in Table 5.6 and the correlation arrays $D_7, D_8$ with correlation vector $(-1, 1, 1, -1)$ in Table 5.9.

The nonuniqueness corresponds to multiple possible explanations, each associated with a different simplex. For example, the correlation could be generated as an equal weight mixture $m_1$ of the common causes represented by the vertices of the tetrahedron $\mathcal{T}_1$ with vertices $\{00; 00\}$, $\{11; 11\}$, $\{00; 11\}$, $\{11; 00\}$, or by an equal weight mixture $m_2$ of the common causes represented by the vertices of the tetrahedron $\mathcal{T}_2$ with vertices $\{01; 01\}$, $\{10; 10\}$, $\{01; 10\}$, $\{10; 01\}$. The vertices represent a pair of tastes for both of Alice's possible peelings $S$ and $T$, and a pair of tastes for both of Bob's possible peelings, $S$ and $T$. The tetrahedron $\mathcal{T}_1$ corresponds to a simulation strategy where Alice generates random pairs of 00 and 11 entries for the items in her $S$-list and $T$-list (0 for $S$ and 0 for $T$, or 1 for $S$ and 1 for $T$), and Bob generates his list by randomly flipping half of Alice's 00 entries to 11 entries, and half of Alice's 11 entries to 00 entries. The tetrahedron $\mathcal{T}_2$ corresponds to a simulation strategy where Alice generates random pairs of 01 and 10 entries for her $S$-list and $T$-list, and Bob generates his list by randomly flipping half of Alice's entry pairs. For each strategy, the marginal probabilities of Alice's two possible tastes, 0 or 1, are both $1/2$, for any peeling, and similarly for Bob. So the two strategies are equivalent for Alice and for Bob separately. The two strategies are also equivalent for Alice and Bob jointly, as reflected in the correlation array of Table 5.7 for bananas that can be peeled either $S$ or $T$, but not both ways. For each strategy, the probabilities of the four possible Alice–Bob pairs of tastes, 00, 01, 10, 11, are equal for any Alice-peeling and Bob-peeling.

A simplex explanation of the correlation assumes access to the tastes for *both* alternative peelings, for Alice and for Bob. From the perspective of the simplex theory, in which the mixtures $m_1$ and $m_2$ are distinct, the local correlation polytope of observational or operational probabilities characterizes a situation where access to the information encoded in the simplex vertices is either not accessible or not accessed for some reason. So something is left out of the story that could, in principle, be added to complete the story as a simplex theory in which the vertices represent the common causes of the correlation. For correlations inside the local polytope, this is always possible, generally in different ways, as the example with the correlation array in Table 5.7 illustrates. For correlations outside the local polytope, there is no theoretical account that provides an explanation of the correlation in terms of common causes.

One could say that correlations in Bananaworld inside $\mathcal{L}$ are completely explained by the bananas having what Einstein called a "being-thus," an independent existence characterized by definite properties, prior to any peeling, that is associated, either deterministically or probabilistically, with a particular taste for a particular peeling:[4]

> If one asks what, irrespective of quantum mechanics, is characteristic of the world of ideas of physics, one is first of all struck by the following: the concepts of physics relate to a real outside world, that is, ideas are established relating to things such as bodies, fields, etc., which claim a "real existence" that is independent of the perceiving subject—ideas which, on the other hand, have been brought into as secure a relationship as possible with sense-data. It is further characteristic of these physical objects that they are thought of as arranged in a physical space-time continuum. An essential aspect of this arrangement of things in physics is that they lay claim, at a certain time, to an existence independent of one another, provided that these objects "are situated in different parts of space." Unless one makes this kind of assumption about the independence of the existence (the "being-thus") of objects which are far apart from one another in space—which stems in the first place from everyday thinking—physical thinking in the familiar sense would not be possible. It is also hard to see any way of formulating and testing the laws of physics unless one makes a clear distinction of this kind.

The combined "being-thus" of a pair of correlated bananas would correspond to a vertex of a simplex in a common cause explanation of the correlation. Correlations outside $\mathcal{L}$ can't be simulated with local resources, so such correlated bananas can't have a "being-thus" in the sense Einstein had in mind. There is no explanation in the sense of a simplex explanation for correlations outside $\mathcal{L}$. To be sure, "physical thinking in the familiar sense" will have to change to accommodate the existence of correlations outside $\mathcal{L}$, but formulating and testing the laws of Bananaworld doesn't present any special problem. The bananas in Bananaworld can be peeled and tasted. The phenomena are objective. We don't know how the universe makes bananas with Bananaworld correlations, but we don't know how the universe makes objects with independent "being-thuses" either. We simply take elementary objects with elementary "being-thuses" as not requiring further explanation in classical physics. What we know is how these elementary objects combine to form complex objects, and how these objects transform dynamically.

Given elementary objects with elementary "being-thuses" as primitive entities, we can understand how classical correlations arise from common causes or direct causes.

The situation in Bananaworld is not really different. We can understand how nonlocal or nonclassical correlations arise in Bananaworld, if we take bananas with indefinite tastes that become definite in an intrinsically random way when peeled as elementary entities not requiring further explanation. In the same way, we can understand how nonlocal correlations arise in our quantum world if we recognize intrinsic randomness as a primitive feature of quantum systems not requiring further explanation.

## The bottom line

- Correlations represented by points inside the local polytope $\mathcal{L}$ in Figure 5.2 represent *observed correlations* that can be simulated by Alice and Bob with classical or local resources.
- A simulation with local resources corresponds to a *common cause explanation*, in the sense that a local simulation generates the observed correlations from a shared random variable whose values represent the possible common causes that explain the observed correlations.
- Geometrically, a local simulation or a common cause explanation of a correlation can be represented by a simplex whose vertices represent the possible common causes. A simplex is a regular convex polytope: a triangle as opposed to a square, or a tetrahedron as opposed to a cube. It is the geometric counterpart of a Boolean algebra.
- The crucial fact about simplices relevant here is that any interior point of a simplex can be represented *uniquely* as a convex combination of the vertices: a linear sum of the vertices, with non-negative coefficients between 0 and 1 that sum to 1. *No other polytope or convex set has this feature.* This uniqueness means that there is no ambiguity in a simplex explanation. The explanation is complete—nothing is left unexplained.
- One could say that correlations in Bananaworld represented by points inside $\mathcal{L}$ are completely explained by the bananas having what Einstein called a "being-thus," an independent existence characterized by definite properties, prior to any peeling, that is associated, either deterministically or probabilistically, with a particular taste for a particular peeling.
- The combined "being-thus" of a pair of correlated bananas would correspond to a vertex of a simplex in a simplex explanation of the correlation. Correlations outside $\mathcal{L}$ can't be simulated with local resources, so such correlated bananas can't have a "being-thus" in the sense Einstein had in mind. There is no explanation in the sense of a simplex explanation for correlations outside $\mathcal{L}$.

## Notes

1. The ideas in this section have their source in work by Itamar Pitowsky, primarily the article "Geometry of quantum correlations," *Physical Review A* 77, 062109 (2008), but also his book *Quantum Probability, Quantum Logic*, Lecture Notes in Physics 321 (Springer, Heidelberg, 1989), and the article "From George Boole to John Bell: The origins of Bell's inequalities," in M. Kafatos (ed.), *Bell's Theorem, Quantum Theory and Conceptions of the Universe* (Kluwer, Dordrecht, 1989), pp. 37–49. As Pitowsky shows, the literature goes back to George Boole's *An Investigation of the Laws of Thought* published in 1854. Other relevant papers by Pitowsky: "The range of quantum probability," *Journal of Mathematical Physics* 27, 1556–1566 (1986) and "Correlation polytopes, their geometry and complexity," *Mathematical Programming A* 50, 395–414 (1991).

   I've also used ideas from Jonathan Barrett, Noah Linden, Serge Massar, Stefano Pironio, Sandu Popescu, and David Roberts, "Nonlocal correlations as an information-theoretic resource," *Physical Review A* 71, 022101 (2005), Jonathan Barrett, "Information processing in generalized probabilistic theories," *Physical Review A* 75, 032304–032325 (2007), and Nicolas Brunner, Daniel Cavalcanti, Stefano Pironio, Valerio Scarani, and Stephanie Wehner, "Bell nonlocality," *Reviews of Modern Physics*, 2014. See also Anupam Garg and N. David Mermin, "Farkas's lemma and the nature of reality: statistical implications of quantum correlations," *Foundations of Physics* 14, 1–39 (1984).
2. Itamar Pitowsky, "Geometry of quantum correlations," *Physical Review A* 77, 062109 (2008).
3. John von Neumann's remark that quantum probabilities are "perfectly new and *sui generis* aspects of physical reality" is from an unpublished manuscript "Quantum Logics (Strict- and Probability-Logics)," reviewed by A. H. Taub in *John von Neumann: Collected Works* (Macmillan, New York, 1962), volume 4, pp, 195–197.
4. The quotation from Einstein that begins "If one asks what, irrespective of quantum mechanics, is characteristic of the world of ideas of physics, . . . " is from the paper: "Quanten-mechanik und wirklichkeit," *Dialectica* 2, 320–324 (1948). The article is reprinted in an English translation by Max Born as "Quantum mechanics and reality," in *The Born–Einstein Letters*, pp. 168–173. The passage quoted is on p. 170.

# 6

# Quantum Magic

The mathematician Andrew Gleason used the term "intertwined" to describe the intricate meshing of commuting and noncommuting observables.[1] In a seminal paper, he proved that it follows from this feature that there is only one way to define probabilities on the structure of quantum properties. I'll follow my colleague Allen Stairs and use the term "entwined." Observables are entwined in the sense that a "yes–no" observable representing a property of a system can belong to different sets of commuting observables that don't commute with each other. Entwinement shows up in the "contextuality" of measurement outcomes, the dependence of an outcome of a measurement on the context defined by what other commuting observables are measured, where the context in this sense can be local or nonlocal.

The "Bananaworld" chapter was all about nonlocal correlations in Bananaworld—correlations between Alice-events and Bob-events that can't be explained by either a direct causal influence between Alice and Bob, or by a common cause.

This chapter is about more quantum magic associated with weird correlations, specifically so-called "pseudo-telepathic" games and the feature of entwinement associated with local contextuality. I'll first consider a tripartite correlation between Alice, Bob, and Clio as an example of a pseudo-telepathic game. Then I'll look at an exotic pentagram correlation between bananas that grow in bunches of ten that combines nonlocality and contextuality. The final section is about contextuality and the Kochen–Specker theorem. I'll show how to simulate these correlations with entangled quantum states in the "More" section at the end of the chapter.

## 6.1 GREENBERGER–HORNE–ZEILINGER BANANAS

The Einstein–Podolsky–Rosen correlation can be simulated with local classical resources, specifically with shared randomness: there is a common cause explanation of the correlation. The Popescu–Rohrlich correlation can't be perfectly simulated with local classical resources, and can't be perfectly simulated with nonlocal quantum resources either. Is there a correlation that can't be simulated with local resources, but can be perfectly simulated with measurements on shared entangled quantum states?

In fact, there is a class of such correlations associated with pseudo-telepathic games. Gilles Brassard, Anne Broadbent, and Alain Tapp introduced the term "pseudo-telepathy" in a 2003 paper:[2]

> We say of a quantum protocol that it exhibits *pseudo-telepathy* if it is perfect provided the parties share prior entanglement, whereas no perfect classical protocol can exist.

They discuss several correlations that can't be simulated with local resources but can be simulated perfectly if the players in the simulation game are allowed to share entangled quantum states before the start of the game. From the perspective of players limited to classical or local resources, the ability to win such games seems inexplicable without telepathic communication between the players—hence the name. Of course, no cheating or telepathy is involved, only quantum magic.

The simulation game for the Greenberger–Horne–Zeilinger correlation, first discussed by Daniel Greenberger, Michael Horne, and Anton Zeilinger,[3] is the simplest pseudo-telepathic game if the responses to the prompts are just 0 or 1 (as opposed to longer strings of 0's and 1's).[4] In Bananaworld, Greenberger–Horne–Zeilinger bananas grow on trees with bunches of three bananas. See Figure 6.1. The tastes of a Greenberger–Horne–Zeilinger banana triple are correlated for certain combinations of peelings (unlike a Popescu–Rohrlich correlation, which is defined for all possible peeling

**Figure 6.1** Alice, Bob, and Clio peeling a bunch of Greenberger–Horne–Zeilinger bananas.

combinations), and the correlation persists if the bananas are separated by any distance, as follows:

- if one banana is peeled $S$ and two bananas are peeled $T$ (so for peelings $STT$, $TST$, $TTS$), an odd number of bananas taste intense (one or three);
- if all three bananas are peeled $S$ (so for the peeling $SSS$), an even number of bananas taste intense (zero or two);
- if one banana is peeled $S$ or $T$, there is an equal probability that the banana tastes ordinary (0) or intense (1), irrespective of whether or not other bananas are peeled, or how they are peeled (so the no-signaling constraint is satisfied).

Consider a simulation game for the correlation of Greenberger–Horne–Zeilinger bananas, where the players are restricted to local resources. The moderator sends separate prompts, $S$ or $T$, randomly to Alice, Bob, and Clio, who are separated and cannot communicate with each other during the game, with the promise that each round of prompts will include two $T$'s or no $T$'s. The conditions for winning a round are:

- if the prompts are $STT$, $TST$, $TTS$, the responses should be 001, 010, 100, or 111 (an odd number of 1's);
- if the prompts are $SSS$, the responses should be 000, 011, 101 or 110 (an even number of 1's);
- the responses are otherwise random (the marginal probabilities for Alice, Bob, and Clio are all $1/2$).

Call the responses of Alice, Bob, and Clio $a, b, c$. To win a round with local resources, Alice, Bob, and Clio must each separately use some rule for choosing 0's or 1's, based on shared lists of random bits generated before the start of the simulation, that satisfies the conditions:

$$
\begin{aligned}
a_S + b_T + c_T &= \text{odd}, \\
a_T + b_S + c_T &= \text{odd}, \\
a_T + b_T + c_S &= \text{odd}, \\
a_S + b_S + c_S &= \text{even},
\end{aligned}
$$

where the subscripts indicate the prompt, $S$ or $T$.

The sum of the terms on the left-hand sides of the four equations is equal to the sum of the terms on the right-hand sides. If you add the terms on the left-hand sides, you get an even number, for any values, 0 or 1, of the terms, because each term occurs twice in the sum, as you can check. But if you add the terms on the right-hand sides, you get an odd number. So there is no assignment of 0's and 1's based on shared random bits that satisfies the correlation, and a local simulation shared randomness is impossible.

If Alice, Bob, and Clio can't simulate the Greenberger–Horne–Zeilinger correlation with local resources, neither can the bananas, which means that they can't have preassigned tastes before they are peeled, nor can their tastes for particular peelings be determined by the values of some variable. As Einstein might put it, a banana in a Greenberger–Horne–Zeilinger triple can't have a "being-thus" that determines its taste for a particular peeling prior to the peeling: the three bananas exhibit nonlocal correlations for which there is no common cause explanation.

The Greenberger–Horne–Zeilinger correlation can't be simulated with local resources, but a perfect simulation of the correlation is possible if Alice, Bob, and Clio are allowed to share many copies of three qubits in a certain entangled state prepared before the start of the simulation. They each store one qubit of the three-particle state, and base their responses to the prompts on appropriate measurements performed on the stored qubits. See the subsection "Simulating Greenberger–Horne–Zeilinger Bananas" in the "More" section at the end of the chapter for the simulation protocol.

## The bottom line

- A pseudo-telepathic game is a game (think of a simulation game for a correlation) for which there is no winning strategy if the players are limited to classical (local) resources, but which can be won by players with quantum resources.
- The three-player simulation game for the correlation of Greenberger–Horne–Zeilinger bananas, which grow on trees with bunches of three bananas, is the simplest pseudo-telepathic game for players who respond to prompts with 0 or 1 (as opposed to longer strings of 0's and 1's).
- Tastes and peelings for a triple of Greenberger–Horne–Zeilinger bananas are correlated as follows:
  - (i) if one banana is peeled $S$ and two bananas are peeled $T$ (so for peelings $STT, TST, TTS$), an odd number of bananas taste intense (one or three);
  - (ii) if all three bananas are peeled $S$ (so for the peeling $SSS$), an even number of bananas taste intense (zero or two);
  - (iii) if one banana is peeled $S$ or $T$, there is an equal probability that the banana tastes ordinary (0) or intense (1), irrespective of whether or not other bananas are peeled, or how they are peeled (so the no-signaling constraint is satisfied).
- The bananas can't have preassigned tastes before they are peeled if they are correlated in this way, so Alice, Bob, and Clio can't win the simulation game if they are limited to local resources.
- The subsection "Simulating Greenberger–Horne–Zeilinger Bananas" in the "More" section at the end of the chapter shows how Alice, Bob, and Clio can win the simulation game with quantum resources.

## 6.2 THE ARAVIND–MERMIN MAGIC PENTAGRAM

Padmanabhan K. Aravind has proposed a variation of a correlation by David Mermin that is particularly informative in revealing structural features of quantum mechanics.[5] Aravind's magic pentagram correlation can be associated with a pseudo-telepathic game and also has a counterpart in Bananaworld.

Aravind–Mermin bananas grow in bunches of ten. The bananas fan out from a central stem in two layers: an outer layer of five large bananas whose tops are arranged like the five vertices 1, 2, 3, 4, 5 of the *pentagram* in Figure 6.2, and an inner layer of five smaller bananas whose tops are at the five vertices 6, 7, 8, 9, 10 of the inner *pentagon* of the pentagram. Although it's not technically correct, I'll use the word "edge" for convenience to refer to a line segment of the pentagram joining two vertices, for example the horizontal line segment labeled $e_1$ joining vertices 3 and 2. Each of the five edges in this sense contains two end nodes corresponding to two vertices, and two inner nodes corresponding to the intersections of the edge with two other edges. The edge $e_1$ contains the inner nodes 9 and 8. The inner node 9 is the intersection of the edge $e_1$ with the edge $e_2$, and the inner node 8 is the intersection of the edge $e_1$ with the edge $e_3$. There are ten nodes in total, and each banana is associated with a node. Any two edges intersect in just one node, or one banana. For any two edges, there are four bananas on each edge, with one banana in common, so seven distinct bananas on any two edges.

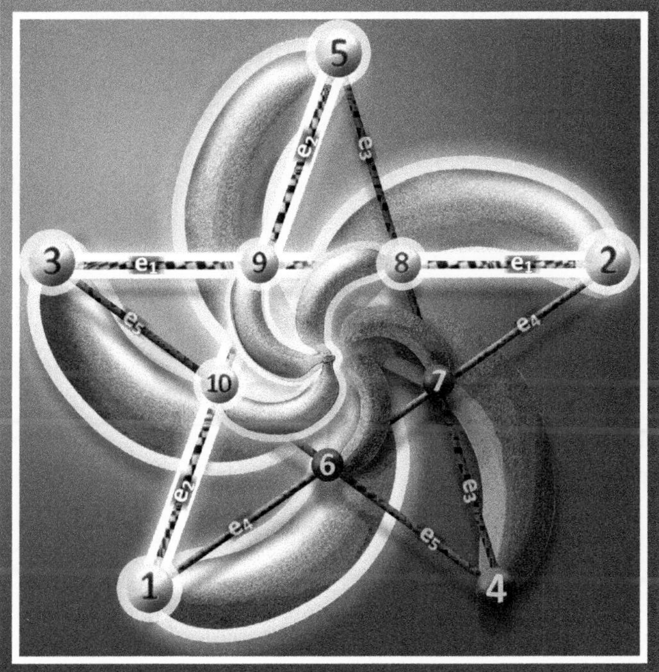


**Figure 6.2** Aravind–Mermin bananas.

The magical thing about these bananas is that if you peel the seven bananas on any two edges (stem end or top end, it doesn't matter), the probability of a particular banana tasting ordinary or intense is $1/2$, but an odd number of bananas on each edge always taste ordinary, either one banana or three bananas, and since there are just four bananas on an edge, an odd number also taste intense. Once you peel the bananas on two edges, the remaining bananas turn out to be inedible. What's remarkable about this is that the bananas can't have definite tastes before they are peeled, and this is an immediate consequence of the odd number property (the "parity," to use a technical term) of an Aravind–Mermin bunch. The illustration in Figure 6.2 shows the bananas on the edges $e_1$ and $e_2$ highlighted as about to be peeled, with banana 9 common to both edges.

Suppose the bananas do all have definite tastes before they are peeled, or some feature that leads to a definite taste when a banana is peeled, regardless of what other bananas are peeled on the two edges to which a banana belongs. All the bananas would have to have this feature and not just the bananas that happen to be peeled, because the choice of which bananas to peel is up to the person peeling the bananas. If, as usual, an ordinary taste is designated by the bit 0 and an intense taste by the bit 1, the sum of these bits for each edge must be an odd number. So if you add up all these sums for the five edges, the total must also be an odd number, because the sum of five odd numbers is an odd number. But—and here's the punch line—each banana occurs twice in this count, because it belongs to two edges, so each taste, 0 or 1, is counted twice, which means that the total must be an even number. Since the assumption that the bananas all have definite tastes before they are peeled leads to a contradiction, the bananas can't have definite tastes before they are peeled.

It follows that Alice and Bob can't simulate the behavior of an Aravind–Mermin bunch of bananas if they are separated and restricted to local resources. In the simulation game, Alice and Bob each receive a different pentagram edge as a prompt, $e_1, e_2, e_3, e_4$, or $e_5$. To win a round of the game, Alice and Bob should each respond with a series of four 0's and 1's assigned to the bananas on the edge corresponding to the prompt so that

(i) they assign the same bit to the common banana (the banana at the node that is the intersection of the two edges);
(ii) Alice's four bits and Bob's four bits should each have an odd number of 0's (and so an odd number of 1's);
(iii) over many rounds of the game, the bit assigned to a banana should be 0 or 1 with equal probability.

I'll call the first rule the "agreement rule," the second rule the "parity rule," because it says that the parity of the sum of the bits assigned to an edge should be odd, and the third rule the "probability rule," because it requires that the probability of a banana tasting ordinary or intense is $1/2$.

They can't pull this off if they are restricted to local resources. The agreement rule can only be satisfied with local resources if Alice and Bob each respond to a prompt by consulting the same shared list with 0's and 1's assigned to the nodes so as to satisfy the parity constraint, for all possible ways of doing this in some random order (to satisfy the equal probability rule). In other words, Alice and Bob would have to share a random list of pentagrams with bits, 0's and 1's, assigned to all the nodes, where the sum of the four bits on each edge is odd. But no such assignment of bits is possible.

There's no way Alice and Bob can win the simulation game with shared randomness, but there is a quantum winning strategy if Alice and Bob share many copies of a certain entangled state, so the game is a pseudo-telepathic game. See the subsection "Simulating Aravind–Mermin Bananas" in the "More" section at the end of the chapter for the simulation protocol.

The interesting thing about the Aravind–Mermin correlation is that it combines contextuality and nonlocality, the structural features of "entwinement" of quantum observables that I mentioned at the beginning of the chapter.

Each node of the magic pentagram is the intersection of two edges. In a quantum simulation of Aravind–Mermin bananas, each node is associated with an observable of a three-qubit system, and each edge is associated with a context defined by a set of compatible or commuting observables that can all be measured together. The observables on any two different edges don't all commute with each other, so an observable associated with a node belongs to two incompatible contexts. The agreement rule forces the taste of a banana to be the same for the two different edges to which it belongs. It's then impossible for the tastes to satisfy the parity rule if all the bananas have definite tastes before they are peeled. So there is no "noncontextual" assignment of tastes to the bananas in an Aravind–Mermin bunch—no assignment in which a banana gets a definite taste that is the same for the two contexts defined by the two edges to which it belongs.

Since it's possible for Alice and Bob to perfectly simulate the Aravind–Mermin correlation with quantum resources, quantum mechanics is contextual—the theory does something that no noncontextual theory can do. The correlation defined by a particular quantum state for the ten observables associated with the ten nodes of a magic pentagram can't arise from measurements of observables with definite noncontextual values before they are measured, which is to say that no noncontextual hidden variable theory can produce the correlation.

## The bottom line

- Aravind–Mermin bananas grow in bunches of ten, five large bananas and five small bananas, radiating outwards from a central stem. The tops of the bananas define the vertices of an outer *pentagram*, and the tops of the small bananas define the vertices of an inter *pentagon* (see Figure 6.2).

*(continued)*

 **The bottom line** *(continued)*

- I use the word "edge" for convenience to refer to a line segment of the pentagram joining two vertices, for example the horizontal line segment labeled $e_1$ joining vertices 3 and 2. If you peel the seven bananas on any two edges (stem end or top end, it doesn't matter), the probability of a particular banana tasting ordinary or intense is 1/2, but an odd number of bananas on each edge always taste ordinary, either one banana or three bananas (and since there are just four bananas on an edge, an odd number taste intense). Once you peel the bananas on two edges, the remaining bananas turn out to be inedible.
- In the simulation game, Alice and Bob each receive a pentagram edge as a prompt and respond with a series of four 0's and 1's assigned to the bananas on the edge in a way that should agree with the correlation. It's an immediate consequence of the odd number property that the bananas can't have definite tastes before they are peeled, so Alice and Bob can't win the simulation game with local resources.
- The interesting thing about the Aravind–Mermin correlation is that it combines contextuality and nonlocality. Each node of the pentagram is the intersection of two edges. In a quantum simulation of Aravind–Mermin bananas, each node is associated with an observable of a three-qubit system, and each edge is associated with a context defined by a set of compatible or commuting observables that can all be measured together.
- The impossibility of winning the simulation game with local resources means that there is no "noncontextual" assignment of tastes to the bananas—no assignment in which a banana gets a definite taste that is the same for the two contexts defined by the two edges to which it belongs. Since it's possible for Alice and Bob to perfectly simulate the Aravind–Mermin correlation with quantum resources (see the subsection "Simulating Aravind–Mermin Bananas" in the "More" section at the end of the chapter), quantum mechanics is contextual: the theory does something that no noncontextual theory can do.

## 6.3 THE KOCHEN–SPECKER THEOREM AND KLYACHKO BANANAS

In the previous section, I showed that there's no noncontextual assignment of tastes to the bananas in an Aravind–Mermin bunch, but it's possible to perfectly simulate the Aravind–Mermin correlation with quantum resources. Simon Kochen and Ernst Paul Specker,[6] and separately John Bell,[7] proved a general version of this "no go" result, now known as the Bell–Kochen–Specker theorem, or just the Kochen–Specker theorem. The theorem says that the observables of a quantum system can't have definite, noncontextual, preexisting values before they are measured, not even for certain finite sets of observables, and it's true for a qutrit, or any quantum system with observables that have three or more possible values. The proof is state independent: it doesn't depend on any particular quantum state. The way certain finite sets of observables are entwined is inconsistent with the observables all having definite values noncontextually. So the question of whether you could recover the probabilities defined by the quantum states of a system if the observables all have definite noncontextual values doesn't arise, because no such assignment of values is even possible.

The theorem doesn't apply to a qubit because there's no entwinement for a qubit—you need observables with at least three possible values. Think of a photon in a state of linear polarization, say vertical polarization in the $z$ direction, denoted by the state $|1\rangle$. There's a unique state of orthogonal polarization, horizontal polarization in the $z$ direction, denoted by the state $|0\rangle$. The two polarization states $|0\rangle$ and $|1\rangle$ define a coordinate system or basis in the two-dimensional state space associated with the observable "polarization in the $z$ direction." For a qutrit, the state space is three-dimensional. Three orthogonal states in the qutrit state space associated with the three possible values of a qutrit observable define a basis. In this case, if you pick a basis state, such as the state $|1\rangle$ on the right in Figure 6.3, there isn't a unique pair of orthogonal states that completes the basis. Rather, there are an infinite number of orthogonal pairs—Figure 6.3 shows two. Any orthogonal pair will do, and each choice for these states is associated with a different three-valued observable that doesn't commute with an observable associated with any other choice. For the qutrit in Figure 6.3, the basis $|1\rangle, |2\rangle, |3\rangle$ is associated with the three possible values of an observable $O$, and the basis $|1\rangle, |2'\rangle, |3'\rangle$ is associated with the three possible values of an observable $O'$. The observables $O$ and $O'$ are entwined, in the sense that the state $|1\rangle$ belongs to both bases or measurement contexts.

Suppose you have a filter that transmits a qutrit in the state $|1\rangle$ and blocks a qutrit in a state orthogonal to $|1\rangle$. The filter is a measuring instrument for a "yes–no" qutrit observable $P$ that has the value "yes" or 1 if the filter transmits the qutrit, and "no" or 0 if the filter blocks the qutrit. The question of contextuality arises for $P$ with respect to the

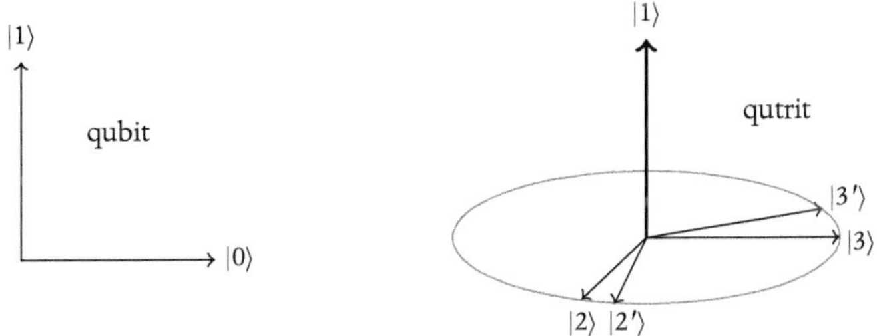


**Figure 6.3** A qubit context, and two qutrit contexts sharing the state vector $|1\rangle$.

observables $O$ and $O'$. Although $O$ and $O'$ are associated with different bases and don't commute, $P$ commutes with both $O$ and $O'$. (Why? Here's the technical explanation. The operator $P$ is just the projection operator onto the state $|1\rangle$. The operator $O$ can be represented as a sum of projection operators onto the state $|1\rangle$ and the orthogonal states $|2\rangle$, $|3\rangle$, with the eigenvalues or possible values of $O$ as coefficients. Similarly, $O'$ can be represented as a sum of projection operators onto $|1\rangle$ and the orthogonal states $|2'\rangle$, $|3'\rangle$, with the eigenvalues of $O'$ as coefficients. I show this for the Pauli operators in the section "The Pauli Operators" in the Supplement, "Some Mathematical Machinery," at the end of the book. Since $P$ commutes with itself and with any orthogonal projection operators, $P$ commutes with both $O$ and $O'$.) So the observable $P$, which can be measured with a $P$-filter with two possible outcomes, can also be measured via a measurement of $O$ with three possible outcomes, $1, 2, 3$, or via a measurement of $O'$ with three possible outcomes, $1, 2', 3'$. If you measure $O$ or $O'$ and get the outcome 1, the qutrit ends up in the state $|1\rangle$, which is also the state of the qutrit after it passes the $P$-filter. So $P$ can be assigned a value "in the context of $O$" via an $O$-measurement, or "in the context of $O'$" via a measurement of $O'$. In the two-dimensional state space of a qubit, there are no observables that are entwined like $O$, $O'$ and $P$, so there's no issue about contextuality.

The Born rule for quantum probabilities guarantees that the outcome 1 for $O$, or for $O'$, or for $P$, has the same probability in any quantum state $|\psi\rangle$, because the probability is the square of the absolute value of the length of the projection of $|\psi\rangle$ onto $|1\rangle$. So quantum *probabilities* are noncontextual. (See the section "The Born Rule" in the Supplement, "Some Mathematical Machinery," at the end of the book.) If quantum observables all had definite noncontextual pre-measurement *values* and measurement merely revealed these values, that would explain the noncontextuality of quantum probabilities. But since different contexts contain noncommuting observables, they can't be measured simultaneously. So there's no reason to think that an observable like $P$, which belongs to different contexts defined by noncommuting observables $O$ and $O'$, would be measured as having the same value, whether $P$ is measured together with $O$ or with $O'$. Putting it differently, if you measure $P$ via a measurement of $O$ and get the outcome 1, there's no justification for saying, counterfactually, that you would have found the corresponding outcome if you had measured $O'$ instead. For all you know, you might have found the outcome $2'$ or $3'$ for $O'$, corresponding to the value 0 for $P$. There's no way to check, because you can't measure $O$ and $O'$ at the same time.

So, does an observable like $P$ have the same value in the context of $O$ as in the context of $O'$, or are the values context-dependent and possibly different in the two contexts? The Kochen–Specker theorem answers the question: a noncontextual assignment of values to the observables of any quantum system at least as complex as a qutrit is impossible. The proof of the Kochen–Specker theorem for a qutrit is equivalent to showing that there's no way to assign 0's and 1's to a particular finite set of points on the surface of a unit sphere, in such a way that every context defined by an orthogonal triple of points gets one 1 and two 0's, where orthogonality is defined for the lines connecting the center of the sphere to the points on the surface.

### The Kochen–Specker theorem

- The Kochen–Specker theorem is a "no go" result for noncontextual hidden variable theories. The theorem says that the observables of a quantum system can't have definite, noncontextual, preexisting values before they are measured, not even for certain finite sets of observables.
- What's shown is that the way certain finite sets of observables are "entwined" is inconsistent with the observables all having definite values noncontextually.
- The theorem is true for a qutrit, or any quantum system with observables that have three or more possible values. There's no entwinement for a qubit—you need observables with at least three possible values.

To get an idea of why the theorem might be true, take all possible Cartesian coordinate systems defined by three orthogonal lines at the center of a sphere. Each triple of orthogonal lines intersects the surface of the sphere at three points and defines a basis or a context associated with a qutrit observable with three possible values. Now assign these points one 1 and two 0's, and try to come up with a scheme for doing this for every triple of points until the entire surface is covered with 0's and 1's. Assigning a 1 to a point as opposed to a 0, independently of the orthogonal triple defining the context to which the point belongs, designates the corresponding value of the observable as the pre-measurement value, or the value you would find if you measured the observable. A particular assignment of 0's and 1's corresponds to a state in the sense of Einstein's "being–thus" for the system, one of the possible ways the observables of the system could have definite values noncontextually. The theorem is the statement that there's no way to do this.

Whenever you assign a 1 to a point $p$ on the surface, you have to assign 0's to all the points on a great circle orthogonal to the line joining the center of the sphere to $p$, not just to two orthogonal points on the great circle, because all these points are orthogonal to $p$. Intuitively, that's too many 0's. If you imagine coloring the points on the surface red for 0 and green for 1, the red area should cover $2/3$ of the surface of the sphere (because each green point should be accompanied by two orthogonal red points), but that seems unlikely if whenever you color a single point on the surface of the sphere green you have to color a whole great circle of points red.

The intuition may be suggestive, but it's not an argument. (There's a continuous infinity of points on the surface of a sphere, and the notion of "counting" infinite sets is not straightforward.) The original Kochen–Specker proof involved 117 directions, or 117 points on the surface of a sphere, associated with 43 orthogonal triples of directions (with some directions belonging to more than one context or orthogonal triple). In effect, they proved that the "coloring problem" for the 117 directions has no solution: it's impossible to color the 117 directions red and green in such a way that every orthogonal triple of directions has one direction colored green and two directions colored red. That's because of the way the 117 directions are entwined, so that some directions belong to more than one orthogonal triple. There are much smaller sets that are now known to yield a state-independent proof of the theorem. There are several proofs with 33 directions for a qutrit, and John Conway and Simon Kochen have a proof with 31 directions.[8]

Curiously, as Roger Penrose pointed out, the polyhedron of three intersecting cubes on top of the left tower of Escher's engraving *Waterfall* in Figure 6.4 contains a representation of the 33 directions in a version of the theorem by Asher Peres![9] The 33 directions of the Peres proof are defined by the lines connecting the center of the three cubes to the vertices and to the centers of the faces and edges. You might say that Escher's *Waterfall* is an illustration of entwinement: some elements of the picture, like the top of the waterfall, belong to two different incompatible perspectives or contexts.

It's not immediately obvious that noncontextuality can be justified as a physically motivated constraint. Following Bohr, Bell pointed out that a measurement outcome

**Figure 6.4** Escher's *Waterfall*.

"may reasonably depend not only on the state of the system (including hidden variables) but also on the complete disposition of the apparatus." The nice thing about the Aravind–Mermin magic pentagram is that noncontextuality is enforced by locality, which is physically motivated. In the simulation game, the only way that Alice and Bob, restricted to local resources, could agree on the value they assign to the observable at the intersection of the edges they receive as prompts is by sharing copies of completely filled out charts of noncontextual value assignments to all ten nodes in the pentagram in some random order. With noncontextuality justified by locality in this way, you get a contradiction if you assume that the observables associated with the nodes on the magic pentagram all have definite values prior to measurement that are simply revealed in a measurement.

To demonstrate the limits of a noncontextual simulation of a correlation, you'd want noncontextuality to be enforced by the conditions of the simulation game. But if enforcing noncontextuality isn't the issue, there's a much simpler way of seeing that certain finite sets of quantum observables can't all have values noncontextually, thanks to an ingenious argument by Alexander Klyachko, M. Ali Can, Sinem Binicioğlu, and Alexander S. Shumovsky.[10] They derive an inequality for any noncontextual assignment of values to five noncommuting observables related in a certain way defined by the five edges of a pentagram. Each edge is associated with a different observable or context, and so each vertex, which belongs to two edges, is associated with a "yes–no" observable that belongs to two contexts. They show that the inequality is violated by the probabilities defined by a specific quantum state.

There's a simple analogue of the inequality in Bananaworld. Klyachko banana trees have bunches of five bananas that splay out from a common stem so that the tops of the bananas correspond to the vertices of a pentagram, as in Figure 6.5. What's peculiar about these Klyachko bananas is that if two bananas connected by an edge of the pentagram are peeled, they always have different tastes: one banana tastes ordinary and the other banana tastes intense. Once two bananas are peeled, the remaining three bananas turn out to be inedible. The probability of any particular banana tasting ordinary or intense is 1/2.

Each of the five vertices of a pentagram belongs to two edges, defining two contexts. Suppose each banana, situated at a pentagram vertex, has a definite taste before being peeled, where the taste is noncontextual in the sense that it doesn't depend on the edge to which the vertex belongs. Then the closest you can get to the correlation of Klyachko bananas is if two bananas that don't belong to the same pentagram edge taste intense (1) and the remaining three bananas taste ordinary (0), or if two bananas that don't belong to the same pentagram edge taste ordinary and the remaining three bananas taste intense. In both these cases, the Klyachko correlation is satisfied for four out of the five pentagram edges. You can check that this is the case for the assignment of tastes in Figure 6.5, where two bananas whose tips are at vertices 3 and 5 are assigned the taste 1. There's no way to

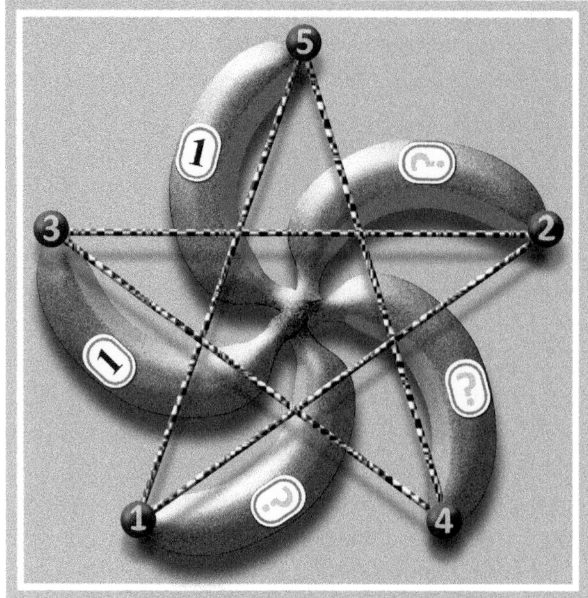


**Figure 6.5** Klyachko bananas.

assign the remaining three bananas definite tastes in such a way that two bananas on an edge have different tastes. Suppose you assign the bananas whose tips are at vertices 1 and 4 the tastes 0. Then assigning the remaining banana whose tip is at vertex 2 either 0 or 1 violates the Klyachko correlation. A 0 conflicts with the assignment of 0 to the banana whose tip is at vertex 1, and a 1 conflicts with the assignment of 1 to the banana whose tip is at vertex 3. You can check other possibilities for the question marks in Figure 6.5, and you'll come to the same conclusion. You can also check that an assignment of tastes with less than two 1's or less than two 0's is consistent with the Klyachko correlation for less than four edges (two edges if there's only one 1 or one 0, and no edges if there are no 1's or no 0's).

So the closest you can get to the Klyachko correlation with noncontextual assignments of tastes to the bananas are assignments in which the maximum number of pentagram edges with different-tasting bananas is four. Since there are five pentagram edges, the maximum probability that the bananas on a pentagram edge have different tastes is $4/5$. The sum of the probabilities for the five edges is therefore less than or equal to $5\times 4/5 = 4$ if the bananas are assigned tastes noncontextually. This is a version of the Klyachko–Can–Binicioğlu–Shumovsky inequality. For Klyachko bananas, the sum of the probabilities is, of course, $5 \times 5/5 = 5$. If tastes are assigned to the bananas via measurements on quantum observables, you can achieve a value of $2\sqrt{5} \approx 4.47$ with a certain quantum state and violate the noncontextuality inequality. I show this in Subsection 6.4.3, "Violating the Klyachko Inequality," in the "More" section at the end of the chapter, following an analogous argument by Klyachko and colleagues.

### The bottom line

- In the Aravind–Mermin simulation game, noncontextuality is enforced by locality. If enforcing noncontextuality isn't the issue, there's a much simpler way of seeing that certain finite sets of quantum observables can't all have values noncontextually, via an argument by Alexander Klyachko and colleagues. They derive an inequality for any noncontextual assignment of values to five observables related in a certain way defined by the five edges of a pentagram, where each edge corresponds to a different observable or measurement context, and they show that the inequality is violated by the probabilities defined by a specific quantum state.
- There's a simple analogue of the inequality in Bananaworld. Klyachko banana trees have bunches of five bananas that splay out from a common stem so that the tops of the bananas correspond to the vertices of a pentagram, as in Figure 6.5. If two bananas connected by an edge of the pentagram are peeled, they always have different tastes. Once two bananas are peeled, the remaining three bananas turn out to be inedible. The probability of any particular banana tasting ordinary or intense is $1/2$.
- The closest you can get to the Klyachko correlation with noncontextual assignments of tastes to the bananas are assignments in which the maximum number of pentagram edges with different-tasting bananas is four. Since there are five pentagram edges, the maximum probability that the bananas on a pentagram edge have different tastes is $4/5$. The sum of the probabilities for the five edges is therefore less than or equal to $5 \times 4/5 = 4$ if the bananas are assigned tastes noncontextually. (For Klyachko bananas, the sum of probabilities is 5.) This is a version of the Klyachko–Can–Binicioğlu–Shumovsky inequality.
- If tastes are assigned to the bananas via measurements on quantum observables, you can achieve a value of $2\sqrt{5} \approx 4.47$ with a certain quantum state and violate the noncontextuality inequality. I show this in Subsection 6.4.3, "Violating the Klyachko Inequality," in the "More" section at the end of the chapter, following an analogous argument by Klyachko and colleagues.

## 6.4 MORE

The quantum simulations of Greenberger–Horne–Zeilinger bananas and Aravind–Mermin bananas are more challenging than the simulations in the "Bananaworld" chapter. If you're lost, look at the Supplement, "Some Mathematical Machinery," at the end of the book, especially the sections "The Pauli Operators," "Noncommutativity and Uncertainty," and "Entanglement".

### 6.4.1 Simulating Greenberger–Horne–Zeilinger Bananas

Here's how Alice, Bob, and Clio can simulate the Greenberger–Horne–Zeilinger correlation. They begin by sharing many copies of three photons in the entangled state $|\Psi\rangle = |000\rangle + |111\rangle$. The state $|000\rangle$ is short for $|0\rangle_A|0\rangle_B|0\rangle_C$, a product state of three photons, labeled *A*, *B*, *C* for Alice, Bob, and Clio, and the state $|111\rangle$ is short for the product state $|1\rangle_A|1\rangle_B|1\rangle_C$. For each entangled three-photon state, Alice stores the photon *A*, Bob the photon *B*, and Clio the photon *C*, and they keep track of their photons as belonging to the first entangled state, the second entangled state, and so on.

The strategy is for each player to respond to the prompt *S* by measuring the observable *X* on his or her qubit (diagonal polarization if the qubit is a photon, or spin in the direction *x* if the qubit is an electron), and to respond to the prompt *T* by measuring the observable *Y* on his or her qubit (circular polarization if the qubit is a photon, or spin in the direction *y* if the qubit is an electron).[11] The measurement outcomes are the possible values (eigenvalues) of the Pauli observables *X* and *Y*, which are $\pm 1$. If the qubit is a photon, the value 1 for the observable *X* represents horizontal polarization, and the value –1 represents vertical polarization. For the observable *Y*, the value 1 represents right circular polarization, and the value –1 represents left circular polarization. (If the qubit is an electron, the values $\pm 1$ represent the two possible spin values, in the direction *x* for an *X*-measurement, and in the direction *y* for a *Y*-measurement.) The response to a prompt is 0 if the measured value of the observable is 1, and 1 if the measured value is –1. (Yes, that's right. Remember that whether you call an outcome $\pm 1$, or 0 or 1, or horizontal or vertical, or up or down, is just a convention.) Alice, Bob, and Clio are allowed to share an unlimited number of copies of the three qubits in the entangled state $|\Psi\rangle$. So for the first round of the simulation game, they measure the qubits in the first copy; for the second round, they measure the qubits in the second copy, and so on.

To see what happens for the prompt *SSS*, when they all measure the observable *X*, express $|\Psi\rangle$ in terms of the *X* eigenstates, which I'll denote here by $|x_+\rangle, |x_-\rangle$ rather than $|+\rangle, |-\rangle$:

$$|x_+\rangle = |0\rangle + |1\rangle,$$
$$|x_-\rangle = |0\rangle - |1\rangle.$$

Then $|0\rangle = |x_+\rangle + |x_-\rangle$ and $|1\rangle = |x_+\rangle - |x_-\rangle$, and if you substitute these expressions for $|0\rangle$ and $|1\rangle$ in $|\Psi\rangle$, you get

$$|\Psi\rangle = |x_+\rangle_A|x_+\rangle_B|x_+\rangle_C + |x_+\rangle_A|x_-\rangle_B|x_-\rangle_C + |x_-\rangle_A|x_+\rangle_B|x_-\rangle_C + |x_-\rangle_A|x_-\rangle_B|x_+\rangle_C,$$

because some of the product states appear with + and – signs and cancel out. If Alice, Bob, and Clio each measure *X* on their qubits, they obtain the outcomes 1, 1, 1 or 1, –1, –1 or –1, 1, –1 or –1, –1, 1 with equal probability 1/4, and so they respond 000, 011, 101, or 110, as required by the correlation (no 1's or two 1's).

To see what happens if the prompts contain one $S$ and two $T$'s, when one player measures $X$ and two players measure $Y$, express one of the qubit states in $|000\rangle$ and $|111\rangle$ as an $X$ eigenstate and two of the qubit states as $Y$ eigenstates. The $Y$ eigenstates are

$$|y_+\rangle = |0\rangle + i|1\rangle,$$
$$|y_-\rangle = |0\rangle - i|1\rangle,$$

where $i = \sqrt{-1}$, so $|0\rangle = |y_+\rangle + |y_-\rangle$ and $|1\rangle = -i|y_+\rangle + i|y_-\rangle$. Substituting as before, say the first qubit state in terms of $X$ eigenstates and the second and third qubit states in terms of $Y$ eigenstates for the prompts $STT$, and simplifying the expression (remember that $i^2 = -1$), you get

$$|\Psi\rangle = |x_-\rangle_A|y_+\rangle_B|y_+\rangle_C + |x_+\rangle_A|y_+\rangle_B|y_-\rangle_C + |x_+\rangle_A|y_-\rangle_B|y_+\rangle_C + |x_-\rangle_A|y_-\rangle_B|y_-\rangle_C.$$

Corresponding substitutions for $TST$ and $TTS$ simply change the order of the $X$ and $Y$ eigenstates. So if two players measure $Y$ and one player measures $X$, they obtain the outcomes $-1, 1, 1$ or $1, 1, -1$ or $1, -1, 1$ or $-1, -1, -1$ with equal probability $1/4$, and so they respond 100, 001, 010, or 111, as required (one 1 or three 1's).

### Simulating Greenberger–Horne–Zeilinger bananas

- To simulate the Greenberger–Horne–Zeilinger correlation, Alice, Bob, and Clio share many copies of the entangled state $|\Psi\rangle = |000\rangle + |111\rangle$. The states $|000\rangle$ and $|111\rangle$ are short for $|0\rangle_A|0\rangle_B|0\rangle_C$ and $|1\rangle_A|1\rangle_B|1\rangle_C$, which are product states of three qubits, labeled $A, B, C$ for Alice, Bob, and Clio.
- If the prompt is $S$, the player measures the observable $X$ on his or her qubit (diagonal polarization if the qubit is a photon). If the prompt is $T$, the player measures the observable $Y$ on his or her qubit (circular polarization if the qubit is a photon). The possible values of $X$ and $Y$ are $\pm 1$. The response to the prompt $S$ is 0 if $X$ is found to have the value 1, representing horizontal polarization in the diagonal direction for a photon, and 1 if $X$ is found to have the value $-1$, representing vertical polarization in the diagonal direction for a photon. (If this is confusing, remember that whether you call an outcome $\pm 1$, or 0 or 1, or horizontal or vertical, or up or down, is just a convention.) The response to the prompt $T$ is similar, with 0 corresponding to the value 1 for $Y$, representing right circular polarization for a photon, and 1 corresponding to the value $-1$ for $Y$, representing left circular polarization.
- The strategy perfectly simulates the Greenberger–Horne–Zeilinger correlation. The discussion explaining why this works refers to the Supplement, "Some Mathematical Machinery," at the end of the book, especially the sections "The Pauli Operators" and "Entanglement."

### 6.4.2 Simulating Aravind–Mermin Bananas

To simulate Aravind–Mermin bananas, Alice and Bob begin by sharing many copies of three pairs of qubits in the state $|\Phi\rangle = |\phi^+\rangle|\phi^+\rangle|\phi^+\rangle$, where $|\phi^+\rangle = |0\rangle_A|0\rangle_B + |1\rangle_A|1\rangle_B$. This is an entangled state prepared as a product of three maximally entangled pairs of qubit states. Alice gets the first qubit of each of the three pairs (indicated by the subscript $A$), and Bob gets the second qubit of each of the three pairs (indicated by the subscript $B$). So for each entangled state $|\Phi\rangle$, Alice gets three qubits and Bob gets three qubits.

If you multiply out the three entangled states (keeping track of the Alice and Bob states, and whether a state belongs to the first, second, or third pair), and express the product states as products of Alice-states and Bob-states, the state $|\Phi\rangle$ can be expressed as

$$|\Phi\rangle = \frac{1}{\sqrt{8}}\big(|000\rangle_A|000\rangle_B + |001\rangle_A|001\rangle_B + |010\rangle_A|010\rangle_B + |011\rangle_A|011\rangle_B$$
$$+|100\rangle_A|100\rangle_B + |101\rangle_A|101\rangle_B + |110\rangle_A|110\rangle_B + |111\rangle_A|111\rangle_B\big).$$

The first state in each pair (for example, the first state $|000\rangle_A$ in the pair $|000\rangle_A|000\rangle_B$) is a state of Alice's three qubits and the second state in each pair is a state of Bob's three qubits.

This is called a "biorthogonal expansion" of the state $|\Phi\rangle$—a linear superposition of correlated orthogonal states for Alice and for Bob. In this case, the biorthogonal expansion includes all the eight orthogonal states of the standard basis defined by the eight possible triples of single-qubit states $|0\rangle$ and $|1\rangle$ in the eight-dimensional state spaces of Alice's three qubits and Bob's three qubits.

It's a property of a biorthogonal expansion that if the values of the coefficients are all equal real numbers (here $1/\sqrt{8}$), then the biorthogonal expansion takes the same form if it's expressed in terms of any other basis states that are related to the standard basis by a transformation with real coefficients (as in the transformation from the standard basis to the basis $|x_+\rangle$, $|x_-\rangle$, where the coefficients are $\pm 1/\sqrt{2}$). So you could equally well express $|\Phi\rangle$ as

$$|\Phi\rangle = \frac{1}{\sqrt{8}}\big(|\alpha_1\rangle_A|\beta_1\rangle_B + |\alpha_2\rangle_A|\beta_2\rangle_B \cdots + |\alpha_8\rangle_A|\beta_8\rangle_B\big),$$

where the states $|\alpha_1\rangle_A, |\alpha_2\rangle_A, \ldots$ are any eight basis states in the state space of Alice's three qubits related to the standard basis states in this way, and the states $|\beta_1\rangle_B, |\beta_2\rangle_B, \ldots$ are the corresponding basis states in the state space of Bob's three qubits.

Here's how Alice and Bob succeed in performing the magic trick required by the rules of the simulation game.[12] If the bananas are labeled as in Figure 6.2, Alice and Bob respond to the prompts labeling edges by measuring the observables associated with the nodes on the edge corresponding to the prompt, according to the scheme in Figure 6.6, where the subscripts label each of the three entangled pairs. They base their response on the measurement outcome, where again the convention is to respond 0 if the measurement outcome on a qubit is 1 and 1 if the measurement outcome is –1.

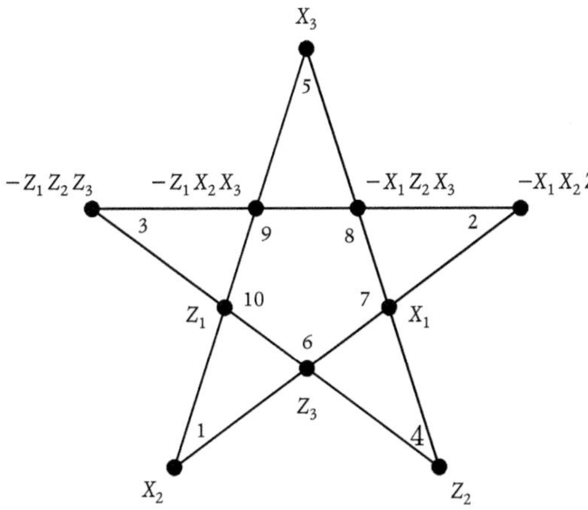


**Figure 6.6** Simulating Aravind–Mermin bananas with quantum measurements.

All the single-qubit observables have possible values $\pm 1$. The product observables have possible values that are products of $\pm 1$'s, so their values are also $\pm 1$. All the observables on an edge of the pentagram commute in pairs, which means that they can all be measured together.

To see this, take the four observables $X_3, -Z_1X_2X_3, Z_1, X_2$ associated with the nodes $5, 9, 10, 1$ on the edge $e_2$. The possible values of $Z_1, X_2$, and $X_3$ are $\pm 1$. Since the observables $Z_1, X_2$, and $X_3$ all commute (because they refer to different qubits), the value of the observable $-Z_1X_2X_3$ is the product of the values of $Z_1, X_2$, and $X_3$, with a minus sign in front, which is also $\pm 1$.

Also, each of the single-qubit observables commutes with $-Z_1X_2X_3$. Take $X_3$ and $-Z_1X_2X_3$, for example. The product of $-Z_1X_2X_3$ with $X_3$ on the right is $-Z_1X_2X_3 \cdot X_3 = -Z_1X_2X_3^2$. The product with $X_3$ on the left is $X_3 \cdot -Z_1X_2X_3 = -Z_1X_2X_3^2$, because $X_3$ and $Z_1$ commute and $X_3$ and $X_2$ commute, so $X_3 \cdot -Z_1X_2X_3 = -Z_1 \cdot X_3 \cdot X_2X_3 = -Z_1X_2 \cdot X_3 \cdot X_3 = -Z_1X_2X_3^2$. The square of a Pauli observable is the identity observable $I$, so $X_3^2 = I$. (See the section "The Pauli Operators" in the Supplement, "Some Mathematical Machinery," at the end of the book.) So you can perform a joint measurement of $Z_1, X_2$, and $X_3$ on the three qubits, with measurement outcomes that yield a sequence of four 1's and –1's for the four observables $X_3, -Z_1X_2X_3, Z_1, X_2$.

The same thing applies to the edges $e_3, e_4$, and $e_5$ with three single-qubit observables and one product observable. See Figure 6.7 for the Aravind–Mermin magic pentagram with edge labels.

What about the edge $e_1$ with four product observables at the nodes $3, 9, 8, 2$? These also commute in pairs. For example, the observables associated with the vertices 3 and 9 are $-Z_1Z_2Z_3$ and $Z_1X_2X_3$. Multiply these together in either order and you get

$$(-Z_1Z_2Z_3) \cdot (-Z_1X_2X_3) = I_1 \cdot -iY_2 \cdot -iY_3 = -I_1Y_2Y_3,$$
$$(-Z_1X_2X_3) \cdot (-Z_1Z_2Z_3) = I_1 \cdot iY_2 \cdot iY_3 \quad = -I_1Y_2Y_3,$$

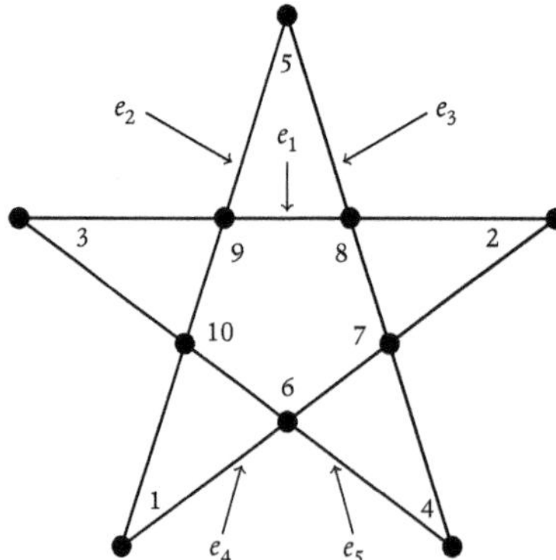


**Figure 6.7** The Aravind–Mermin magic pentagram with edges labeled $e_1, e_2, e_3, e_4, e_5$.

because $ZZ = I$, $XZ = iY$, and $ZX = -iY$. (See the section "Noncommutativity and Uncertainty" in the Supplement, "Some Mathematical Machinery," at the end of the book for the products of Pauli observables.)

If Alice or Bob gets a prompt that labels one of the edges $e_2, e_3, e_4, e_5$ with only one product observable, they simply measure the single-qubit observables on their three qubits, which determines the value of the product observable (as the product of the values of the single-qubit observables multiplied by −1). For the edge $e_1$ with four product observables, the situation is trickier, but since the four product observables commute, they can in fact all be measured together in a measurement on the composite three-qubit system, without assigning values to the single-qubit observables in the products.[13]

Why will a simultaneous measurement of the four observables associated with the vertices of an edge produce an odd number of 1's and an odd number of −1's? Take the edge $e_2$ again, with nodes 5, 9, 10, 1. The product of these four observables, $X_3 \cdot -Z_1X_2X_3 \cdot Z_1 \cdot X_2$, can be reordered as $-Z_1^2 \cdot X_2^2 \cdot X_3^2$. The square of any of these observables is $I$, where $I$ is the identity observable, so the product of the four observables is $-I_1I_2I_3$. The identity observable is like the number 1 (if you multiply any observable $Q$ by $I$, you get $Q$) and has the value 1 in any quantum state. So $-I_1I_2I_3$ has the value $-1 \cdot 1 \cdot 1 = -1$, which means that the product of the measurement outcomes of these four observables, the product of the sequence of 1's and −1's, must be −1. It follows that an odd number of these measurement outcomes will be 1 and an odd number −1, corresponding to an odd number of bananas tasting ordinary and an odd number tasting intense.

Similarly, the product of the four observables associated with one of the edges $e_3, e_4, e_5$ with one triple product observable is $-I_1I_2I_3$. That's because the triple product observable is just a triple product of the three single-qubit observables on the three remaining nodes of the edge, multiplied by −1, and the product of any observable with itself is $I$.

For the horizontal edge $e_1$ with four triple product observables, the product of these observables is also $-I_1I_2I_3$. To see this, take the products in pairs (the order doesn't matter because the observables all commute). The product of the observables for the nodes 3 and 9 is $-Z_1Z_2Z_3 \cdot -Z_1X_2X_3 = I_1 \cdot iY_2 \cdot iY_3 = -I_1Y_2Y_3$. The product of this with the observable associated with node 8 is $-I_1Y_2Y_3 \cdot -X_1Z_2X_3 = X_1 \cdot iX_2 \cdot -iZ_3 = X_1X_2Z_3$. The product of $X_1X_2Z_3$ with the observable $-X_1X_2Z_3$ associated with node 2 is $-I_1I_2I_3$. So the product of the values of the four observables is $-1$.

By sharing the entangled state $|\Phi\rangle$, Alice and Bob ensure that if they both receive the same prompts and so measure the same observables on their respective qubits, they will agree. If they get different prompts associated with different edges that intersect at a node, the form of the biorthogonal expansion ensures that they will agree on the measurement outcome of the observable corresponding to the common node. See the section "The Biorthogonal Decomposition Theorem" in the Supplement, "Some Mathematical Machinery," at the end of the book for an explanation.

**Simulating Aravind–Mermin bananas**

- To simulate the Aravind–Mermin correlation, Alice and Bob share many copies of the state $|\Phi\rangle = |\phi^+\rangle|\phi^+\rangle|\phi^+\rangle$, where $|\phi^+\rangle = |0\rangle_A|0\rangle_B + |1\rangle_A|1\rangle_B$. Alice gets the first qubit of each of the three pairs (indicated by the subscript $A$), and Bob gets the second qubit of each of the three pairs (indicated by the subscript $B$). So for each entangled state $|\Phi\rangle$ Alice gets three qubits and Bob gets three qubits.
- In the simulation game, the prompts label edges in Figure 6.6, as indicated in Figure 6.7. Alice and Bob respond to the prompts by measuring the observables associated with the nodes on the edge corresponding to the prompt according to the scheme in Figure 6.6, where the subscripts label the three entangled pairs. They base their responses on the measurement outcomes.
- The discussion explaining why this strategy perfectly simulates the Aravind–Mermin correlation refers to the Supplement, "Some Mathematical Machinery," at the end of the book, especially the sections "The Pauli Operators," "Noncommutativity and Uncertainty," and "Entanglement."

## 6.4.3 Violating the Klyachko Inequality

In Section 6.3, "The Kochen–Specker Theorem and Klyachko Bananas," I showed that the closest you can get to the Klyachko correlation in a simulation with noncontextual assignments of tastes to the bananas is a simulation in which the maximum number of pentagram edges with different-tasting bananas is four. Since there are five pentagram edges, the maximum probability that the bananas on a pentagram edge have different tastes is $4/5$. So the sum of the probabilities for the five edges is less than or equal to $5 \times 4/5 = 4$ if the bananas are assigned tastes noncontextually. This is a version of the Klyachko–Can–Binicioğlu–Shumovsky inequality. If tastes are assigned to the bananas via measurements on quantum observables, you can violate the inequality and achieve a value of $2\sqrt{5} \approx 4.47$ with a certain quantum state, closer to the value 5 for Klyachko bananas than the sum of probabilities for any noncontextual assignment of tastes. The argument below is an exercise in three-dimensional Euclidean geometry, following an analogous argument by Klyachko and colleagues.[14]

Consider a unit sphere and imagine a circle $\Sigma_1$ on the equator of the sphere with an inscribed pentagon and pentagram, with the vertices of the pentagram labeled in order $1, 2, 3, 4, 5$, as in Figure 6.8.

The angle subtended at the center $O$ by adjacent vertices of the *pentagon* (for example, vertices 1 and 4) is $2\pi/5$, so the angle subtended at $O$ by adjacent vertices defining an edge of the *pentagram* (for example, vertices 1 and 2) is $\theta = 2 \times 2\pi/5 = 4\pi/5$, which is greater than $\pi/2$. If the radii linking $O$ to the vertices are pulled upwards towards the north pole of the sphere, the circle with the inscribed pentagon and pentagram will move up on the sphere towards the north pole. Since $\theta = 0$ when the radii point to the north pole (and the circle vanishes), $\theta$ must pass through $\pi/2$ before the radii point to the north pole. So it must be possible to draw a circle, $\Sigma_2$, on the sphere at some point between the equator and the north pole, with an inscribed pentagon and pentagram, such that the lines joining $O$ to the two vertices of an edge of the pentagram, for example the edge with vertices 1 and 2, are orthogonal. Call the centre of this circle $P$—see Figure 6.9.

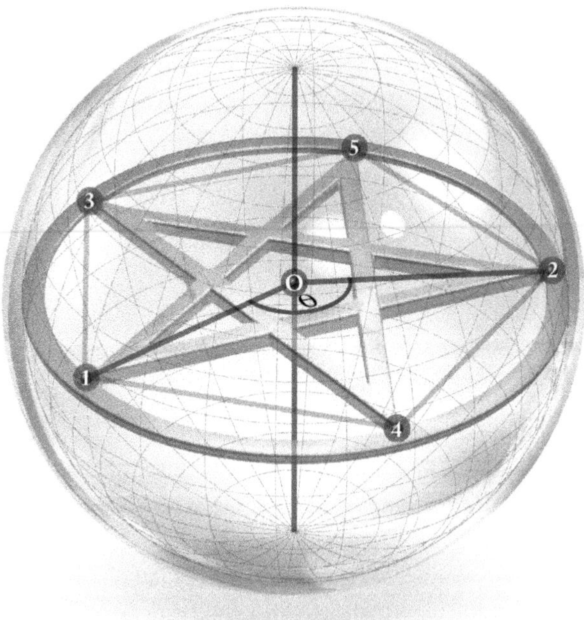


**Figure 6.8** Circle $\Sigma_1$ on the equator of a unit sphere with an inscribed pentagram.

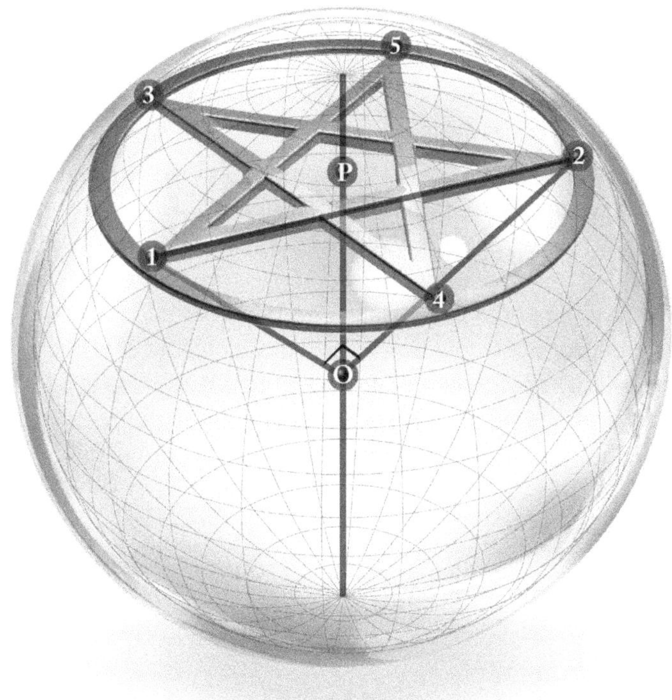


**Figure 6.9** Circle $\Sigma_2$ with an inscribed pentagram.

It's possible, then, to define five orthogonal triples of unit vectors, corresponding to five bases in the three-dimensional Hilbert space or state space of a qutrit, representing the eigenstates of five different noncommuting observables or measurement contexts:

$$\begin{aligned} &|1\rangle,\ |2\rangle,\ |v\rangle \\ &|2\rangle,\ |3\rangle,\ |w\rangle \\ &|3\rangle,\ |4\rangle,\ |x\rangle \\ &|4\rangle,\ |5\rangle,\ |y\rangle \\ &|5\rangle,\ |1\rangle,\ |z\rangle. \end{aligned}$$

Here, $|v\rangle$ is orthogonal to $|1\rangle$ and $|2\rangle$, $|w\rangle$ is orthogonal to $|3\rangle$ and $|4\rangle$, and so on, and each vector $|1\rangle, |2\rangle, |3\rangle, |4\rangle, |5\rangle$ from $O$, the center of the sphere, to the points $1, 2, 3, 4, 5$ on the surface of the sphere, belongs to two different contexts. For example, $|1\rangle$ belongs to the context defined by $|1\rangle, |2\rangle, |v\rangle$ and to the context defined by $|5\rangle, |1\rangle, |z\rangle$. The vectors $|u\rangle, |v\rangle, |x\rangle, |y\rangle, |z\rangle$ play no role in the rest of the argument, so a context is defined by one of the edges $12, 23, 34, 45, 51$ of the pentagram in the circle $\Sigma_2$.

Now consider assigning probabilities to the vertices of the pentagram by a quantum state $|\psi\rangle$ defined by a unit vector that passes through the north pole of the sphere, and so through the point $P$ in the center of the circle $\Sigma_2$. According to the Born rule, the probability assigned to a vertex, say the vertex 1, by $|\psi\rangle$, is

$$|\langle 1|\psi\rangle|^2 = \cos^2\phi,$$

where $|1\rangle$ is the unit vector defined by the radius from $O$, the center of the unit sphere, to the vertex 1, and $\phi$ is the angle between $|\psi\rangle$ and $|1\rangle$. See Figure 6.10. (To remind you: the Born rule says that the probability is the square of the length of the projection of $|\psi\rangle$ onto $|1\rangle$.)

Here's how to work out $\cos^2\phi$. The line from $O$ to 1 is a radius of the unit sphere, so length 1, and the line from $O$ to $P$ is orthogonal to the circle $\Sigma_2$ and so orthogonal to the line from 1 to $P$. The cosine of $\phi$ is the length of the base of the right-angled triangle $\{P, 1, O\}$ divided by the length of the hypotenuse. The base of the triangle is the line from $O$ to $P$ with length $r$, and the hypotenuse is the line from $O$ to 1 with length 1 (the right angle is the angle at $P$). So $\cos\phi = r$. Also, $r^2 + s^2 = 1$ by Pythagoras's theorem, where $s$ is the length of the line from 1 to $P$, so

$$\cos^2\phi = r^2 = 1 - s^2.$$

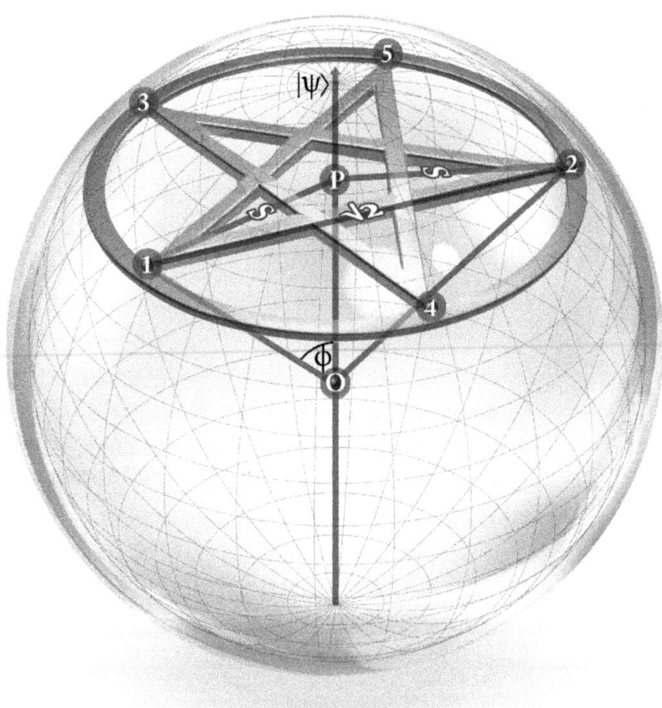


**Figure 6.10** Pentagram with probabilities defined by a quantum state through the center.

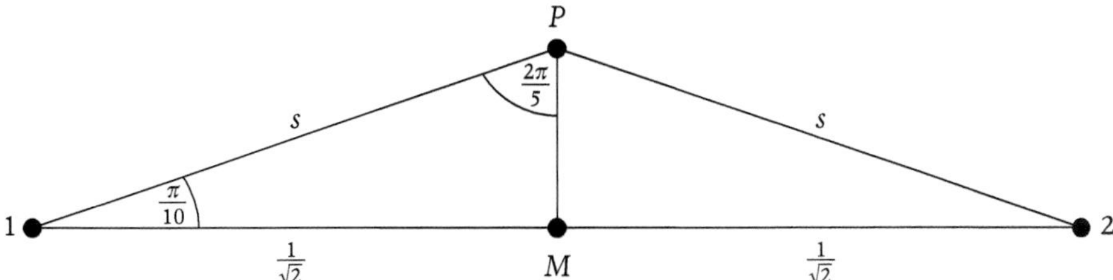


**Figure 6.11** Calculating $s$.

You're almost there. To find the length $s$, look at the triangle defined by the points $\{P, 1, 2\}$ in Figure 6.10, reproduced on its own in Figure 6.11. The angle subtended at $P$ by the lines joining $P$ to the two vertices of an edge is $4\pi/5$. Since the lines from the center $O$ of the sphere to the vertices of an edge of the pentagram in $\Sigma_2$ are orthogonal radii of length 1, the length of the edge joining the vertices labeled 1 and 2 is $\sqrt{2}$, by Pythagoras's theorem. If you draw a vertical line from $P$ to the midpoint, $M$, of the edge from vertex 1 to vertex 2, the length of the line from vertex 1 to $M$ is $\frac{\sqrt{2}}{2} = \frac{1}{\sqrt{2}}$. The right-angled triangle with the vertices $P, 1, M$ has an angle $\frac{1}{2} \cdot \frac{4\pi}{5} = \frac{2\pi}{5}$ at $P$ and $\frac{\pi}{2} - \frac{2\pi}{5} = \frac{\pi}{10}$ at vertex 1. See Figure 6.11. So $s \cos \frac{\pi}{10} = \frac{1}{\sqrt{2}}$, which gives $s = \frac{1}{\sqrt{2} \cos \frac{\pi}{10}}$. From the previous expression for $s$, you have $\cos^2 \phi = 1 - s^2 = 1 - \frac{1}{2\cos^2 \frac{\pi}{10}}$.

For any angle $\theta$, $1 + \cos\theta = 2\cos^2 \frac{\theta}{2}$ (see the subsection "Some Useful Trigonometry" in the "More" section of the "Bananaworld" chapter), so

$$1 - \frac{1}{2\cos^2 \dfrac{\pi}{10}} = 1 - \frac{1}{1 + \cos \dfrac{\pi}{5}} = \frac{\cos \dfrac{\pi}{5}}{1 + \cos \dfrac{\pi}{5}}.$$

Since $\cos \pi/5 = \dfrac{1}{4}(1 + \sqrt{5})$, this expression becomes

$$\frac{\dfrac{1}{4}\left(1 + \sqrt{5}\right)}{1 + \frac{1}{4}\left(1 + \sqrt{5}\right)} = \frac{1}{\sqrt{5}}.$$

This is the probability assigned to a vertex by the state $|\psi\rangle$. The probability of a vertex is the probability, in the state $|\psi\rangle$, that the vertex is selected or assigned a 1 in a measurement of the observable corresponding to an edge containing the vertex. The probability that the two vertices of an edge get assigned different values is the probability that one or the other of the vertices is selected and gets assigned a 1 in a measurement of the observable corresponding to the edge. That's $1/\sqrt{5} + 1/\sqrt{5} = 2/\sqrt{5}$. So the sum of these probabilities for all five edges is $5 \times 2/\sqrt{5} = 2\sqrt{5} \approx 4.47$, which is closer to the value 5 for Klyachko bananas than the maximum noncontextual value 4.

##  Violating the Klyachko inequality

- In Section 6.3, "The Kochen–Specker Theorem and Klyachko Bananas," I showed that the sum of the probabilities for the five pentagram edges is less than or equal to four in any noncontextual assignment of tastes to the bananas. Klyachko and colleagues show that if tastes are assigned to the bananas via measurements on quantum observables, you can achieve a value of $2\sqrt{5} \approx 4.47$ with a certain quantum state and violate the noncontextuality inequality.
- It's possible to draw a circle on a unit sphere between the equator and the north pole, with an inscribed pentagon and pentagram such that the lines joining the center of the sphere to the two vertices of an edge of the pentagram are orthogonal. See Figure 6.9.
- The vertices on each edge of the pentagram can therefore be associated with two of the three possible values of a qutrit observable (the lines joining the center of the sphere to the vertices are the orthogonal eigenstates of the observable). Each edge is associated with a different noncommuting observable. Since a vertex belongs to two edges, each vertex is associated with two measurement contexts.
- The quantum state $|\psi\rangle$ that passes through the north pole of the sphere assigns each vertex a probability of $1/\sqrt{5}$. Each edge, with two vertices, is assigned a probability $1/\sqrt{5} + 1/\sqrt{5} = 2/\sqrt{5}$. So the sum of the probabilities for all five edges is $5 \times 2/\sqrt{5} = 2\sqrt{5} \approx 4.47$, which is greater than 4, the maximum noncontextual value.

# Notes

1. The term "intertwined" was used by A.N. Gleason in a seminal paper, "Measures on the closed subspaces of Hilbert space," *Journal of Mathematics and Mechanics* 6, 885–893 (1957), to refer to the way in which observables in quantum mechanics are related; see p. 886. In general, an observable representing a property of a system belongs to different sets of commuting observables that don't commute with each other. I'll follow my colleague, Allen Stairs, and use the term "entwinement."
2. The term "pseudo-telepathy" was introduced by Gilles Brassard, Anne Broadbent, and Alain Tapp in their paper "Multi-party pseudo-telepathy," Proceedings of the 8th International Workshop on Algorithms and Data Structures in *Lecture Notes in Computer Science* 2748, 1–11 (2003). See also Gilles Brassard, Anne Broadbent, and Alain Tapp, "Quantum pseudo-telepathy," *Foundations of Physics* 35, 1877–1907 (2005); Anne Broadbent and André Allan Méthot, "On the power of non-local boxes," *Theoretical Computer Science* C 358, 3–14 (2006).
3. The Greenberger–Horne–Zeilinger (GHZ) correlation is from Daniel Greenberger, Michael Horne, and Anton Zeilinger, "Going beyond Bell's theorem," in M. Kafatos (ed.), *Bell's*

*Theorem, Quantum Theory, and Conceptions of the Universe* (Kluwer Academic, Dordrecht, 1989), pp. 69–72.

4. The proof that there is no two-party pseudo-telepathic game if the responses to the prompts are 0 or 1 is by Richard Cleve, Peter Hoyer, Ben Toner, and John Watrous, "Consequences and limits of nonlocal strategies," arXiv:quant-ph/0404076v2.
5. The Aravind–Mermin correlation is from the paper by P. K. Aravind, "Variation on a theme by GHZM," arXiv:quant-ph/0701031. Aravind's correlation was inspired by a similar pentagram correlation discussed by David Mermin in the paper "Hidden variables and the two theorems of John Bell," *Reviews of Modern Physics* 65, 803–814 (1993). Mermin's pentagram was in turn inspired by the Greenberger–Horne–Zeilinger correlation. Mermin's version of the pseudo-telepathic game is often referred to as the Mermin–GHZ game. Mermin comments (p. 810): "That the three-spin form of the Greenberger–Horne–Zeilinger version of Bell's Theorem could be reinterpreted as a version of the Bell-KS Theorem was brought to my attention by A. Stairs." See also Mermin's paper "Simple unified form for the major no-hidden-variables theorems," *Physical Review Letters* 65, 3373–3376 (1990).
6. The Kochen–Specker theorem: Simon Kochen and E. P. Specker, "On the problem of hidden variables in quantum mechanics," *Journal of Mathematics and Mechanics*, 17, 59–87 (1967).
7. Bell's version of the Kochen–Specker theorem: John Stuart Bell, "On the problem of hidden variables in quantum mechanics," *Review of Modern Physics* 38, 447–452 (1966). Reprinted in John Stuart Bell, *Speakable and Unspeakable in Quantum Mechanics* (Cambridge University Press, Cambridge, 1989).
8. The Conway-Kochen proof is unpublished. See also Sixia Yu and C.H. Oh, "State-independent proof of Kochen-Specker theorem with 13 rays," *Physical Review Letters* 108, 030402 (2012). While there exist noncontextual assignments of 0's and 1's to the thirteen rays such that one and only one ray in each orthogonal triple is assigned a 1, Oh and Yu show that every such assignment contradicts a prediction of quantum mechanics.
9. Asher Peres, "Two simple proofs of the Kochen–Specker theorem," *Journal of Physics A: Math. Gen* 24, L175–L178 (1991). Peres mentions Roger Penrose's curious observation in his book *Quantum Theory: Concepts and Methods* (Kluwer, Dordrecht, 1993), p. 212. Escher's *Waterfall* and *Ascending and Descending* were inspired by "impossible" objects designed, respectively, by Roger Penrose and his father, the geneticist Lionel Penrose, and reported in a paper published in the *British Journal of Psychology* in 1958. The Penroses sent a copy of the paper to Escher. David Mermin also mentions the connection between Escher and Penrose in "Hidden variables and the two theorems of John Bell," *Reviews of Modern Physics* 65, 803–815 (1993), p. 809. See also John Horgan, "The artist, the physicist, and the waterfall," *Scientific American* 268, February, 1993, p. 30.
10. The Klyachko version of the Kochen–Specker theorem: Alexander A. Klyachko, "Coherent states, entanglement, and geometric invariant theory," arXiv: quant-ph/0206012 (2002), and Alexander A. Klyachko, M. Ali Can, Sinem Binicioğlu, and Alexander S. Shumovsky, "Simple test for hidden variables in spin-1 systems," *Physical Review Letters* 101, 020403 (2008). I first learned of the Klyachko–Can–Binicioğlu–Shumovsky inequality and the connection with the Kochen–Specker theorem when Ben Toner sketched the argument in the air and on a scrap of paper at a party at my house in 2008 for our annual conference "New Directions in the Foundations of Physics." Ben Toner and David Bacon proved the surprising result that Alice and Bob can simulate the correlation of a maximally entangled state like $|\phi^+\rangle$ if they are allowed one bit of communication at each round of the simulation game: Ben F. Toner and David

Bacon, "The communication cost of simulating Bell correlations," *Physical Review Letters* 90, 157904 (2003).

11. The quantum strategy for simulating the Greenberger–Horne–Zeilinger correlation is a modified version of the strategy in Gilles Brassard, Anne Broadbent, and Alain Tapp, "Quantum pseudo-telepathy," *Foundations of Physics* 35, 1877–1907 (2005).
12. The quantum strategy for simulating Aravind–Mermin bananas is from P. K. Aravind, "Variation on a theme by GHZM," arXiv:quant-ph/0701031.
13. See Christoph Simon, Marek Zukowski, Harald Weinfurter, and Anton Zeilinger, "Feasible 'Kochen–Specker' experiment with single particles," *Physical Review Letters* 85, 1782 (2000), for how to measure the product observables $-Z_1Z_2Z_3$, $-Z_1X_2X_3$, $-X_1Z_2X_3$, $-Z_1X_2X_3$ on the edge $e_1$ of the Aravind–Mermin magic pentagram in a quantum simulation of Aravind–Mermin bananas, without measuring the single-qubit observables (so the measurement outcomes provide values for the product observables without providing values for the single-qubit observables).
14. The quantum strategy for violating the Klyachko inequality is based on the discussion in Alexander A. Klyachko, M. Ali Can, Sinem Binicioğlu, and Alexander S. Shumovsky "Simple test for hidden variables in spin-1 systems," *Physical Review Letters* 101, 020403 (2008).

# 7
# Quantum Secrets

This chapter is about sharing secrets, and the difference between Bananaworld, the quantum world, and the classical world about what you can and can't do.

Suppose Alice and Bob are separated and want to communicate a secret message, without revealing any information to Eve, the evil eavesdropper. There are standard, publicly available schemes for encoding messages as sequences of bits by representing letters of the alphabet, spaces, and punctuation symbols as binary numbers. In a classical world, using some such scheme, they can communicate with complete security if they share what's called a "one-time pad," a cryptographic key represented by a sequence of random bits at least as long as the number of bits required to communicate the message. In fact, this is the only way to achieve perfect security in a classical world. To send a message to Bob, Alice communicates which bits in the key Bob should flip. The resulting sequence of bits is the message. Even if Alice's communication about which bits are flipped is a public announcement over the Internet, Eve has no information about the secret message as long as she has no information about the key.

Here's another way to think about this. The "flipping" message that Alice sends Bob—the "cyphertext"—is a sequence of 0's and 1's, where 0 means "don't flip" and 1 means "flip." Suppose the key (a sequence of bits in the one-time pad) is 01001 and the message Alice wants to communicate—the "plaintext"—is 10011. You get the cyphertext by adding the key to the plaintext bitwise, or mod 2 (so that adding 1 to 1 takes you back to 0):

$$\begin{array}{rr} \text{plaintext} & 10011 \\ \text{key} & \underline{01001} \\ \text{cyphertext} & 11010 \end{array}$$

To extract the plaintext from the cyphertext, you simply add the key to the cyphertext:

$$\begin{array}{rr} \text{cyphertext} & 11010 \\ \text{key} & \underline{01001} \\ \text{plaintext} & 10011 \end{array}$$

The problem is that messages communicated in this way are only secret if Alice and Bob use a different sequence of random bits for each message, in other words if they really do use a *one-time* pad. If they use the same random key for several messages, Eve will be able to figure out the correspondence between letters of the alphabet and subsequences of bits in the key by relating statistical features of the messages to the way words are composed of letters. If Alice and Bob share a very long key they can use this for several messages, but eventually the key will run out and they'll need to share a new key, which means relying on trusted couriers or some similar method to distribute the key. Classical physics can't guarantee the security of a key distribution procedure. If Eve does gain access to the key, perhaps by bribing the couriers, she can copy it without revealing her intervention to Alice and Bob, who would then be in the disastrous position of continuing to use the key without suspecting that their messages are being read by Eve. That's the key distribution problem.

Classical physics also can't guarantee that a procedure for storing a key is completely secure, and can't guarantee that breaching the security and copying a key will always be detected. So apart from the key distribution problem, there's a key storage problem.

Quantum entanglement provides a way of solving these problems because entanglement is "monogamous." An entanglement between Alice's qubit and Bob's qubit can't be shared with Eve's qubit, and any attempt by Eve to eavesdrop on communications between Alice and Bob can always be revealed by exploiting Bell inequalities. In fact, Alice and Bob can share secrets even if Eve is from an advanced extraterrestrial civilization and uses some post-quantum theory to compromise the security of the devices Alice and Bob use to communicate, so long as Eve's post-quantum theory is a no-signaling theory. One could say that the possibility of *privacy* in this sense is what distinguishes correlations in the quantum world, or any nonclassical no-signaling world, from classical correlations.

The next section is about how Alice and Bob can share secrets with unconditional security—security guaranteed by the laws of physics—in a nonclassical no-signaling world.

## 7.1 REVEALING EVE IN BANANAWORLD

In Bananaworld, privacy is possible because Alice and Bob can generate secret bits that only they know by peeling and tasting Popescu–Rohrlich bananas. To generate a sequence of secret bits when they are separated, they begin by picking lots of pairs of bananas from a Popescu–Rohrlich tree. Alice keeps one banana from each pair, and Bob the other, and they label the bananas in order so they can keep track of the different pairs when they separate. To share a secret bit, Alice peels and tastes a banana, and Bob does the same with his paired banana. They publicly announce the way they peel their bananas, either from the top end or the stem end, but they keep the tastes private. Since Alice knows how Bob peeled his banana, and Bob knows how Alice peeled her banana, they now share the same secret bit. If the peelings are *SS*, *ST*, or *TS*, the tastes are the same, so they share 0 if the bananas taste ordinary or 1 if the bananas taste intense. If the peeling is *TT*, the tastes are opposite. In this case, Alice (but not Bob) flips the bit corresponding to her taste to share the same bit as the bit corresponding to the taste of Bob's banana. To share more secret bits, they peel and taste more bananas.

There's no storage problem, because Alice and Bob store bananas, not the random key. If they run out of bananas, they pick some more from a Popescu–Rohrlich tree. Since they are distributing bananas and not a secret key, there's no distribution problem. The key is generated *when needed* by peeling and tasting bananas. There's no information Eve can gain even if she has access to Alice's or Bob's stored bananas, because the Popescu–Rohrlich correlation is monogamous, as I showed in Section 4.2, "Really Random Bananas," in Chapter 4, "Really Random," and can only be shared between Alice and Bob, and not a third party.

If Eve could share the correlation by cloning Alice's bananas, she would be able to use the public information about how Alice peels her bananas to monitor the key as it's generated by peeling and tasting her cloned bananas, and that would deprive Bob of his free choice to peel his banana by the stem end or the top end, or violate the no-signaling principle. If Eve peels her cloned banana differently from the way Alice peels her banana, so one banana is peeled *S* and the other *T*, and these events occur before Bob peels his banana, then Bob must peel *S* if the tastes of Alice's banana and the cloned banana are the same. Otherwise, if Bob peels *T*, the taste of his banana would have to be the same as the taste of the banana peeled *S* but different from the taste of the banana peeled *T*, which is impossible since the tastes of these bananas are the same. If the tastes of Alice's banana and the cloned banana are different, then Bob must peel *T*. Otherwise, if he peels *S*, the taste of his banana would have to be the same as the taste of the banana peeled *S* and the taste of the banana peeled *T*, which is impossible since the tastes of these bananas are different. Alternatively, if Bob peels before Alice and Eve, the no-signaling principle is violated.

Free choice here, as in the chapter "Really Random," means that the choice is independent of any events that are not in the future light cone of the choice, which is to say that the choice is independent of anything that could not in principle have been caused by the choice. It's an interesting further question to consider how much free choice you need to guarantee completely private secret communication—whether partial rather than total independence is sufficient. Jonathan Barrett and Nicolas Gisin have investigated this question.[1]

Suppose it's inconvenient for Alice and Bob to pick their own bananas and they rely on someone, who turns out to be in the pay of Eve, to supply them with pairs of bananas. Alice and Bob can check that the bananas are genuine Popescu–Rohrlich pairs, and not fakes that Eve can monitor or control, by peeling and tasting a randomly chosen fair sample of the bananas and publicly communicating the way they peeled their bananas and the tastes. If the peelings and tastes don't satisfy the Popescu–Rohrlich correlation, they know there's something wrong with the bananas and they look for a new supplier and start over. If the peelings and tastes satisfy the Popescu–Rohrlich correlation for the sample, and they are confident that the sample is a fair sample, they use the remaining bananas to generate a secret key. The monogamy of the correlation guarantees that Eve has no information about the key.

So if Alice and Bob use Popescu–Rohrlich bananas to communicate in Bananaworld, secret communication is guaranteed by the monogamy of the correlation and free choice or no signaling. But privacy is also possible even if the banana pairs are supplied by an adversary and are not in fact Popescu–Rohrlich pairs, because eavesdropping by Eve can always be revealed in a nonclassical, no-signaling world even if the correlation is not monogamous. Depending on the extent to which the bananas have been compromised, Alice and Bob can still distill a secret key from the correlations by a procedure called "privacy amplification." See the subsection "Revealing Eve in a Device-Independent Scenario" later in this section. The same sort of thing is possible if the currently known correlation phenomena are as described by quantum mechanics, like the correlations of pairs of photons in the entangled quantum state $|\phi^{+}\rangle = |0\rangle|0\rangle + |1\rangle|1\rangle$, but quantum mechanics has been replaced by some (no-signaling) post-quantum theory, where the correlations aren't necessarily monogamous.

I'll show how this can be done below, but first I'll consider how Alice and Bob could share a secret in a world in which quantum mechanics is true.

**The bottom line**

- To share a secret bit in Bananaworld, Alice peels and tastes a Popescu–Rohrlich banana, and Bob does the same with his paired banana. They publicly announce the way they peel their bananas, but they keep the tastes private.
- Since Alice knows how Bob peeled his banana, and Bob knows how Alice peeled her banana, they now share the same secret bit. If the peelings are *SS*, *ST*, or *TS*, the tastes are the same, so they share the bit 0 if the bananas taste ordinary or 1 if the bananas taste intense. If the peeling is *TT*, the tastes are opposite. In this case, Alice (but not Bob) flips the bit corresponding to her taste to share the same bit as the bit corresponding to the taste of Bob's banana. To share more secret bits, they peel and taste more bananas.
- There's no storage problem, because Alice and Bob store bananas, not the random key. If they run out of bananas, they pick some more from a Popescu–Rohrlich tree. Since they are distributing bananas and not a secret key, there's no distribution problem. The key is generated when needed by peeling and tasting bananas.
- There's no information an eavesdropper, Eve, can gain without giving herself away even if she has access to Alice's or Bob's stored bananas, because the Popescu–Rohrlich correlation is monogamous and can only be shared by Alice and Bob and not a third party.

## 7.2 REVEALING EVE IN THE QUANTUM WORLD

Here's a way Alice and Bob could share a secret key in a quantum world.[2] Alice has a collection of photons, one for each entangled pair in the state $|\phi^+\rangle$, and Bob has a collection of paired photons. Alice measures the polarization of her photons in order, choosing randomly from the directions $0, \pi/8, 2\pi/8$ with respect to some direction $z$ they agree on in advance, and Bob measures the polarizations of his photons in order, choosing randomly from the directions $\pi/8, 2\pi/8, 3\pi/8$. They communicate the directions of their polarization measurements publicly, but not the outcomes, and they divide the measurements into two sets: one set when they both measured polarization in the direction $\pi/8$, or when they both measured polarization in the direction $2\pi/8$, and one set when Alice measured polarization in directions 0 or $2\pi/8$ and Bob measured polarization in directions $\pi/8$ or $3\pi/8$. For the first set, when they measured polarization in the same direction, the outcomes are random but perfectly correlated in the state $|\phi^+\rangle$, so they share these random bits. The second set is not used for the key, but to reveal an eavesdropper.

The correlation array for the measurements in the second set is given by Table 4.6, reproduced here as Table 7.1, where $A, A'$ represent polarizations in directions $0, 2\pi/8$ for Alice, and $B, B'$ represent polarizations in directions $\pi/8, 3\pi/8$ for Bob.

From the correlation array you can read off that

$$p(\text{same}\,|\,AB)+p(\text{same}\,|\,A'B)+p(\text{same}\,|\,A'B')+p(\text{different}\,|\,AB') = 4\cos^2\frac{\pi}{8} = 2+\sqrt{2} \approx 3.414,$$

since $\cos^2\frac{\pi}{8} = \frac{2+\sqrt{2}}{4}$, one of the useful trigonometric facts listed in Subsection 3.5.1, "Some Useful Trigonometry," in the "More" section of the "Bananaworld" chapter, or look at the table of approximate values in Table 4.7, reproduced here as Table 7.2.

In the "Bananaworld" chapter, in Section 3.2, "Popescu–Rohrlich Bananas and Bell's Theorem," I derived an expression for an equivalent sum of probabilities as $4 \times \frac{1}{2}(1 + \frac{K}{4})$.

**Table 7.1** Quantum correlation array for revealing Eve in the key distribution protocol with entangled states $|\phi^+\rangle$.

| | Alice | $A$ | | $A'$ | |
|---|---|---|---|---|---|
| Bob | | 0 | 1 | 0 | 1 |
| $B$ | 0 | $\frac{1}{2}\cos^2\frac{\pi}{8}$ | $\frac{1}{2}\sin^2\frac{\pi}{8}$ | $\frac{1}{2}\cos^2\frac{\pi}{8}$ | $\frac{1}{2}\sin^2\frac{\pi}{8}$ |
| | 1 | $\frac{1}{2}\sin^2\frac{\pi}{8}$ | $\frac{1}{2}\cos^2\frac{\pi}{8}$ | $\frac{1}{2}\sin^2\frac{\pi}{8}$ | $\frac{1}{2}\cos^2\frac{\pi}{8}$ |
| $B'$ | 0 | $\frac{1}{2}\sin^2\frac{\pi}{8}$ | $\frac{1}{2}\cos^2\frac{\pi}{8}$ | $\frac{1}{2}\cos^2\frac{\pi}{8}$ | $\frac{1}{2}\sin^2\frac{\pi}{8}$ |
| | 1 | $\frac{1}{2}\cos^2\frac{\pi}{8}$ | $\frac{1}{2}\sin^2\frac{\pi}{8}$ | $\frac{1}{2}\sin^2\frac{\pi}{8}$ | $\frac{1}{2}\cos^2\frac{\pi}{8}$ |

**Table 7.2** Approximate values for the probabilities in Table 7.1.

| Bob \ Alice | | A: 0 | A: 1 | A′: 0 | A′: 1 |
|---|---|---|---|---|---|
| B | 0 | .427 | .073 | .427 | .073 |
| | 1 | .073 | .427 | .073 | .427 |
| B′ | 0 | .073 | .427 | .427 | .073 |
| | 1 | .427 | .073 | .073 | .427 |

The optimal quantum value of $K$, the Tsirelson bound, is $K_Q = 2\sqrt{2}$, so the optimal quantum value of the sum of probabilities is $\frac{4}{2}(1 + \frac{2\sqrt{2}}{4}) = 2 + \sqrt{2}$, as given by the correlation array in Table 7.1. The optimal value of $K$ for classical or common cause correlations is 2, so the optimal value of the sum of probabilities is $\frac{4}{2}(1 + \frac{2}{4}) = 3$.

If the sum of probabilities for the measurements in the second set of polarization measurements is $2+\sqrt{2}$, and this set of measurements can be assumed to be a fair sample of the total set of measurements, then Alice and Bob infer that the correlation is indeed the correlation of the Bell state $|\phi^+\rangle$. Since this correlation is monogamous, like the Popescu–Rohrlich correlation, they can be confident that Eve has no information about the key they abstracted from the perfectly correlated first set of measurements. Any attempt by Eve to gain information about the shared key will involve tampering with the photons in some way and destroying the entangled state. In that case, the sum of probabilities in the second set can't be more than the optimal classical value of 3, because only an entangled quantum state can yield a value greater than 3. This quantum key distribution protocol was first proposed by Artur Ekert in a 1991 paper in *Physical Review Letters*.

## The bottom line

- In a quantum world, the correlation of a pair of qubits in the entangled state $|\phi^+\rangle$—the closest you can get in a quantum world to the correlation of a pair of Popescu–Rohrlich bananas—is monogamous. The Ekert protocol shows how to exploit the monogamy of entanglement to allow Alice and Bob to share a secret key.
- To share a secret key, Alice and Bob share many pairs of photons in the state $|\phi^+\rangle$. Alice measures the directions of her photons in order, choosing randomly from the directions $0, \pi/8, 2\pi/8$ with respect to some direction $z$ they agree on in advance, and Bob measures the polarizations of his photons in order, choosing randomly from the directions $\pi/8, 2\pi/8, 3\pi/8$. They communicate the directions of their polarization measurements publicly, but not the outcomes, and they divide the measurements into two sets: one set when they both measured polarization in the same direction, $\pi/8$ or $2\pi/8$, and

*(continued)*

 **The bottom line** *(continued)*

one set when Alice measured polarization in directions 0 or $2\pi/8$ and Bob measured polarization in directions $\pi/8$ or $3\pi/8$. The outcomes of polarization measurements in the same direction on two qubits in the state $|\phi^+\rangle$ are perfectly correlated, so the first set is used to establish the secret key.

- The second set is used to reveal eavesdropping by Eve. Any attempt by Eve to obtain information about the shared key will involve tampering with the photons in the entangled state $|\phi^+\rangle$. The monogamy of the entangled state ensures that Alice and Bob will be able to detect any eavesdropping.

## 7.3 REVEALING EVE IN A DEVICE-INDEPENDENT SCENARIO

Ekert and Renner have an insightful way of representing general no-signaling correlations in terms of a parameter $\varepsilon$ that measures the amount of deviation from the Popescu–Rohrlich correlation.[3] Table 7.1 can expressed as Table 7.3, with the optimal quantum value of $\varepsilon$ as $\sin^2 \frac{\pi}{8}$, which is approximately $2 \times .073 = .146$, as you can read off from Table 7.2. For the Popescu–Rohrlich correlation corresponding to this correlation array, $\varepsilon = 0$. That's the Popescu–Rohrlich correlation $a \oplus b = (A \oplus 1) \cdot B$ with the correlation array in Table 4.5, reproduced here as Table 7.4.

The parameter $\varepsilon$ is related to the Clauser–Horne–Shimony–Holt quantity $K$, or to the parameter $E = K/4$. The sum of probabilities $p(\text{same}\,|\,AB)+p(\text{same}\,|\,A'B)+p(\text{same}\,|\,A'B')+$ $p(\text{different}\,|\,AB')$ is equal to $4 \times \frac{1}{2}(1 + \frac{K}{4}) = 4 \times \frac{1}{2}(1 + E)$, and from the correlation array in

**Table 7.3** No-signaling correlation array.

| Bob | Alice | $A$ 0 | $A$ 1 | $A'$ 0 | $A'$ 1 |
|---|---|---|---|---|---|
| $B$ | 0 | $\frac{1-\varepsilon}{2}$ | $\frac{\varepsilon}{2}$ | $\frac{1-\varepsilon}{2}$ | $\frac{\varepsilon}{2}$ |
| | 1 | $\frac{\varepsilon}{2}$ | $\frac{1-\varepsilon}{2}$ | $\frac{\varepsilon}{2}$ | $\frac{1-\varepsilon}{2}$ |
| $B'$ | 0 | $\frac{\varepsilon}{2}$ | $\frac{1-\varepsilon}{2}$ | $\frac{1-\varepsilon}{2}$ | $\frac{\varepsilon}{2}$ |
| | 1 | $\frac{1-\varepsilon}{2}$ | $\frac{\varepsilon}{2}$ | $\frac{\varepsilon}{2}$ | $\frac{1-\varepsilon}{2}$ |

**Table 7.4** The Popescu–Rohrlich correlation corresponding to $\varepsilon = 0$.

| Bob \ Alice | | $S$ 0 | $S$ 1 | $T$ 0 | $T$ 1 |
|---|---|---|---|---|---|
| $S$ | 0 | 1/2 | 0 | 1/2 | 0 |
| | 1 | 0 | 1/2 | 0 | 1/2 |
| $T$ | 0 | 0 | 1/2 | 1/2 | 0 |
| | 1 | 1/2 | 0 | 0 | 1/2 |

Table 7.3 you can read off the sum of probabilities as $4 \times (1-\varepsilon)$. So $4 \times (1-\varepsilon) = 4 \times \frac{1}{2}(1+E)$, from which it follows that $\varepsilon = \frac{1}{2}(1 - E)$.

In the "Bananaworld" chapter, I showed that correlations for which $E$ is less than or equal to $1/2$ are classical or local because they can be simulated with local resources. That amounts to saying that the variables $A, A'$ and $B, B'$ have definite preassigned values, because a successful local simulation requires Alice and Bob to share a list where each item on the list corresponds to an assignment of values to the four variables $A, A', B, B'$. Since $\varepsilon = \frac{1}{2}(1 - E)$, the condition $E \leq 1/2$ for classical or local correlations is equivalent to $\varepsilon \geq 1/4$. So for $\varepsilon < 1/4$, the correlations are nonclassical or nonlocal and exclude the possibility of definite preassigned values to the observables.

In Section 4.3, "Really Random Qubits," in the chapter "Really Random," I used an argument by Ekert and Renner to show that the probability of Eve successfully predicting Alice's measurement outcomes for the correlation in Table 7.1, based on any information held by Eve about events before the preparation of the correlated systems, in any reference frame, can be no better than a random guess. The argument is independent of quantum mechanics and holds for any no-signaling theory. Specifically, I showed that if Eve has a device that is designed to output values $z$ that are the same as the outcomes for Alice's $A$-measurement, then the probability that Alice's measurement outcome is $z$ is less than or equal to $\frac{1}{2}(1 + P_{2-})$. Since $P_{2-} + P_{2+} = 4$ and $P_{2+} = 4(1 - \varepsilon)$, it follows that $P_{2-} = 4\varepsilon$. So, in terms of $\varepsilon$, the probability that the outcome of Alice's $A$-measurement is tracked by $z$ is less than or equal to $\frac{1}{2}(1 + 4\varepsilon)$. The same goes for the outcome of Alice's $A'$-measurement, or the outcomes of any of Bob's measurements. For pairs of photon polarization measurements separated by an angle $\pi/8$, this probability is $\frac{1}{2}(1 + 4\sin\frac{\pi}{8}) \approx .793$. So, whatever her technological abilities, Eve's guess will be wrong more than 20% of the time.

Suppose Alice and Bob follow the Ekert protocol described in the previous subsection to share a secret key. Alice measures polarization in directions $0, \pi/8, 2\pi/8$, and Bob measures polarization in directions $\pi/8, 2\pi/8, 3\pi/8$. Suppose that in the future the currently known correlation phenomena are as *described* by quantum mechanics, but the phenomena are no longer *explained* by quantum mechanics, but by some no-signaling post-quantum theory, where entanglement isn't necessarily monogamous. Perhaps an adversary has already discovered this and come up with such a theory. Or suppose Alice and Bob suspect that the photon pairs they are using for the protocol might have come from Eve and so might not in fact be entangled photons in the state $|\phi^+\rangle$. From the polarization measurements in directions $\pi/8, 2\pi/8$, Alice and Bob can establish a "raw" key. They can't assume that the correlations are monogamous. So they can't confirm that Eve knows nothing about the key even if they calculate $\varepsilon$ as .146, the quantum value for the entangled state $|\phi^+\rangle$, from the polarization measurements in directions $0, 2\pi/8$ for Alice and $\pi/8, 3\pi/8$ for Bob. But they can use the value of $\varepsilon$ to put a bound on how much information Eve has about the raw key—even an Eve from a galaxy far, far away with amazing technological abilities using a post-quantum theory superior to quantum mechanics (provided the post-quantum theory satisfies the no-signaling principle).

For $\varepsilon = .146$, the value for the entangled state $|\phi^+\rangle$, Eve's probability of correctly guessing a bit in the raw key is close to 80%. Alice and Bob can distill a shorter secret key from the raw key by privacy amplification. Here's the idea. Suppose Alice and Bob know that Eve knows at most one of two bits. If they add the two bits together mod 2, the resulting bit will be secret. For Eve, the operation amounts to adding a random bit to the bit she knows, so her probability of guessing the secret bit goes from 80% to 50%, which is no better than her ability to guess the outcome of a random coin toss. The cost to Alice and Bob is that a raw key of two bits is reduced by half. This is the simplest example of a variety of privacy amplification operations that can be applied to distill a secret key from a raw key. The technique works because Alice and Bob can use Bell inequalities to reveal *how much* Eve knows about the raw key.

If Alice measures polarization in directions $0, 2\theta, 4\theta, \ldots, (2n-2)\theta$, so for even multiples of $\theta$, and Bob measures polarization in directions $\theta, 3\theta, 5\theta, \ldots, (2n-1)\theta$, so for odd multiples of $\theta$, then, as in Section 4.3, "Really Random Qubits," in the chapter "Really Random," a chain argument shows that the probability of Eve correctly predicting the outcome of Alice's measurement outcome, for any of her measurements, is at most $\frac{1}{2}(1 + P_{n-})$.

Now $P_{n-} = 2n - P_{n+}$, and $P_{n+}$ is

$$\begin{aligned} P_{n+} &= p(\text{outcomes same}\,|\,A = 0, B = \theta) + p(\text{outcomes same}\,|\,A = 2\theta, B = \theta) + \cdots \\ &\quad + p(\text{outcomes different}\,|\,A = 0, B = (2n-1)\theta). \end{aligned}$$

In the case $n = 2$, when Alice and Bob each measure one of two observables, at most three out of four of these probabilities can be 1 if the correlations are classical or local (as I showed in the proof of Bell's theorem in Section 3.2, "Popescu–Rohrlich Bananas and Bell's Theorem" in the "Bananaworld" chapter). In the general case, each of these $2n$ probabilities can be 1, except the last (or, more generally, at most $2n - 1$ of these probabilities can be 1). So $P_{n+}$ can be at most $2n - 1$, or equivalently $P_{n-} \geq 1$. Eve's probability of correctly predicting the outcomes of Alice's measurements is less than or equal to $\frac{1}{2}(1 + P_{n-})$, which means that for classical correlations there is no constraint on Eve's ability to do this.

In a world in which the known phenomena, such as polarization correlations, are as described by quantum theory, you can have correlations for which $P_{n-} = 2n \sin^2 \frac{\pi}{4n}$. So $P_{n-} \leq 2/n$, since $\sin^2 \frac{\pi}{4n} \leq 1/n^2$, and $P_n$ can be as small as you like. In the limit as $n$ tends to infinity and $\theta$ tends to zero, $P_{n-}$ tends to zero. The probability that Eve correctly predicts the outcome of Alice's $A$-measurement tends to $1/2$, which is no better than a random guess.

So what's the upshot of all this? Private secret communication is possible in a quantum world because any attempt by an eavesdropper, Eve, to obtain information about the communication between Alice and Bob, or to influence the communication, can be revealed by exploiting the monogamy of entanglement correlations, assuming that Alice and Bob can freely choose what they measure. Classical correlations are not monogamous because they hold between variables with definite preassigned values which can be copied an unlimited number of times. Nothing prevents Alice, Bob, and Eve, or anyone else, from sharing the same preassigned values, or functionally related values, so any correlation between Alice and Bob could also hold between Alice and Eve, or between Bob and Eve, or between an unlimited number of parties.

Remarkably, Alice and Bob can distill a secret key from a raw key by privacy amplification if they can put a bound on what Eve knows about the raw key. They can reveal what Eve might know by exploiting Bell correlations, even if they suspect that the devices they use have been compromised by Eve. So as long as correlations are no-signaling and Alice and Bob are free to choose what they measure, device-independent cryptography is possible even if the explanation of phenomena now described by quantum mechanics is through a (no-signaling) post-quantum theory.[4]

## The bottom line

- The ability of Alice or Bob to share a secret key in Bananaworld or in a quantum world can be extended to a device-independent scenario. Alice and Bob can share a secret key even if they suspect that the devices they use to prepare and measure entangled quantum states ate insecure and might have been supplied by an adversary.
- Alice and Bob can establish a "raw key" by measuring polarization in the same directions on pairs of photons they assume to be in the entangled state $|\phi^+\rangle$. By exploiting the chained Bell inequality in Section 4.3, "Really Random Qubits," in the chapter "Really Random," they can put a bound on how much information an eavesdropper, Eve, has about the raw key.

(continued)

 **The bottom line** *(continued)*

- If they know how much Eve knows, they can distill a shorter secret key from the raw key by privacy amplification. As a simple example, suppose Alice and Bob know that Eve knows at most one of two bits. If they add the two bits together mod 2, the resulting bit will be secret. For Eve, the operation amounts to adding a random bit to the bit she knows, so her probability of guessing the secret bit is 1 / 2, which is no better than a random guess. The cost to Alice and Bob is that a raw key of two bits is reduced by half.

## Notes

1. How much free choice do Alice and Bob need to infer nonlocality from a violation of Bell's inequality? Jonathan Barrett and Nicolas Gisin investigate this question in "How much measurement independence is needed in order to demonstrate nonlocality?" *Physical Review Letters*, 106, 100406 (2011).
2. Artur Ekert proposed this key distribution scheme based on quantum entanglement in "Quantum cryptography based on Bell's theorem," *Physical Review Letters* 67, 661–663 (1991).
3. My discussion follows the *Perspective* article by Artur Ekert and Renato Renner, "The ultimate physical limits of privacy," in *Nature* 507, 443–447 (2014).
4. Jonathan Barrett, Lucien Hardy, and Adrian Kent showed that key distribution could be secure in a device-independent, no-signaling scenario in "No signaling and quantum key distribution," *Physical Review Letters* 95, 010503 (2005).

# 8

# Quantum Feats

This chapter is about three amazing quantum feats: using a quantum computer to manipulate entangled quantum states to solve certain problems dramatically faster than a classical computer running any known classical algorithm, exploiting quantum entanglement to teleport information, and exploiting entanglement to transmit two classical bits with one qubit. Building a practically useful quantum computer is still a quantum engineer's dream, but quantum teleportation is a reality that several groups have implemented.

## 8.1 QUANTUM COMPUTATION

As you may know, a quantum computer offers the possibility of a potentially massive speedup over any classical computer. Popular accounts generally describe the speedup as the result of "quantum parallelism"—the supposed ability of a quantum computer to perform a huge number of computations at the same time by encoding information as qubits rather than bits, on the grounds that a qubit can somehow represent 0 and 1 at the same time while a bit is either 0 or 1. In fact, a quantum computer achieves a speedup over known classical algorithms by being able to decide whether a disjunction, an "or" statement, "*P* or *Q* or *R* or . . .," is true or false, without having to work out whether the individual components of the "or" statement, like *P*, are either true or false. So rather than achieving a speedup over classical computers by performing multiple computations simultaneously, a quantum computer can compute a solution to a problem in fewer steps than a classical computer, because it can do so without working out the answers to questions that aren't asked, but that a classical computer needs to answer to solve the problem.[1]

If you ask a classical computer whether "*P* or *Q* or *R* or . . ." is true or false, the computer has to first figure out whether *P* is true (or begin with some other component and figure out whether that is true). The computer then has to store the information in computer memory as a preliminary step in answering the question about the "or" statement as a whole. If *P* is false, the computer has to check *Q*. If *Q* is false, it has to check *R*, and so on. In the worst case, the computation will involve as many steps as there are components in the "or" statement, where each step is a mini-computation.

Now, the question is whether the "or" statement *as a whole* is true or false, not which components of the "or" statement are true and which are false. Is there a way of getting the answer to this question without first going to all the work of answering the question about $P$, and then about $Q$, and then about $R$, ..., and storing the information in computer memory? In fact, this is impossible with a classical computer, because the only way a classical computer can decide whether an "or" statement is true is by checking the truth or falsity of its component statements until it finds one that's true, or finds that none are true. By contrast, a quantum computer can decide the question without going through all the steps required to show that at least one of the components, $P, Q, R, \ldots$, is true or that none are.

I'll illustrate how this works by going through the simplest quantum algorithm, proposed by David Deutsch.[2] Suppose you have a function that takes 0 or 1 as the only possible input values and produces output values that are also either 0 or 1. A function with inputs and outputs that are 0's and 1's (or sequences of 0's and 1's) is called a Boolean function. You could imagine different functions that involve combinations of Boolean sums and products, but these can all be divided into two sorts:

- functions that produce the same output for the two possible inputs 0 and 1: $f(0) = f(1) = 0$, *or* $f(0) = f(1) = 1$;
- functions that produce different outputs for the two possible inputs 0 and 1: $f(0) = 0$ and $f(1) = 1$, *or* $f(0) = 1$ and $f(1) = 0$.

Deutsch calls Boolean functions that produce the same output for the two possible inputs "constant," and Boolean functions that produce different outputs for the two possible inputs "balanced." To say that a function is constant is to say that "$f(0) = f(1) = 0$ *or* $f(0) = f(1) = 1$." Similarly, to say that a function is balanced is to say that "$f(0) = 0$ and $f(1) = 1$ *or* $f(0) = 1$ and $f(1) = 0$." So the problem of figuring out whether a Boolean function $f$ is constant or balanced amounts to figuring out which of these two "or" statements is true.

The only way to do this, if you don't have a quantum computer, is to evaluate $f$ twice, once for each of the two possible inputs 0 or 1, and compare the output values you get. That's two steps for 0 and 1, or two mini-computations, as well as further steps to store and compare the results of the mini-computations.

Now suppose you have a quantum computer, and you know how to program the computer so as to represent the function $f$ dynamically as a transformation of quantum states. (To understand how this works, you'll need to know something about how quantum states evolve in time. See the section "Quantum Dynamics" in the Supplement, "Some Mathematical Machinery," at the end of the book.)

A quantum computer has an input register for quantum states representing the input values of $f$, and an output register for quantum states representing the output values. These are one-qubit registers that are initially set to the quantum state $|0\rangle_i|0\rangle_o$. For example, the states $|0\rangle_i$ and $|0\rangle_o$ could be the quantum states of two photons both linearly polarized in some direction. The first state in the product with subscript $i$ is the state of the input register and the second state with subscript $o$ is the state of the output register. I'll drop the subscripts $i$ and $o$ in the following—the order (input first, output second) indicates which state represents the input state and which state represents the output state.

A transformation called a Hadamard transformation is applied to the input register. A Hadamard transformation takes the state $|0\rangle$ to the state $|0\rangle + |1\rangle$ and the state $|1\rangle$ to the state $|0\rangle - |1\rangle$. (Reminder: the convention is to omit the two equal coefficients of the states $|0\rangle$ and $|1\rangle$, which are both $1/\sqrt{2}$ here.) For a photon, a Hadamard transformation transforms an initial state of linear polarization, horizontal or vertical in some direction $z$, to a state of diagonal polarization, horizontal or vertical in a direction 45° to $z$. So the first step in Deutsch's algorithm is:

$$|0\rangle|0\rangle \xrightarrow{H} (|0\rangle + |1\rangle)|0\rangle = |0\rangle|0\rangle + |1\rangle|0\rangle,$$

where the $H$ above the arrow indicates that the arrow represents a Hadamard transformation.

The second step is to apply the unitary transformation $U_f$ corresponding to the function $f$ to both registers. The transformation $U_f$ will transform the state $|0\rangle|0\rangle$ of the input and output registers to the state $|0\rangle|f(0)\rangle$ (because the function $f$ takes the input 0 to the output $f(0)$, where $f(0)$ is either 0 or 1), and it will transform the state $|1\rangle|0\rangle$ of the input and output registers to the state $|1\rangle|f(1)\rangle$ (because the function $f$ takes the input 1 to the output $f(1)$, where $f(1)$ is either 0 or 1). So $U_f$ will transform a linear superposition of these states into a linear superposition of $|0\rangle|f(0)\rangle$ and $|1\rangle|f(1)\rangle$:

$$|0\rangle|0\rangle + |1\rangle|0\rangle \xrightarrow{U_f} |0\rangle|f(0)\rangle + |1\rangle|f(1)\rangle.$$

If $f$ is constant, $f(0)$ and $f(1)$ are both 0, or both 1. So the final combined state of the input and output registers is one of the two states:

$$|c_1\rangle = |0\rangle|0\rangle + |1\rangle|0\rangle, \tag{8.1}$$

$$|c_2\rangle = |0\rangle|1\rangle + |1\rangle|1\rangle, \tag{8.2}$$

depending on whether $f$ takes both input values to 0, or whether $f$ takes both input values to 1. The states $|c_1\rangle$ and $|c_2\rangle$ are just product states: $|c_1\rangle = (|0\rangle + |1\rangle)|0\rangle$ and $|c_2\rangle = (|0\rangle + |1\rangle)|1\rangle$. The state of the input register is the same in both cases, while the state of the output register is $|0\rangle$ in the case of $|c_1\rangle$ and $|1\rangle$ in the case of $|c_2\rangle$, and these states are orthogonal. So $|c_1\rangle$ and $|c_2\rangle$ are orthogonal.

The orthogonal states $|c_1\rangle$ and $|c_2\rangle$ "span" a plane in the four-dimensional state space of the qubits. (The state spaces of the input and output qubits are two-dimensional, so the combined state space of the two qubits is four-dimensional.) Think of the state vectors as represented by two orthogonal lines. To say that the orthogonal states $|c_1\rangle$ and $|c_2\rangle$ span a plane in this state space is to say that any vector in the plane can be represented as a linear sum or superposition of these two state vectors. Call this plane the "constant plane."

If you add these two state vectors with equal coefficients $1/\sqrt{2}$ (to ensure that you end up with a state vector of unit length) you get a state vector halfway between the two (at an angle 45° to the two state vectors):

$$|c_1\rangle + |c_2\rangle = |0\rangle|0\rangle + |0\rangle|1\rangle + |1\rangle|0\rangle + |1\rangle|1\rangle.$$

(Reminder: the equal coefficients here are all $1/\sqrt{4} = 1/2$.)

If the function is balanced, the final combined state of the input and output registers is one of the two states:

$$|b_1\rangle = |0\rangle|0\rangle + |1\rangle|1\rangle, \tag{8.3}$$

$$|b_2\rangle = |0\rangle|1\rangle + |1\rangle|0\rangle, \tag{8.4}$$

depending on whether $f$ takes 0 to 0 and 1 to 1, or $f$ takes 0 to 1 and 1 to 0. You can check that the state vectors $|b_1\rangle$ and $|b_2\rangle$ are orthogonal. (In this case, the calculation is not so obvious because $|b_1\rangle$ and $|b_2\rangle$ are entangled states. You'll need to show that the scalar product is zero. See the sections "Dirac's Ingenious Idea" and "The Pauli Operators" in the Supplement, "Some Mathematical Machinery," at the end of the book.) The two balanced state vectors span a plane called the "balanced plane."

Again, if you add these two state vectors with equal coefficients you get a state vector halfway between the two (at an angle 45° to the two state vectors), which is *the same state vector* as the linear superposition of $|c_1\rangle$ and $|c_2\rangle$:

$$|b_1\rangle + |b_2\rangle = |0\rangle|0\rangle + |0\rangle|1\rangle + |1\rangle|0\rangle + |1\rangle|1\rangle.$$

In other words, this state vector belongs to both planes, which is to say that the planes intersect in the line spanned by this state vector (the line defined by any multiple of this state vector).

If, instead of the original computational basis states $|0\rangle$, $|1\rangle$ for the input and output registers, you use the transformed basis states $|0'\rangle = H|0\rangle = |0\rangle + |1\rangle$ and $|1'\rangle = H|1\rangle = |0\rangle - |1\rangle$, then the state vector in the intersection of the constant and balanced planes is just $|0'\rangle|0'\rangle$:

$$|0'\rangle|0'\rangle = (|0\rangle + |1\rangle)(|0\rangle + |1\rangle) = |0\rangle|0\rangle + |0\rangle|1\rangle + |1\rangle|0\rangle + |1\rangle|1\rangle.$$

For a photon, this corresponds to expressing everything in terms of diagonally polarized states instead of the original linearly polarized states.

The state vectors $|0'\rangle|0'\rangle$ and $|0'\rangle|1'\rangle$ are rotated 45° to $|c_1\rangle$ and $|c_2\rangle$ (because $|0'\rangle|0'\rangle = |c_1\rangle + |c_2\rangle$ and $|0'\rangle|1'\rangle = (|0\rangle + |1\rangle)\cdot(|0\rangle - |1\rangle) = |0\rangle|0\rangle + |1\rangle|0\rangle - |0\rangle|1\rangle - |1\rangle|1\rangle) = |c_1\rangle - |c_2\rangle$). So these state vectors also span the constant plane (any vector in the plane can also be expressed as a linear superposition of these orthogonal state vectors). Similarly, the balanced plane is spanned by the state vectors $|0'\rangle|0'\rangle$ and $|1'\rangle|1'\rangle$. These state vectors are rotated 45° to $|b_1\rangle$ and $|b_2\rangle$ (because $|0'\rangle|0'\rangle = |b_1\rangle + |b_2\rangle$ and $|1'\rangle|1'\rangle = (|0\rangle - |1\rangle)\cdot(|0\rangle - |1\rangle) = |b_1\rangle - |b_2\rangle$).

To sum up, the state vector of the input and output registers ends up, after the unitary transformation corresponding to the function $f$, either in the constant plane or the balanced plane, depending on whether $f$ is constant or balanced. The constant plane is spanned by the state vectors $|0'\rangle|0'\rangle$, $|0'\rangle|1'\rangle$, and the balanced plane is spanned by the state vectors $|0'\rangle|0'\rangle$, $|1'\rangle|1'\rangle$.

So to find out whether the function $f$ is constant or balanced, you could simply measure an observable with two possible values ("eigenvalues") associated with the states ("eigenstates") $|0'\rangle$, $|1'\rangle$ of the input register, and the same observable of the output register. For two photons, that would involve measuring the diagonal polarization of each photon—passing the photons through two beamsplitters with their axes of polarization in the diagonal direction and seeing where each photon ends up. If the state is $|c_1\rangle$ or $|c_2\rangle$ in the constant plane, you'll get one of the outcomes $|0'\rangle|0'\rangle$ or $|0'\rangle|1'\rangle$ with probability $1/2$, since $|c_1\rangle$ and $|c_2\rangle$ are both at an angle 45° to $|0'\rangle|0'\rangle)$ and to $|0'\rangle|1'\rangle$. If the state is $|b_1\rangle$ or $|b_2\rangle$ in the balanced plane, you'll get one of the outcomes $|0'\rangle|0'\rangle$ or $|1'\rangle|1'\rangle$ with probability $1/2$, since $|b_1\rangle$ and $|b_2\rangle$ are both at an angle 45° to $|0'\rangle|0'\rangle$ and to $|1'\rangle|1'\rangle$.

If you get the outcome $|0'\rangle|0'\rangle$, which occurs with probability $1/2$ in both cases, whether $f$ is constant or balanced, you can't distinguish the two cases, and you'll need to repeat the procedure. But if you don't get the outcome $|0'\rangle|0'\rangle$, the algorithm answers the question about which "or" statement is true, corresponding to whether $f$ is constant or balanced. If the outcome is $|0'\rangle|1'\rangle$, $f$ is constant, and if the outcome is $|1'\rangle|1'\rangle$, $f$ is balanced. When the algorithm succeeds, it does so without storing any information about whether the components of the "or" statement are true or false (whether in the constant case $f$ takes both possible inputs to 0, *or* whether $f$ takes both possible inputs to 1, and in the balanced case, whether $f$ takes 0 to 0 and 1 to 1, *or* whether $f$ takes 0 to 1 and 1 to 0).

Deutsch's algorithm has an even probability of failing, but there is a variation of the algorithm by Richard Cleve, Arthur Ekert, Christina Macchiavello, and Michelle Mosca that avoids this feature.[3] I won't go into the details of this variation because it's a little more complicated, and the point of going through Deutsch's algorithm was simply to show how a quantum computation can work out whether an "or" statement is true of a function without working out whether the components of the "or" statement are true or false.

Daniel Simon's algorithm for finding the period of a function, and Peter Shor's factorization algorithm, the "killer app" for quantum computation, work in a similar way. See the discussion in the subsection "Two Quantum Algorithms" in the "More" section at the end of the chapter.

Computer scientists classify decision problems into different complexity classes depending on how hard it is to solve the problem. A decision problem is in complexity class **P** if a classical computer can solve the problem in polynomial time, which is to say that a classical computer can solve the problem in a number of computational steps that is a polynomial function of the size of the input. (A polynomial is a function that can be expressed as a sum of powers of one or more variables multiplied by coefficients. For example, $5x^4 + 2x^3 + x + 1$ is a polynomial function of the variable $x$. If the problem is, say, to decide whether $p$ is a factor of $N$, the size of the input to the problem is the number of bits required to store the integers $p$ and $N$.)[4]

A problem is in complexity class **NP** if a classical computer running a nondeterministic algorithm can solve the problem in polynomial time. (A nondeterministic algorithm allows the computer to sometimes make random choices between alternative computational steps.) This complexity class is equivalent to the class of decision problems for which a proposed solution can be *verified* in polynomial time by a classical computer running a deterministic algorithm. For example, the problem of factoring a large positive integer $N$ into its prime factors is in **NP**, but it's relatively easy to check whether a given integer is a factor by dividing it into $N$ and seeing whether there is a remainder, using a deterministic algorithm that can be carried out in polynomial time. An **NP**-complete problem is a problem in **NP** that any problem in **NP** can be transformed into in polynomial time. So an **NP**-complete problem is at least as hard as any other problem in **NP**. A problem is in **PSPACE** if it can be solved by a classical computer using a polynomial amount of memory.

The complexity class **P** is included in **NP** which is included in **PSPACE**, but it's still an open problem—*the* big problem in complexity theory—to show that these complexity classes are in fact distinct. Problems in **NP** are regarded as "hard" relative to problems in **P**, but for all we know, the two complexity classes could turn out to coincide (although no one working in the field thinks that this is so).

In popular accounts of quantum computation, there are often claims that a quantum computer would be "exponentially faster" than any classical computer. Rather, what's been shown is that some quantum algorithms, like Shor's algorithm, are exponentially faster than any *known* classical algorithm. It's possible to prove that a quantum computer running a quantum algorithm could solve certain problems exponentially faster than a classical computer, if both computers had access to an "oracle" that provided a particular solution to an unsolved problem. Relative to an oracle, you can prove that **P** $\neq$ **NP**, but the open problem in complexity theory is whether, in the unrelativized sense, these complexity classes are distinct. So, for all we know, there could be an as yet undiscovered classical factorization algorithm that runs on a classical computer that is just as efficient as Shor's quantum algorithm.

## The bottom line

- A quantum computer offers the possibility of a potentially massive speedup over any classical computer. Popular accounts generally describe the speedup as the result of "quantum parallelism"—the supposed ability of a quantum computer to perform a huge number of computations at the same time by encoding information as qubits rather than bits, on the grounds that a qubit can somehow represent 0 and 1 at the same time while a bit is either 0 or 1.

(continued)

 **The bottom line** *(continued)*

- In fact, a quantum computer achieves a speedup over known classical algorithms by being able to solve a decision problem about whether a disjunction, an "or" statement, "$P$ or $Q$ or $R$ or . . .," is true or false, without having to work out whether the individual components of the "or" statement, like $P$, are either true or false. Rather than achieving a speedup over classical computers by performing multiple computations simultaneously, a quantum computer can compute a solution to a problem in fewer steps than a classical computer, because it can do so without working out the answers to questions that aren't asked, but that a classical computer needs to answer to solve the problem.
- The discussion illustrates this by going through the simplest quantum algorithm, proposed by David Deutsch. Simon's algorithm for finding the period of a function, and Shor's factorization algorithm, the "killer app" for quantum computation, work in a similar way. See the "More" section at the end of the chapter for a discussion.

## 8.2 QUANTUM TELEPORTATION

Here's a quantum feat you can do with quantum correlations but not with classical correlations. Suppose Alice and Bob, who are separated, share a pair of qubits in the maximally entangled state $|\phi^+\rangle_{AB} = |0\rangle_A|0\rangle_B + |1\rangle_A|1\rangle_B$. You hand Alice a qubit $Q$ in some quantum state $\alpha|0\rangle_Q + \beta|1\rangle_Q$, unknown to Alice, and ask her to send the state to Bob, without actually carrying the qubit to him. Alice can perform whatever manipulations she likes on $Q$, but she is only allowed to communicate classical information to Bob. It turns out that it takes just two bits of information, 00, 01, 10, or 11, to tell Bob, who could be any distance at all from Alice, how to manipulate his qubit so that it ends up in the state $\alpha|0\rangle_Q + \beta|1\rangle_Q$. That's remarkable, because there's an infinite set of qubit states, so it would take an infinite number of bits to specify the state precisely (for example, to single out the direction of linear polarization of a photon among all the infinite possible directions).[5]

Now, Alice doesn't know the state $|\psi\rangle_Q$, and there is no way she can find out the identity of an unknown quantum state by measuring a single qubit. If she had many qubits, all in the same state, she could measure some qubit observables and infer the state from the measurement statistics with an accuracy that depends on the number of qubits. Specifically, she could divide a sample of qubits into three sets and measure one of the three noncommuting Pauli observables $X, Y, Z$ on the qubits in each set to find the state. (The Pauli observables are "informationally complete." See the section "Informationally Complete Observables" in the Supplement, "Some Mathematical Machinery," at the end of the book.) But she can only measure one observable on a single qubit $Q$ (because a measurement changes the state $|\psi\rangle_Q$ randomly), and that will produce one of the two possible outcomes, for any state of $Q$, which provides no information about the state.

In fact, it wouldn't help if Alice knew the state of $Q$, because that would leave her with the problem of transmitting an infinite number of bits to Bob, but there is a rather simple procedure, at least conceptually, by which Alice can accomplish this feat by *teleporting* the unknown state of $Q$ to Bob.

Here's how she does it. Alice implements the teleportation by performing a certain joint measurement on the qubit $Q$ and her qubit $A$. The measurement produces one of four entangled states of $Q$ and $A$, with equal probability, and leaves Bob's qubit $B$ in a pure quantum state that is no longer entangled with Alice's qubit $A$. Alice can tell Bob which outcome she found by sending him two bits. This information enables Bob to select one of four operations that transforms the state of his qubit to the initial state of $Q$.

At the start, the three-qubit state of the combined system, the qubit $Q$, and the entangled pair $AB$ is the product of the state $\alpha|0\rangle_Q + \beta|1\rangle_Q$ and the entangled state $|\phi^+\rangle_{AB}$. This product state can be expressed as a linear superposition, with equal coefficients $1/2$ (the square root of $1/4$), of the four product states

$$
\begin{aligned}
|\phi^+\rangle_{QA} &\otimes (\alpha|0\rangle_B + \beta|1\rangle_B),\\
|\phi^-\rangle_{QA} &\otimes (\alpha|0\rangle_B - \beta|1\rangle_B),\\
|\psi^+\rangle_{QA} &\otimes (\alpha|1\rangle_B + \beta|0\rangle_B),\\
|\psi^-\rangle_{QA} &\otimes (\alpha|1\rangle_B - \beta|0\rangle_B),
\end{aligned}
$$

where the subscript $QA$ indicates a state of the composite two-bit system, $Q$, and Alice's qubit, $A$, and the subscript $B$ indicates a state of Bob's qubit. The "$\otimes$" here is the symbol for a product of two quantum states, which I insert here for clarity (technically, it's the "tensor product"; see the section "Entanglement" in the Supplement, "Some Mathematical Machinery," at the end of the book).

The states $|\phi^+\rangle_{QA}$, $|\phi^-\rangle_{QA}$, $|\psi^+\rangle_{QA}$, $|\psi^-\rangle_{QA}$ are entangled states of the qubit $Q$ and Alice's qubit $A$, the four Bell states (see p.247). This expression for the three-qubit product state is just another way of writing the state—no manipulations of the qubits are involved. It's a little tedious, but you can check this by writing out the expressions for each of the Bell states and multiplying everything out explicitly.

The Bell states are mutually orthogonal, so they represent a basis in the four-dimensional state space of the composite system $QA$, associated with a joint observable of $QA$ that Alice can measure. If Alice measures this Bell observable, she will get one of four outcomes that I'll label $00, 01, 10, 11$ with equal probability $1/4$. For the outcome $00$, the system $QA$ ends up in the entangled Bell state $|\phi^+\rangle_{QA}$ and Bob's qubit ends up in the state $\alpha|0\rangle_B + \beta|1\rangle_B$, which is just the state of $Q$ that Alice is supposed to send to Bob. The measurement of the Bell observable on the composite system $QA$ produces the entangled state of $QA$ and leaves the qubits $A$ and $B$ unentangled. For the three other possible outcomes, $01$, $10$, or $11$, the system $QA$ will be in the Bell state $|\phi^-\rangle_{QA}$, $|\psi^+\rangle_{QA}$, or

$|\psi^-\rangle_{QA}$ and Bob's qubit will be the corresponding state $\alpha|0\rangle_B - \beta|1\rangle_B$, $\alpha|1\rangle_B + \beta|0\rangle_B$, or $\alpha|1\rangle_B - \beta|0\rangle_B$, which Bob can transform to the required state by a quantum operation.

So to complete the teleportation, Alice sends Bob two bits of information that correspond to the outcome of her measurement on *QA*. If Bob receives the message 00, he does nothing: the state of his qubit *B* is the same as the state of *Q*. If Bob receives the message 01, he applies the transformation *Z* to his qubit, which leaves $|0\rangle_B$ unchanged and takes $|1\rangle_B$ to $-|1\rangle_B$, so $\alpha|0\rangle_B - \beta|1\rangle_B \xrightarrow{Z} \alpha|0\rangle_B + \beta|1\rangle_B$. If Bob receives the message 10, he applies the transformation *X* to his qubit, which flips the states $|0\rangle_B$ and $|1\rangle_B$, so $\alpha|1\rangle_B + \beta|0\rangle_B \xrightarrow{X} \alpha|0\rangle_B + \beta|1\rangle_B$. If Bob receives the message 11, he applies the transformation *X* to his qubit, followed by the transformation *Z*. So $\alpha|1\rangle_B - \beta|0\rangle_B \xrightarrow{X} \alpha|0\rangle_B - \beta|1\rangle_B \xrightarrow{Z} \alpha|0\rangle_B + \beta|1\rangle_B$. If the qubits are photons, Bob can implement these transformations by passing his photon through an optical device that alters the direction of linear polarization. (For the transformations *X* and *Z*, see the section "Quantum Dynamics" in the Supplement, "Some Mathematical Machinery," at the end of the book. Note that the state $\alpha|1\rangle_B - \beta|0\rangle_B$ is equivalent to the state $-(\alpha|1\rangle_B - \beta|0\rangle_B) = \beta|0\rangle_B - \alpha|1\rangle_B$. The state vectors point in opposite directions but lie along the same ray, which is all that matters for the quantum state. See the section "The Pauli Operators" in the Supplement.)

There's a different procedure for implementing a teleportation protocol that avoids the technical difficulty of measuring a joint observable of the composite system *QA*, which is rather more challenging than simply measuring a single qubit observable, like the linear polarization of a photon.[6] Begin, as before, with the three-qubit state of the combined system *Q* and the entangled pair *AB*, the product of the state $|\psi\rangle_Q = \alpha|0\rangle_Q + \beta|1\rangle_Q$ and the entangled state $|\phi^+\rangle_{AB}$. This product state can also be expressed as

$$|\psi\rangle_Q \otimes |\phi^+\rangle_{AB} = \alpha|0\rangle_Q \otimes (|0\rangle_A|0\rangle_B + |1\rangle_A|1\rangle_B) + \beta|1\rangle_Q \otimes (|0\rangle_A|0\rangle_B + |1\rangle_A|1\rangle_B).$$

In the first part of the teleportation procedure, Alice sends her two qubits through a controlled-not (CNOT) gate, with *Q* as the control qubit and *A* as the target. A CNOT gate does nothing if the state of the control is $|0\rangle$, but flips the state of the target if the control is $|1\rangle$ (hence the term "controlled-not"; see the section "Quantum Dynamics" in the Supplement, "Some Mathematical Machinery," at the end of the book). After the transformation by the CNOT gate, the three-qubit state is

$$\alpha|0\rangle_Q \otimes (|0\rangle_A|0\rangle_B + |1\rangle_A|1\rangle_B) + \beta|1\rangle_Q \otimes (|1\rangle_A|0\rangle_B + |0\rangle_A|1\rangle_B).$$

The state of Alice's qubit *A* stays the same in the part of the state correlated with the state $|0\rangle_Q$ of *Q*. What's changed is that the state of *A* is flipped ($|0\rangle_A \to |1\rangle_A$ and $|1\rangle_A \to |0\rangle_A$) in the part of the state correlated with the state $|1\rangle_Q$ of *Q*.

In the second part of the teleportation procedure, Alice applies a Hadamard transformation to the qubit *Q* that takes the qubit state $|0\rangle_Q$ to $|0\rangle_Q + |1\rangle_Q$ and the qubit state $|1\rangle_Q$ to $|0\rangle_Q - |1\rangle_Q$. After the Hadamard transformation, the three-qubit state is

$$\alpha(|0\rangle_Q + |1\rangle_Q) \otimes (|0\rangle_A|0\rangle_B + |1\rangle_A|1\rangle_B) + \beta(|0\rangle_Q - |1\rangle_Q) \otimes (|1\rangle_A|0\rangle_B + |0\rangle_A|1\rangle_B),$$

and this can be expressed (after multiplying everything out) as a linear superposition, with equal coefficients $1/2$, of the four states

$$
\begin{aligned}
&|0\rangle_Q \otimes |0\rangle_A \otimes (\alpha|0\rangle_B + \beta|1\rangle_B),\\
&|0\rangle_Q \otimes |1\rangle_A \otimes (\alpha|1\rangle_B) + \beta|0\rangle_B),\\
&|1\rangle_Q \otimes |0\rangle_A \otimes (\alpha|0\rangle_B) - \beta|1\rangle_B),\\
&|1\rangle_Q \otimes |1\rangle_A \otimes (\alpha|1\rangle_B) - \beta|0\rangle_B).
\end{aligned}
$$

The three-qubit state started out as a product of the state of $Q$ and the entangled state of Alice's qubit $A$ and Bob's qubit $B$. After Alice's manipulation, the three-qubit state is an entangled state that is a linear superposition of product states, each of which is a product of a state of $Q$, a state of Alice's qubit $A$, and a state of Bob's qubit $B$. The qubits $Q$ and $A$ are in Alice's possession. If Alice now measures the $Q$-observable associated with the basis $|0\rangle_Q$, $|1\rangle_Q$ for $Q$, and the same $A$-observable associated with the basis $|0\rangle_A$, $|1\rangle_A$ for $A$ (so she measures linear polarization if the qubits are photons, by passing them through two beamsplitters), she will get one of four outcomes with equal probability $1/4$ corresponding to the states $|0\rangle_Q|0\rangle_A$, $|0\rangle_Q|1\rangle_A$, $|1\rangle_Q|0\rangle_A$, $|1\rangle_Q|1\rangle_A$. For each of these outcomes, Bob's qubit $B$ will be in the corresponding state $\alpha|0\rangle_B + \beta|1\rangle_B, \alpha|1\rangle_B + \beta|0\rangle_B, \alpha|0\rangle_B - \beta|1\rangle_B$, or $\alpha|1\rangle_B - \beta|0\rangle_B$.

So, as before, to complete the teleportation, Alice sends Bob two bits of information that correspond to the outcomes, 0 or 1, of her measurements on the qubits $Q$ and $A$. If Bob receives the bits 00, his qubit is in the same state as the qubit $Q$. If he receives the bits 01, he applies the transformation $X$ to transform the state of his qubit to the state of $Q$. If he receives the bits 10, he applies the transformation $Z$, and if he receives the bits 11, he applies the transformations $X$ followed by $Z$, in that order, to transform the state of his qubit to the state of $Q$.

Alice's feat exploits a remarkable feature of entangled correlations that Schrödinger called remote "steering" in a two-part paper he published in 1935 and 1936 as a response to the Einstein–Podolsky–Rosen argument. If Alice and Bob share an entangled state like the state $|\phi^+\rangle$, the reduced state of Bob's system—the state of Bob's system alone—is a completely random mixed state that represents, equivalently, *any* equal weight mixture of a pair of orthogonal qubit states, for example the states $|0\rangle_B$, $|1\rangle_B$, or *any* one of an infinite set of mixtures of nonorthogonal states, for example the equal weight mixture of the four states $\alpha|0\rangle_B + \beta|1\rangle_B, \alpha|0\rangle_B - \beta|1\rangle_B, \alpha|1\rangle_B + \beta|0\rangle_B, \alpha|1\rangle_B - \beta|0\rangle_B$ in the teleportation protocol. These mixtures are all indistinguishable: they have the same outcome probabilities for any measurements. By suitably entangling an ancilla or auxiliary particle with her qubit (the word "ancilla" comes from the Latin for maidservant) and measuring an observable of the ancilla and her qubit, Alice can "steer" Bob's qubit into any one of these mixtures that she chooses. For the state $|\phi^+\rangle$ and without an ancilla, Alice can steer Bob's qubit into different mixtures of orthogonal states by an appropriate measurement on his qubit. To steer Bob's qubit into a mixture of nonorthogonal states (as in the teleportation protocol), Alice entangles her qubit with an ancilla qubit $Q$ in an appropriate way, and she measures an observable of her qubit and the ancilla, perhaps following some preliminary manipulation of her qubits as in the second teleportation protocol. The different possible outcomes of her measurement will be correlated with different states of Bob's qubit. So after her measurement, Alice will be in a position to tell Bob that his qubit is in a certain quantum state, or how to manipulate his state to the initial state of $Q$.

## Quantum teleportation

- Alice and Bob, who are separated, share a pair of qubits in the maximally entangled state $|\phi^+\rangle_{AB} = |0\rangle_A|0\rangle_B + |1\rangle_A|1\rangle_B$. Alice receives a qubit $Q$ in some state $|\psi\rangle_Q = \alpha|0\rangle_Q + \beta|1\rangle_Q$, unknown to Alice. Quantum teleportation is a procedure in which Alice measures the two qubits in her possession, $Q$ and $A$, to get one of four possible outcomes, representing two bits of information. These two bits, communicated by Alice to Bob, are sufficient for Bob to be able to manipulate his qubit $B$ so that the quantum state of $B$ ends up as $\alpha|0\rangle_B + \beta|1\rangle_B$, the same as the original state of $Q$.
- The feat exploits a remarkable feature of entangled correlations that Schrödinger called remote "steering." If Alice and Bob share an entangled state like the state $|\phi^+\rangle$, the reduced state of Bob's system—the state of Bob's system alone—is a completely random mixed state that represents, equivalently, *any* equal weight mixture of a pair of orthogonal qubit states, or *any* one of an infinite set of mixtures of nonorthogonal states. Alice can remotely "steer" Bob's qubit into any mixture she chooses by performing an appropriate measurement on her qubit, or on her qubit and a suitable auxiliary particle. To steer Bob's qubit into a mixture of nonorthogonal states (as in the teleportation protocol), Alice entangles her qubit with an auxiliary particle in an appropriate way, and she measures an observable of her qubit and the auxiliary particle. The different possible outcomes of her measurement will be correlated with different states of Bob's system. So after her measurement, Alice will be in a position to tell Bob that his qubit is in a certain quantum state, or how to manipulate his qubit to achieve a certain quantum state.

That's what's going on here. After Alice's measurement on her qubit and the qubit $Q$, she can tell Bob that his qubit is in a particular state belonging to an equal weight mixture of four states $|t_{00}\rangle_B = \alpha|0\rangle_B + \beta|1\rangle_B$, $|t_{01}\rangle_B = \alpha|0\rangle_B - \beta|1\rangle_B$, $|t_{10}\rangle_B = \alpha|1\rangle_B + \beta|0\rangle_B$, $|t_{11}\rangle_B = \alpha|1\rangle_B - \beta|0\rangle_B$. (This equal weight mixture is represented by the completely random mixed state $\frac{1}{2}I$, where $I$ is the identity operator. If you represent the states $|t_{00}\rangle_B, |t_{01}\rangle_B, |t_{10}\rangle_B, |t_{11}\rangle_B$ as operators $P_{00}, P_{01}, P_{10}, P_{11}$, you can can check that $\frac{1}{2}I = \frac{1}{4}(P_{00} + P_{01} + P_{10} + P_{11})$. See the section "Mixed States" in the Supplement, "Some Mathematical Machinery," at the end of the book.)

Now this is the same mixed state as Bob's qubit *at the start*, before Alice's CNOT operation and Hadamard operation. So what Alice chooses to do in her location (whether she implements a CNOT transformation followed by a Hadamard transformation or not) doesn't, in itself, change the quantum state of Bob's qubit. But if Alice measures her qubits, she gets a definite outcome, and this outcome is correlated with a particular pure state in a particular mixture of nonorthogonal states: one of the states $|t_{00}\rangle_B$, $|t_{01}\rangle_B$, $|t_{10}\rangle_B$, $|t_{11}\rangle_B$, with equal probability of $1/4$. So it looks like Alice's measurement does bring about an instantaneous remote change in Bob's qubit: something Alice does puts Bob's qubit into one of four pure states, with equal probability. Since the outcome of Alice's measurement is random and not under her control, she can't control the particular pure state that Bob's qubit ends up in. Nevertheless, Alice's measurement puts Bob's qubit into *a particular mixture* in the set of equivalent mixtures corresponding to the quantum state of the qubit (the completely random mixed state).

This is how Schrödinger put it in the second part of a two-part paper on entanglement correlations between separated systems, motivated by the Einstein–Podolsky–Rosen argument:[7]

> If for a system which consists of two entirely separated systems the representative (or wave function) is known, then current interpretation of quantum mechanics obliges us to admit *not only* that by suitable measurements, taken on *one* of the two parts only, the state (or representative or wave function) of the *other* part can be determined without interfering with it, *but also* that, in spite of this non-interference, the state arrived at *depends* quite decidedly on *what* measurements one chooses to take—not only on the results they yield. So the experimenter, even with this indirect method, which avoids touching the system itself, controls its future state in very much the same way as is well-known in the case of a direct measurement. In this paper it will be shown that the control, with the indirect method, is in general not only *as* complete but even more complete. For it will be shown that *in general* a sophisticated experimenter can, by a suitable device which does *not* involve measuring non-commuting variables, produce a non-vanishing probability of driving the system into any state he chooses; whereas with the ordinary direct method at least the states orthogonal to the original one are excluded.

What Schrödinger meant by the last comment is that an experimenter can produce any polarization state of a photon, say, by passing it through a beamsplitter or other optical device for measuring the polarization, except for a state orthogonal to the original state of the photon, which has zero probability as the outcome. But if the photon is part of an entangled state like $|\phi^+\rangle = |0\rangle|0\rangle + |1\rangle|1\rangle$, an experimenter can produce any state of polarization at all for the photon (with non-zero probability) by a suitable measurement on the remote paired photon and an ancilla. He comments later that an arbitrary state "can be given a non-vanishing probability of its turning up in the *first* system by a suitable treatment of the second one. Since it has a finite chance of turning up, it will certainly turn up, if precisely the same experiments are repeated sufficiently often." Schrödinger found this so disturbing that he conjectured that entanglement would rapidly dissipate as two entangled systems separated. Schrödinger's pessimistic conjecture has turned out to be false. Teleportation has been experimentally confirmed, and entanglement has been demonstrated over fairly large distances (notably by Anton Zeilinger's group between La Palma and Tenerife, two of the Canary Islands, which are more than 140 kilometers apart).

Are we really forced to conclude that what Alice does to her qubits at her location in a teleportation protocol somehow instantaneously changes the qubit at Bob's location, which could be as remote from Alice's qubits as you like, even in a distant galaxy? Of course, there is no violation of the no-signaling principle, because without Alice's message—the two classical bits that Alice sends Bob in some conventional way—Bob has no way of knowing how his qubit has changed: any measurement he carries out on his qubit will produce a random outcome before or after Alice's measurement on her qubits. But Alice knows that before her measurement the state of Bob's qubit was a completely random mixed state, and after her measurement, depending on what she chooses to measure and the outcome of the measurement, Bob's qubit is in a particular pure state. If Alice knows this, then the universe knows this too, as Nicolas Gisin puts it,[8] which would mean that there is some objective change in Bob's qubit, caused instantaneously by Alice's measurement.

In a teleportation procedure, Alice carries out a measurement on two qubits in her possession (or she carries out a measurement after some operations on the qubits), one of which is entangled with Bob's qubit. Bob's qubit, which is originally in a mixed state, then ends up in a particular pure state correlated with Alice's measurement outcome. Now, this is pretty much what happens if Alice and Bob share an entangled state like $|\phi^+\rangle$ and Alice measures her qubit. If she finds the outcome $|0\rangle$, Bob's qubit, which was originally in a completely random mixed state for which any polarization measurement would produce either outcome with equal probability, is in the pure state $|0\rangle$. If she finds the outcome $|1\rangle$, Bob's qubit is in the pure state $|1\rangle$. And this is very much like what happens if Alice and Bob share a pair of Popescu–Rohrlich bananas and Alice peels her banana by the top end ($T$) and finds that the taste is ordinary. Before Alice peels and tastes her banana, the probability of Bob's banana tasting intense if he peels $T$ is $1/2$, but after Alice peels and tastes her banana, the probability of Bob's banana tasting intense if he peels $T$ is 1. It would seem that there has been an objective change in Bob's banana caused, instantaneously, by Alice peeling her banana $T$ and finding the taste ordinary. The probability of Bob's banana tasting intense if he peels $T$ has gone from $1/2$ to 1, so it's now certain that Bob's banana will taste intense if he peels $T$, while it could have tasted ordinary before.

Since the same worry arises for teleportation or steering as for the correlations of entangled states or Popescu–Rohrlich bananas, the issue boils down to getting some things straight about Popescu–Rohrlich correlations that haven't come up before.[9]

The change here, for teleportation or entanglement or Popescu–Rohrlich bananas, is a change in probability, and probability is a tricky business. One standard way of understanding probability in the subjective sense is in terms of fair betting odds. If you estimate the probability of Lucky Jim winning a horse race as 1/2 or 50%, what you mean is that you think the odds are even that he'll win. In other words, you'd be willing to bet a dollar that Lucky Jim will win the race, on the expectation of winning a dollar from me if I accept the bet and he wins, or losing a dollar to me if he loses. If you think the probability is 3/4 or 75%, then you think the odds are 3 to 1 that Lucky Jim will win the race, which means that you'd be willing to bet three dollars and expect to win one dollar from me if I accept the bet and he wins. If he loses, I get three dollars from you. If you estimate the probability of Lucky Jim winning as 3/4 for many different races, then in roughly 3/4 of the races you expect to win a dollar, and in 1/4 of the races you expect to lose three dollars to me, so we'll come out even, which is to say that the odds are fair betting odds.

For subjective Bayesians, all probabilities are subjective in this sense, but there are other probabilities that don't seem to be subjective at all. (Bayesians take the name from Thomas Bayes, the 18th-century theologian and mathematician, who made significant contributions to the theory of statistical inference.) The probability that a photon in a state of linear polarization passes or is blocked by a polarization filter doesn't seem to have anything to do with anyone's estimate of fair betting odds, but rather with a stable feature of the world, something objective.

A photon, horizontally polarized in a direction $z$, has a probability $\cos^2\theta$ of passing an analyzer with its axis of polarization at an angle $\theta$ to $z$. One way of understanding what this means is to think of the probability as a property of the photon, a tendency or disposition or propensity of the photon to pass the analyzer, or in general to behave in a certain way when put in an experimental setting with alternative outcomes.

There's another way of thinking about objective probability or objective chance, without committing to the idea of a propensity or some sort of "oomph" associated with quantum systems, following the minimalist view of David Hume, the 18th-century Scottish philosopher, as developed by David Lewis. Lewis refers to the totality of events, past, present, and future as the "Humean mosaic."[10] Rather than regarding quantum mechanics as a theory of propensities, you can think of quantum probabilities as objective probabilities referring to statistical patterns in the mosaic. These objective probabilities are the probabilities in the best candidate system of generalizations about the statistical patterns. The "best candidate system" is understood as best in terms of simplicity (in how the generalizations about chances are formulated), strength (or informativeness with respect to what the chances are in various scenarios), and fit (with respect to how well the chances fit the statistical facts).

Lewis specifies the relation between subjective probabilities or "credences" and these objective probabilities or objective "chances" in terms of what he calls the "principal principle." The principal principle says that your credences should conform to the objective chances. So if the objective chance of some event occurring is 1/2, your subjective probability or credence should also be 1/2 if you want to avoid losing bets against nature. On this minimalist Humean or Lewisian way of thinking about objective probabilities, it makes sense to talk about objective probabilities even in a deterministic universe. If, like Laplace's demon, you could take everything into account, all the "hidden variables," you wouldn't need probabilities to describe events, but there could still be statistical patterns in a deterministic universe, like the fact that tossing a fair coin results in a "head" in roughly half the cases in a sufficiently long run of coin tosses.

Suppose Alice and Bob share a pair of Popescu–Rohrlich bananas. Alice peels her banana $T$ and finds the taste ordinary. Before she peeled and tasted her banana, her subjective probability about the taste of Bob's banana tasting intense if he peeled $T$ was $1/2$. Now her subjective probability is 1, because the tastes are different if they both peel $T$. Does this mean that Alice's peeling and tasting her banana has caused a change in Bob's banana—a change that would have to be instantaneous, since the correlations are independent of how far apart Alice and Bob are, or who peels and tastes first? As my colleague Allen Stairs emphasizes (referring to Popescu–Rohrlich boxes rather than bananas):[11]

> To give the matter a Humean turn, there's nothing inconsistent in assuming that the correlations are just basic patterns in the mosaic, supervening on nothing more than themselves. The point isn't to insist on Humeanism; rather it's that the Humean picture serves as a consistency check. It represents one way of describing a world where the correlations hold but what's done to one box doesn't bring about what happens to the other.

The pattern in the mosaic in this case is just the Popescu–Rohrlich correlation that if both bananas are peeled $T$, they have different tastes. Expressed in terms of objective probabilities, this is $p(\text{different} \mid TT) = 1$. If Alice peels $T$ and finds the taste ordinary, she can infer, with certainty, that if Bob also peeled his banana $T$, then the taste of his banana was intense. This is Alice's subjective conditional probability (conditional on both Alice and Bob peeling $T$ and on Alice's banana tasting ordinary). It goes along with a corresponding objective conditional probability, which could be expressed as $p(1_B \mid T_A T_B 0_A)$. These conditional probabilities follow from the objective probability expressing the correlation. There's no need to suppose that after Alice peels and tastes her banana there is, instantaneously, a property of Bob's banana that wasn't there before, a propensity to taste intense if the banana is peeled $T$.

As Stairs points out, if Bob *in fact* peeled $T$, then (given that Alice peeled $T$ and found the taste ordinary) the taste of Bob's banana was certainly intense, since the tastes of the two bananas are always different if Alice and Bob both peeled $T$. But there are no grounds for saying that if, counterfactually, Bob had peeled $S$ then his banana would have tasted ordinary, the same as the taste of Alice's banana. If Alice's peeling and tasting her banana is responsible for bringing about a property of Bob's banana, then one could make this sort of inference about what would happen or would have happened in a counterfactual situation, because this would be guaranteed by the existence of the property. But if there is no such property caused by Alice's action, then all you have is the pattern in the mosaic, the correlation, and nothing follows about what would happen in a counterfactual situation. One might just as well say that if, counterfactually, Bob had peeled $S$ rather than $T$ and found the taste intense, then both his banana and Alice's banana would have tasted intense. If the Alice events and Bob events are spacelike separated in the relativistic sense, there's no fact of the matter about who peeled and tasted first, so whether the Alice events caused the Bob events or the Bob events caused the Alice events depends on the reference frame.

Stairs doesn't claim that the probabilities of Popescu–Rohrlich bananas or quantum probabilities *aren't* propensities, and that what Alice does in her part of the universe doesn't instantaneously change the propensities in Bob's part of the universe. Rather, the argument is that there's another option: you could think of these nonclassical probabilistic correlations as patterns in the Humean mosaic, in which case what Alice does has no instantaneous effect in Bob's part of the universe. Bananaworld is a possible world. There's no contradiction in supposing that the tastes of two bananas could be correlated in a nonclassical way depending on how they are peeled. You could assume that the correlation arises because of action at a distance, and that peeling and tasting a banana causes an instantaneous change in a remote paired banana, so that the remote banana ends up with an instruction set relating peelings and tastes that it didn't have before. But it's equally consistent to take a minimal Humean or Lewisian view that the correlation is just a brute fact about bananas that grow on certain trees in Bananaworld, and on this view there needn't be any action at a distance.

Parallel arguments can be made for entangled quantum states and for quantum teleportation. In quantum teleportation, Alice and Bob exploit a nonclassical correlation to do something that would be impossible with a classical correlation. This is an amazing feat from a classical point of view, but it doesn't require thinking that Alice's action causes an instantaneous change to Bob's qubit. It's consistent to take a minimalist view about correlations in our quantum world too.

**The bottom line**

- Should we conclude that what Alice does to her qubits at her location in a teleportation protocol somehow instantaneously changes the qubit at Bob's location, which could be as remote from Alice's qubits as you like, even in a distant galaxy? The same question arises for teleportation or steering as for the correlations of entangled states or Popescu–Rohrlich bananas.
- You could assume that the correlation arises because of action at a distance, and that peeling and tasting a banana causes an instantaneous change in a remote paired banana, so that the remote banana ends up with an instruction set relating peelings and tastes that it didn't have before. But it's equally consistent to take a minimal Humean or Lewisian view that the correlation is just

(*continued*)

 **The bottom line** *(continued)*

a brute fact about bananas that grow on certain trees in Bananaworld, and on this view there needn't be any action at a distance.

- Parallel arguments can be made for entangled quantum states and for quantum teleportation. In quantum teleportation, Alice and Bob exploit a nonclassical correlation to do something that would be impossible with a classical correlation. This is an amazing feat from a classical point of view, but it doesn't require thinking that Alice's action causes an instantaneous change to Bob's qubit. It's consistent to take a minimalist view about correlations in our quantum world too.

## 8.3 SUPERDENSE CODING

Here's one final quantum feat, another example of entanglement-assisted communication that follows immediately from the possibility of quantum teleportation. Alice can exploit a shared entangled state to send two classical bits to Bob with one qubit.[12]

Suppose Alice and Bob share the entangled state $|\phi^+\rangle = |0\rangle|0\rangle + |1\rangle|1\rangle$. Depending on the two bits she wants to communicate, Alice does some quantum processing on her qubit. If Alice wants to send Bob the bits 00, she does nothing. To send the bits 01, she applies the transformation $Z$ to her qubit, which leaves $|0\rangle$ unchanged and takes $|1\rangle$ to $-|1\rangle$. This transforms the entangled state to the state $|\phi^-\rangle = |0\rangle|0\rangle - |1\rangle|1\rangle$. To send the bits 10, she applies the operation $X$ to her qubit, which flips $|0\rangle$ to $|1\rangle$ and $|1\rangle$ to $|0\rangle$, and so transforms the entangled state to $|\psi^+\rangle = |0\rangle|1\rangle + |1\rangle|0\rangle$. To send the bits 11, she applies the transformation $X$, which produces the entangled state to $|0\rangle|1\rangle + |1\rangle|0\rangle$, followed by Z, which changes the state to $|\psi^-\rangle = |0\rangle|1\rangle - |1\rangle|0\rangle$.

After performing these operations on her qubit, she sends the qubit to Bob. Bob now has two qubits in his possession, which are in one of the four orthogonal entangled Bell states $|\phi^+\rangle$, $|\phi^-\rangle$, $|\psi^+\rangle$, $|\psi^-\rangle$. Bob can measure an observable of the qubit pair with four possible outcomes corresponding to the four Bell states. The outcome of Bob's measurement reveals Alice's operation on her qubit, and so reveals the two bits that distinguish the four operations.

What's interesting about superdense coding is that Alice can send two bits to Bob by manipulating and transmitting just one qubit, by exploiting the entanglement between her qubit and Bob's qubit. So information could be transmitted this way between different parts of a quantum computer with stored entangled qubits shared by the two locations. That's impossible to do with a bit, even by exploiting shared bits.[13]

 **The bottom line**

- Superdense coding is another example of entanglement-assisted communication that follows immediately from the possibility of quantum teleportation. Alice can send two bits to Bob by manipulating and transmitting just one qubit, by exploiting the entanglement between her qubit and Bob's qubit.

## 8.4 MORE

### 8.4.1 Two Quantum Algorithms

Daniel Simon's algorithm for finding the period of a periodic Boolean function works in a similar way to Deutsch's algorithm.[14] The possible inputs to the function are sequences of 0's and 1's. Suppose the inputs are two-bit sequences: 00, 01, 10, or 11. A periodic function takes the same output value for input values that are separated by a certain interval, the period of the function. In this case the period can be 01 or 10 or 11. If the period is 01, the function takes 00 to the same output value as 01, and it takes 10 to the same output value as 11, and similarly for the other two possible periods. (The notation 00 and 01 is shorthand for the two-bit sequences $(0, 0)$ and $(0, 1)$, and $(0, 0) \oplus (0, 1) = (0, 1)$, because adding the first bits in each sequence gives 0 and adding the second bits gives 1. Similarly, $(1, 0) \oplus (0, 1) = (1, 1)$.)

So each period corresponds to a particular way of dividing up the input values into subsets, where the function takes the input values in each subset to the same output value (different output values for different subsets). The period 01 is associated with the subsets $\{00, 01\}$ and $\{10, 11\}$, the period 10 is associated with the subsets $\{00, 10\}$ and $\{01, 11\}$, and the period 11 is associated with the subsets $\{00, 11\}$ and $\{01, 10\}$.

Simon's algorithm associates a subspace in the state space of the input register with each period and the corresponding way of dividing up the input values into subsets. In the case where the inputs are two-bit sequences, the subspace associated with a period is a plane. Two of the planes correspond to the constant and balanced planes in Deutsch's algorithm (the plane for the period 10 corresponds to the constant plane, and the plane for the period 11 corresponds to the balanced plane). A third plane is associated with the period 01 and, as in Deutsch's algorithm, all three planes intersect in a line. After the unitary transformation that takes the input/output registers to the final state, an appropriate measurement of the input register produces an outcome associated with the particular subspace corresponding to the period of the function. There is a chance of finding the outcome corresponding to the intersection, in which case the algorithm fails. If that happens, you repeat the algorithm until you get a result that distinguishes the periods.

In Deutsch's problem, one "or" statement is associated with the two alternative ways in which a function can be constant, and the other "or" statement is associated with the two alternative ways in which a function can be balanced. The problem is to determine which of these two "or" statements is true of a given function. Deutsch's algorithm enables you to do this without finding out whether the components in the "or" statement are true or false. In Simon's problem, each of several "or" statements is associated with one of the alternative possible periods of a periodic function, $f$. In the simple case when the inputs to the function are two-bit sequences, the "or" statement for the period 01 says: "the input value is in the subset $\{00, 01\}$, in which case the output value is $f(00) = f(01)$, *or* the input value is in the subset $\{10, 11\}$, in which case the output value is $f(10) = f(11)$." Two other "or" statements are associated with the periods 10 and 11. The problem is to determine which of the three "or" statements is true of the function. Simon's algorithm enables you to do this without finding out whether or not the components in the "or" statement are true or false.

The "killer app" in quantum computation is Peter Shor's quantum algorithm for factoring a positive integer $N = pq$ into its prime factors $p$ and $q$. It's relatively easy to multiply two large prime numbers $p$ and $q$ and get the product $N$, but hard to preform the reverse operation and find the prime factors or divisors $p$ and $q$ from $N$. If $N$ is a 130-digit integer, and you had access to a network of hundreds of computers running the most efficient classical factorization algorithm known, it would take about a month to find the two 65-digit factors. For a 400-digit integer, it would take about $10^{10}$ years, which is roughly the age of the universe!

Banks and financial networks currently exploit the difficulty of factoring large integers to encrypt their communications via the RSA public key encryption protocol, named after the inventors Ron Rivest, Adi Shamir, and Leonard Adelman. If Alice wants to send a message $m$ (in binary digits) to Bob, she uses Bob's public key, which is available to everyone, to encrypt the message. Bob's public key consists of two large integers, $s$ and $N$. Alice encrypts $m$ as the cyphertext $c = m^s \mod N$. Bob decrypts the message as $m = e^t \mod N$, where $e$ is the base of the natural logarithm, using an integer $t$ that depends on $s$ and the prime factors $p, q$ of $N$. Bob, of course, knows the value of $t$, but for anyone else to figure out $t$ from $s$ and $N$ would require computing the factors $p$ and $q$, which could take a very long time if $N$ is a large enough integer. (The fact that the RSA protocol works for particular values of $s, t, p, q$ depends on some number theory, which I won't go into here.) Shor's algorithm would allow anyone with access to a quantum computer to break RSA, which would have devastating consequences for our financial system and any institutions that rely on RSA encryption for privacy. This explains the race to build a quantum computer, and the urgency to find alternative quantum or device-independent cryptographic protocols that are immune to attack by a quantum computer.

Shor's algorithm exploits some number theory, specifically the fact that the two prime factors $p, q$ of a positive integer $N = pq$ can be found by determining the period of a function $f(x) = a^x \mod N$ for any $a < N$ coprime to $N$. (A number $a$ is coprime to $N$ if $a$ has no common factors with $N$ other than 1.) Once you know the period, you can factor $N$ if $r$ is even and $a^{r/2} \neq -1 \mod N$. This will be the case with probability greater than 1/2 if you choose $a$ randomly. If it's not, you choose another value of $a$. The factors of $N$ are the greatest common factors of $a^{r/2} \pm 1$ and $N$—the largest integer that divides $a^{r/2} + 1$ and $N$ without remainder, and the largest integer that divides $a^{r/2} - 1$ and $N$ without remainder. There's an efficient algorithm, Euclid's algorithm, for finding the greatest common factor of two positive integers. So the problem of finding the two prime factors of a positive integer $N$ reduces to the problem of finding the period of a certain periodic function. For a quantum computer, this involves deciding between different "or" statements corresponding to the different possible periods of the function, which can be computed efficiently without going through the steps to determine if the components of the "or" statements are true or false.

## Notes

1. The discussion of quantum computation follows the treatment in my paper "Quantum computation from a quantum logical perspective," *Quantum Information and Computation* 7, 281–296 (2007).
2. Deutsch proposed the problem (the "XOR problem") and the quantum algorithm as a solution in David Deutsch, "Quantum theory, the Church–Turing principle and the universal quantum computer," *Proceedings of the Royal Society of London A*, 400, 97–117 (1985).
3. For the improved version of Deutsch's algorithm, see Richard Cleve, Arthur Ekert, Christina Macchiavello, and Michelle Mosca, "Quantum algorithms revisited," *Proceedings of the Royal Society A* 454, 339–354 (1998).
4. For complexity and the **P** versus **NP** problem, see Scott Aaronson's entertaining book *Quantum Computers Since Democritus* (Cambridge University Press, Cambridge, 2013). The title page of his blog *Shtetl-Optimized* at http://www.scottaaronson.com/blog/ used to begin with the motto: "Quantum computers are not known to be able to solve NP-complete problems in polynomial time, and can be simulated classically with exponential slowdown." The motto now reads: "If you take just one piece of information from this blog: Quantum computers would not solve hard search problems instantaneously by simply trying all the possible solutions at once."
5. Quantum teleportation was first proposed in a 1993 paper by Charles H. Bennett, Gilles Brassard, Claude Crépeau, Richard Jozsa, Asher Peres, and William K. Wootters, "Teleporting an unknown quantum state via dual classical and Einstein–Podolsky–Rosen channels," *Physical Review Letters* 70, 1895–1899 (1993). It was first demonstrated experimentally by two groups in 1997, one in Innsbruck and one in Rome: Dik Bouwmeester, Jian-Wei Pan, Klaus Mattle, Manfred Eibl, Harald Weinfurter, and Anton Zeilinger, "Experimental quantum teleportation," *Nature* 390, 575–579 (1997), and D. Boschi, S. Branca, Francesco De Martini, Lucien Hardy, and Sandu Popescu, "Experimental realization of teleporting an unknown pure quantum state via dual classical and Einstein–Podolsky–Rosen channels," *Physical Review Letters* 80,

1121–1125 (1998). For recent long-distance teleportation, see Ivan Marcikic, Hugues de Riedmatten, Wolfgang Tittel, Hugo Zbinden, and Nicolas Gisin, "Long-distance teleportation of qubits at telecommunication wavelengths," *Nature* 421, 509–513 (2003), for an implementation of quantum teleportation with photons over a distance of 55 meters, and Xiao-Song Ma, Thomas Herbst, Thomas Scheidl, Daqing Wang, Sebastian Kropatschek, William Naylor, Bernhard Wittmann, Alexandra Mech, Johannes Kofler, Elena Anisimova, Vadim Makarov, Thomas Jennewein, Rupert Ursin, and Anton Zeilinger, "Quantum teleportation over 143 kilometres using active feed-forward," *Nature* 489, 269–273 (2012) for an implementation with photons over 143 kilometres between the two Canary Islands of La Palma and Tenerife.

6. The second procedure for a teleportation protocol is from Michael A. Nielsen and Isaac L. Chuang, *Quantum Computation and Quantum Information* (Cambridge University Press, Cambridge, 2000), p. 26.
7. Erwin Schrödinger's two-part paper in which he introduces the notion of remote quantum "steering": "Discussion of probability relations between separated systems," *Mathematical Proceedings of the Cambridge Philosophical Society,'* 31, 555–563 (1935) "Probability relations between separated systems," *Mathematical Proceedings of the Cambridge Philosophical Society* 32, 446–452 (1936). The quotation is from p. 446. The subsequent comment that an arbitrary state 'can be given a non-vanishing probability of its turning up in the *first* system ...' is on p. 451.
8. Nicolas Gisin's remark that if Alice knows something, then "the universe knows it too" is from private correspondence.
9. My discussion of how to make sense of quantum teleportation (and quantum entanglement in other scenarios), without assuming that some sort of action at a distance is involved, follows Allen Stairs, "A loose and separate certainty: Caves, Fuchs and Schack on quantum probability one," *Studies in History and Philosophy of Modern Physics* 42, 158–166 (2011).
10. On David Lewis and the Humean mosaic, see Roman Frigg and Carl Hoefer, "Determinism and chance from a Humean perspective," in D. Dieks, W. Gonzalez, S. Harmann, M. Weber, F. Stadler, and T. Eubel (eds.), *The Present Situation in the Philosophy of Science* (Springer, Berlin, 2010). Stairs cites Frigg and Hoefer in his paper on Caves, Fuchs, and Schack.
11. Allen Stairs, "A loose and separate certainty: Caves, Fuchs and Schack on quantum probability one," *Studies in History and Philosophy of Modern Physics* 42, 158–166 (2011), p. 162.
12. Superdense coding was first proposed by Charles H. Bennett and Stephen J. Wiesner, "Communication via one- and two-particle operators on Einstein–Podolsky–Rosen states," *Physical Review Letters*, 69, 2881–2884 (1992). For more on superdense coding, see the section on superdense coding in Section 2.3, pp. 97–98, of Michael A. Nielsen and Isaac L. Chuang, *Quantum Computation and Quantum Information* (Cambridge University Press, Cambridge, 2000).
13. An experiment confirming superdense coding: Klaus Mattle, Harald Weinfurter, Paul G. Kwiat, and Anton Zeilinger, "Dense coding in experimental quantum communication," *Physical Review Letters* 76, 4656–4659 (1996).
14. The discussion of Simon's algorithm and Shor's algorithm follows the treatment in my paper "Quantum computation from a quantum logical perspective," *Quantum Information and Computation* 7, 281–296 (2007) and the article by Adriano Barenco, "Quantum computation: an introduction" in Hoi-Kwong Lo, Sandu Popescu, and Tim Spiller, *Introduction to Quantum Computation and Information* (World Scientific Publishing Company, Singapore, 1998). See also Michael A. Nielsen and Isaac L. Chuang, *Quantum Computation and Quantum Information* (Cambridge University Press, Cambridge, 2000). See Nielsen and Chuang for references to Simon's algorithm and Shor's algorithm.

# 9

# Why the Quantum?

The title of this chapter comes from one of John Wheeler's "Really Big Questions."[1] One could ask why the world is quantum rather than classical, but the question here is: why is the world quantum rather than superquantum? Popescu and Rohrlich posed this question in the paper in which they showed that there are superquantum correlations that don't violate the no-signaling principle. In particular, as an extremal case, PR box correlations don't violate the no-signaling principle. So why are there no Popescu–Rohrlich bananas in our world? What principle excludes Popescu–Rohrlich bananas, or Popescu–Rohrlich correlations between particles?

As I showed in the "Bananaworld" chapter, the probability of perfectly simulating Popescu–Rohrlich bananas is $1/2(1 + K/4)$, where $K$ is the Clauser–Horne–Shimony–Holt quantity (a sum of expectation values). For a simulation with classical or local resources, the optimal value of $K$ is $K_L = 2$, so the optimal probability is no more than $3/4$, and for a simulation with quantum resources the optimal value of $K$ is $K_Q = 2\sqrt{2}$, the Tsirelson bound, so the optimal probability is approximately .85. If you could use Popescu–Rohrlich bananas or PR boxes, the probability of simulating the correlations of Popescu–Rohrlich bananas would, of course, be 1, so the value of $K$ for PR boxes is $K_{PR} = 4$, which is also the optimal value for no-signaling correlations. In terms of $E = K/4$, these quantities are $E_C = 1/2$ for the optimal classical value, $E_Q = 1/\sqrt{2}$ for the quantum Tsirelson bound, and $E_{PR} = 1$ for PR boxes.

Popescu and Rohrlich pointed out that correlations for which $E$ lies between the Tsirelson bound $1/\sqrt{2}$ and the PR box value 1 don't violate the no-signaling principle. In an ingenious argument using PR boxes in a 2009 *Nature* paper, Marcin Pawłowski, Tomasz Patarek, Dagomir Kaszlikowski, Valerio Scarani, Andreas Winter, and Marek Żukowski show that the Tsirelson bound follows from an information-theoretic principle that they call "information causality." They conclude the *Nature* article with the comment:[2]

> In conclusion, we have identified the principle of Information Causality, which precisely distinguishes physically realized correlations from nonphysical ones (in the sense that quantum mechanics cannot reach them). It is phrased in operational terms and in a theory-independent way and therefore we suggest it is at the same foundational level as the no-signaling condition itself, of which it is a generalization.

> The new principle is respected by all correlations accessible with quantum physics while it excludes all no-signaling correlations, which violate the quantum Tsirelson bound. Among the correlations that do not violate that bound it is not known whether Information Causality singles out exactly those allowed by quantum physics. If it does, the new principle would acquire even stronger status.

In fact, it has been shown subsequently that information causality fixes part of the boundary between quantum correlations and superquantum correlations for two qubits, except for a small set of non-quantum correlations below the Tsirelson bound.[3] These cases are still unresolved: it's not known whether or not they can be excluded by information causality. (As I mentioned in Chapter 5, "The Big Picture," the set of quantum correlations has a complicated boundary. The Tsirelson bound is the *maximum* value of the Clauser–Horne–Shimony–Holt correlation for quantum systems. The boundary goes below this bound, so there are non-quantum correlations below the Tsirelson bound.) For three qubits, however, it is now known that the principle excludes all but one extremal no-signaling correlation, but this superquantum correlation can't be excluded by a bipartite principle like information causality.[4] Information causality also can't explain the quantum bound in violations of noncontextuality inequalities like the Klyachko–Can–Binicioğlu–Shumovsky inequality in Section 6.3, "The Kochen–Specker Theorem and Klyachko Bananas," in the "Quantum Magic" chapter, and in Subsection 6.4.3, "Violating the Klyachko Inequality" in the "More" section of that chapter. So the principle of information causality by itself does not answer Wheeler's question. Nevertheless, it's quite astonishing how the Tsirelson bound, $1/\sqrt{2}$, emerges from the principle of information causality. I show this in Subsection 9.3.4, "The Tsirelson Bound from Information Causality," in the "More" section at the end of this chapter.

In a doctoral dissertation presented to Oxford University and in a later paper, Wim van Dam showed that communication complexity would be trivial or vacuous in a world with PR boxes (or Popescu–Rohrlich bananas).[5] Communication complexity is about how much communication you would need between Alice and Bob, who are in separate places and each have a data set of a certain number of bits, if they want to compute a function of these bits. Since they are in different locations, they need to exchange some bits to compute the function. How many bits they need to exchange depends on the specific function and on the available resources, for example, whether or not Alice and Bob share entangled quantum states. At least one bit has to be exchanged in the computation, or else the no-signaling principle would be violated (because Alice could signal instantaneously to Bob if the value of the function depended on Alice's data set, without any communication from Alice to Bob). The computational complexity of a function is regarded as trivial if just one bit of communication between Alice and Bob is all that's needed to compute a value of the function for a pair of inputs. Van Dam showed that with access to PR boxes, Alice and Bob could solve any communication problem of this sort with just one bit of communication. So the information-theoretic principle "communication complexity is nontrivial," in the sense that more than one bit should be exchanged to complete a task distributed between Alice and Bob, rules out PR boxes. In fact, the principle rules out more than this. Gilles Brassard, Harry Buhrman, Noah Linden, André Allan Méthot, Alain Tapp, and Falk Unger showed that communication complexity would become trivial if you could approximate PR boxes with a probability greater than 90% (specifically, greater than $\frac{3+\sqrt{6}}{6} \approx .908$).[6] It's a significant sea change in the foundations of physics that information-theoretic principles of this sort are investigated as possible constraints on physical processes.

Other principles have been proposed from which it is possible to derive the Tsirelson bound. Miguel Navascués and Harald Wunderlich derive the Tsirelson bound from a principle they call "macroscopic locality."[7] Macroscopic locality is, roughly, the requirement that a fundamental physical theory should reduce to classical physics for macroscopic systems, in the sense that the coarse-grained correlations you get from limited precision measurements on systems composed of an enormous number of microsystems should satisfy any Bell inequality and be recoverable from a local hidden variable theory. Macroscopically local correlations can violate information causality (that's the title of a paper by Daniel Cavalcanti, Alejo Salles, and Valerio Scarani).[8] For no-signaling theories, it's known that a broader class of correlations than quantum correlations satisfy macroscopic locality. Paul Skrzypczyk, Nicolas Brunner, and Sandu Popescu introduce the idea of a "coupler" for PR boxes that allows the boxes to interact, and they derive the Tsirelson bound from features of the interaction between noisy or imperfect PR boxes.[9]

Information-theoretic principles like information causality and van Dam's principle about communication complexity exclude some superquantum correlations, so they go some way to answering the question raised in the title of this chapter, but it's known that they don't provide a complete answer. An intriguing principle, developed by Adán Cabello as the "exclusivity principle" or "Specker's principle,"[10] and separately by Tobias Fritz, Ana Belén Sainz, Remigiusz Augusiak, Jonatan Bohr Brask, Rafael Chaves, Anthony Leverrie, and Antonio Acín as the principle of "local orthogonality," [11] rules out correlations beyond the quantum limit for Klyachko bananas (see the "Quantum Magic" chapter) and many other cases where contextuality or nonlocality is relevant.

Section 9.1, "Information Causality," is a general discussion about the motivation for the principle of information causality as an information-theoretic constraint. I give a more precise definition of information causality in the "More" section of this chapter, where I'll show how the Tsirelson bound follows from the principle. In Section 9.2, "The Parable of the Over-Protective Seer", I'll say something about the exclusivity principle, which Cabello traces to Ernst Specker's story about a seer from Nineva.

## 9.1 INFORMATION CAUSALITY

Information causality is a generalization of the no-signaling principle, a constraint that limits the amount of information that a receiver of information can gain about a sender's data set.[12] If Alice is the sender and Bob the receiver, the principle says that, given all Bob's local resources and $M$ classical bits communicated by Alice, Bob's information gain is at most $M$ bits. The no-signaling principle is just information causality for $M = 0$: if Alice communicates nothing to Bob, then there is no information in the statistics of Bob's outputs about Alice's data set. Classical and quantum correlations satisfy the constraint, but Popescu–Rohrlich correlations and a large class of superquantum correlations don't.

There's a sense in which Popescu–Rohrlich bananas are just "too good to be true." As Wim van Dam showed, communication complexity would become trivial with PR boxes. So in Bananaworld, with access to Popescu–Rohrlich bananas, Alice and Bob could solve any communication problem with just one bit of communication, and that seems too good to be true.

But there are other remarkable feats that would be possible with Popescu–Rohrlich bananas. Suppose Alice has a data set consisting of two bits, and she is allowed to send Bob one bit of information. She would like her one-bit message to somehow encode enough information about her data set so that Bob can receive just one of these bits, depending on his choice, and which bit he chooses to receive depends on how he decodes this information. Oblivious transfer is a hypothetical protocol that allows Alice and Bob to implement this information-theoretic task.[13] At the end of the protocol, Bob ends up knowing the value of one of Alice's bits, whichever bit he chooses to receive, and Alice does not know which bit—the protocol leaves her oblivious of Bob's choice. There's no way to do this with a 100% success rate in a classical world or a quantum world, but it is possible in Bananaworld.

Think of the task as a guessing game where Alice and Bob are separated and can communicate with a moderator during the game, but not with each other, except for one bit that Alice is allowed to send to Bob at each round of the game. The moderator chooses a pair of bits, $x_0, x_1$, for each round, announces them to Alice, and asks Bob to guess a particular bit, either the first bit in the pair, or the second bit. Alice can send Bob one bit of information, but she doesn't know which bit he's supposed to guess. If Alice and Bob have a supply of shared Popescu–Rohrlich bananas, they can win the game. Bob will be able to guess the designated bit correctly, using the one bit communicated by Alice.

To remind you, peelings and tastes of a pair of Popescu–Rohrlich bananas are correlated so that the tastes are the same for the three combinations of peelings $SS, ST, TS$, with equal probability of either banana tasting ordinary (0) or intense (1), but the tastes are different if Alice and Bob both peel $T$, again with equal probability for the two possible tastes. Using Boolean addition $\oplus$, or addition modulo 2 for 0 and 1, where multiples of 2 are set to 0, the correlation can be expressed as:

$$a \oplus b = A \cdot B.$$

Here, $A$ represents Alice's peeling, 0 for $S$ or 1 for $T$, $B$ represents Bob's peeling, 0 for $S$ or 1 for $T$, and $a$ and $b$ represent the tastes of Alice's banana and Bob's banana, 0 for ordinary and 1 for intense.

Here's how they do it. Alice peels a banana from the stem end if the bits she receives from the moderator are the same, and from the top end if the bits are different. (That's the same as saying that she peels $S$ if $x_0 \oplus x_1 = 0$, and $T$ if $x_0 \oplus x_1 = 1$.) Suppose the taste of her banana is $a$. She sends the one-bit message $m = 0$ to Bob if the taste is the same as her first bit, and $m = 1$ if the taste is different from her first bit. (That's the same as saying that she sends Bob the message $m = x_0 \oplus a$.) If the bit Bob has to guess is $x_0$, he peels his paired banana from the stem end. If the bit he has to guess is $x_1$, he peels his banana from the top end. Suppose the taste of his banana is $b$. Bob guesses the designated bit as 0 if the taste of his banana is the same as the message bit $m$, and he guesses 1 if it's different from $m$. (That's the same as saying that he guesses $m \oplus b$.)

Here's why the strategy works. If Bob is supposed to guess Alice's first bit $x_0$, he peels his banana from the stem end. Then the taste of his banana is the same as the taste of Alice's banana, however she peels her banana. The message bit $m$ is 0 or 1 depending on whether the taste of Alice's banana is the same as Alice's first bit ($m = 0$), the bit Bob has to guess, or different from Alice's first bit ($m = 1$). Since the taste of Bob's banana is the same as the taste of Alice's banana, the message bit tells Bob whether or not the taste of his banana is the same as the bit he has to guess ($m = 0$), or different from the bit he has to guess ($m = 1$). So he guesses the bit as the same as the taste of his banana if $m = 0$, and different from the taste of his banana if $m = 1$.

Putting it formally, if Bob is supposed to guess Alice's first bit, he peels his banana from the stem end, which is represented by $B = 0$. Then $A \cdot B = 0$ so $a \oplus b = 0$. Bob guesses the bit as $m \oplus b$. Since $m = x_0 \oplus a$, in this case $m \oplus b$ is $x_0 \oplus a \oplus b = x_0$. If Bob is supposed to guess Alice's second bit $x_1$, he peels his banana from the top end, which is represented by $B = 1$. Then $A \cdot B = A$ so $a \oplus b = A$. As before, Bob computes his guess as $m \oplus b = x_0 \oplus a \oplus b$, which in this case is $x_0 \oplus A$. Since $A = x_0 \oplus x_1$, this is $x_0 \oplus x_0 \oplus x_1$, which is equal to $x_1$, because $x_0 \oplus x_0 = 0$. In both cases, Bob guesses correctly.

Figure 9.1 illustrates the guessing game when Alice has two bits, $x_0 = 1, x_1 = 0$, and Bob has to guess the first bit $x_0$. Since Alice's bits are different, she peels $T$. Bob peels $S$ because he is supposed to guess Alice's first bit. The tastes of their bananas must be the same in this case. The message bit $m$ is 0 or 1, depending on whether the taste of Alice's banana is the same as the taste of her first bit, or different from the taste of her first bit. In this case, the taste of Alice's banana is different from her first bit, so $m = 1$. Bob guesses 0 or 1, depending on whether the taste of his banana is the same as $m$ or different from $m$. Since the taste of Bob's banana is 0 and $m = 1$, he guesses 1.

Suppose Bob has to guess Alice's second bit $x_1 = 0$. The only difference between the two cases is that in the first case Bob peels $S$, and in the second case he peels $T$. What Alice does must be the same in the two cases, because she doesn't know which bit the moderator asks Bob to guess. Since they both peel $T$ now, the tastes of their bananas must be different. If Alice's banana tastes 0 and Bob's banana tastes 1, the message bit $m$ is 1 (because the taste of Alice's banana is different from the first bit). So Bob guesses 0 because the taste of his banana and $m$ are both 1. If Alice's banana tastes 1 and Bob's banana tastes 0, the message bit $m$ is 0 (because the taste of Alice's banana is the same as the first bit). So again Bob guesses 0 because the taste of his banana and $m$ are both 0.

The guessing game gets a lot more complicated if Alice gets more than two bits at each round of the game and Bob has to guess a particular bit chosen by the moderator. See Subsection 9.3.1, "The Guessing Game," in the "More" section at the end of the chapter.

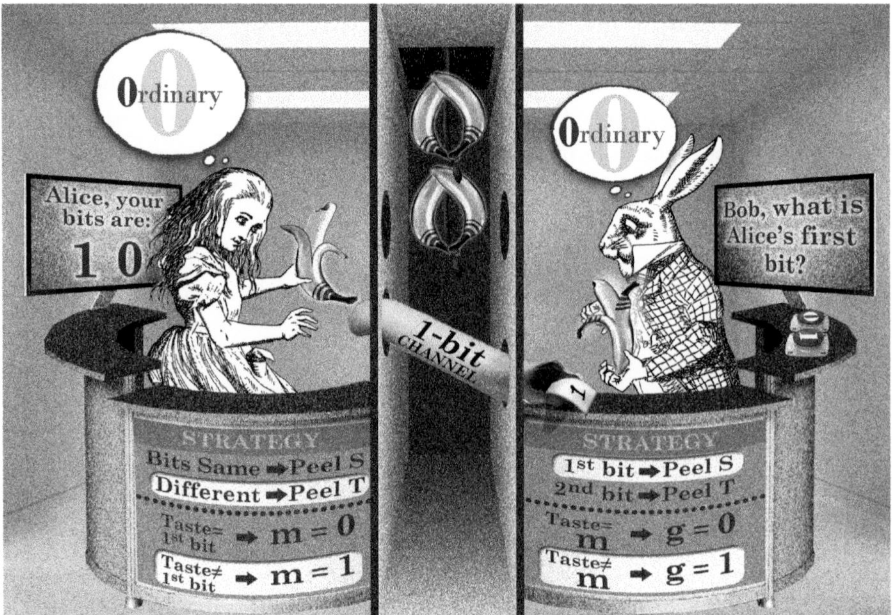


**Figure 9.1** The guessing game if Alice has two bits and Bob has to guess a designated bit. Alice's message bit is represented by *m*, and Bob's guess by *g*.

Here's something else they could do with Popescu–Rohrlich bananas. Alice and Bob want to go on a date, but only if they both like each other. In other words, they would like to compute a function that takes the value 1 or "yes, go on a date" if they both like each other (if both inputs to the function are 1 for "like"), but takes the value 0 or "no" if at least one party does not like the other (if the inputs are both 0 for "don't like," or one input is 0 and the other input is 1). In effect, they want to compute a classical conjunction, an AND function, which outputs 1, or "yes," if and only if both inputs are 1 or "yes."

Now, a device like a calculator that computes an AND function doesn't guarantee the privacy of Alice's input and Bob's input. Unlike a pair of Popescu–Rohrlich bananas, where Alice's banana is in a location that is secure for Alice and Bob's banana is in a location that is secure for Bob, a local device with an input by both parties isn't private for either party. Even if Alice and Bob input 0 or 1 into the device without anyone else being able to see the input, once the input information is in the device in the circuits that compute the function, it's no longer private. There's no principle of physics that prevents someone, in particular Alice or Bob, from accessing this information.

In a classical or quantum world, there's no way Alice and Bob can solve the dating problem without risking revealing information that they both want to keep private. Alice doesn't want Bob to know that she likes him if he doesn't like her, and similarly for Bob. In Bananaworld, they can compute this function, while keeping private the information they want to keep private. Alice and Bob pick a pair of Popescu–Rohrlich bananas. They take one banana each and go off to different parts of the island to peel their bananas privately, peeling $S$ for input 0 or "no," or $T$ for input 1 or "yes." After peeling the bananas, they get together and announce their tastes. If the tastes are different, they know that both inputs were 1, so they happily go on a date. In this case, of course, Alice knows that Bob likes her, and Bob knows that Alice likes him, but that's fine. If the tastes are the same, they know only that either Alice did not like Bob, or that Bob did not like Alice, or that the dislike was mutual. While Bob can infer that Alice doesn't like him if he likes her, this knowledge is private, and so is his pain. So Bob avoids any humiliation, and similarly for Alice.

If this sounds trivial, the same problem arises if Alice and Bob represent two multinational companies that are contemplating a merger. Each company is interested in a merger only if the other company's assets are greater than a certain amount. They would like to be able to decide the issue, but in a way that keeps information about their assets private if the merger doesn't go through. You'll be able to think of other scenarios for which a supply of Popescu–Rohrlich bananas would be really helpful.

Equipped with Popescu–Rohrlich bananas, Alice and Bob could achieve an even more spectacular feat than one-out-of-two oblivious transfer. At each round of a game, a moderator sends Alice $N$ random and independent bits $x_0, x_1, \ldots, x_{N-1}$ and asks Bob, who is separated from Alice, to guess a particular bit that the moderator chooses. Alice is allowed to send Bob a one-bit message at each round. As usual, Alice and Bob are allowed to communicate and plan a strategy before the game starts, but once the game starts the only communication between them is the one-bit message that Alice is allowed to send to Bob at each round of the game. They win a round if Bob guesses correctly, and they win the game if Bob always guesses correctly over any number of rounds. The problem they have to solve is that Alice has to decide on the message bit she sends to Bob at each round of the game independently of the bit that Bob has to guess, which is unknown to Alice.

Pawłowski and colleagues show that there is a strategy that will allow Alice and Bob to win the game with shared PR boxes (equivalently, with shared Popescu–Rohrlich bananas). That is, for any round, and for any bit the moderator chooses, Bob will be able to correctly guess the value of the designated bit in Alice's list. There's also a winning strategy if Bob is required to correctly guess any designated subset of Alice's data set, where the moderator chooses which bits in Alice's data set Bob has to guess at each round. For this game, Alice is allowed to send Bob $M$ bits, where $M < N$ is the number of bits that Bob has to guess. The subsection "The Guessing Game" in the "More" section at the end of the chapter shows how this is done with PR boxes.

So a PR box or a pair of Popescu–Rohrlich bananas would be an unreasonably powerful device in an information-theoretic sense. The motivation for information causality as an information-theoretic principle is that it excludes feats like these which, intuitively, seem "too good to be true". I'll formulate the principle of information causality more precisely and show how it works in Subsection 9.3.3, "How Information Causality Works," in the "More" section at the end of the chapter.

### The bottom line

- Why are there no Popescu–Rohrlich bananas in our world? What principle excludes Popescu–Rohrlich bananas? I showed in the "Bananaworld" chapter that the probability of perfectly simulating Popescu–Rohrlich bananas is $1/2(1 + K/4)$, where $K$ is the Clauser–Horne–Shimony–Holt quantity (a sum of expectation values). Popescu and Rohrlich pointed out that correlations for which $E = K/4$ lies between the Tsirelson bound $1/\sqrt{2}$ and the PR box value 1 don't violate the no-signaling principle. So the question is, in part, why the Tsirelson bound?
- There's a sense in which a pair of Popescu–Rohrlich bananas is an unreasonably powerful device in an information-theoretic sense. For example, with a supply of Popescu–Rohrlich bananas, Alice and Bob could compute the value of any Boolean function of their private data sets with just one bit of communication. One would expect that harder problems require more communication than easier problems. As another example, they could win a guessing game where a moderator gives Alice a list of two (or more) bits and asks Bob to guess a particular bit in the list. If Alice is allowed to send Bob one bit of information at each round of the game, they could win the game with supply of shared Popescu–Rohrlich bananas.
- In a 2009 *Nature* paper, Marcin Pawłowski and colleagues show that the Tsirelson bound follows from an information-theoretic principle that they call "information causality." The motivation for information causality as an information-theoretic principle is that it excludes feats like these that are "too good to be true."
- The principle is formulated more precisely and discussed further in Subsection 9.3.3, "How Information Causality Works," in the "More" section at the end of the chapter.

## 9.2 THE PARABLE OF THE OVER-PROTECTIVE SEER

This section is about an alternative principle proposed as an answer to the question "why the quantum?" The exclusivity principle, or the principle of local orthogonality, is variously stated as "events that are mutually exclusive pairwise are jointly exclusive," or "the sum of the probabilities of pairwise mutually exclusive events is less than or equal to 1." Adán Cabello credits a 1960 paper by Ernst Specker, "Die Logik nicht gleichzeitig entscheidbarer Aussagen" ("The logic of propositions that are not simultaneously decidable"),[14] as the source of the exclusivity principle.[15] In the 1960 paper, Specker motivates a related principle that Cabello calls Specker's principle—"pairwise decidable propositions are simultaneously decidable"—with a story about an ancient seer from Nineva, who requires his daughter's suitors to pass a test that involves predicting, for each of three boxes, whether the box holds a gem or not, when there could be as few as zero gems or as many as three gems. After a suitor's prediction, two boxes, which the suitor declares to be both empty or both full, are opened. The prediction game can't be won because the boxes have a magical property: for any two selected boxes that are opened (and only two boxes can be opened), one contains a gem and one is empty. That's impossible if each of the three boxes either has a gem or not before being opened, because then there must be two boxes that are either both empty (if there are no gems in any of the boxes, or one gem in one of the boxes), or two boxes that both contain a gem (if there are two or three gems, distributed one to a box).

Here's the story, dubbed "The Parable of the Over-Protective Seer" in a paper on Specker's principle by Yeong-Cherng Liang, Robert Spekkens, and Howard Wiseman.[16] The translation, from the Liang–Spekkens–Wiseman paper, is an amalgam of separate translations by Allen Stairs and by Michiel Seevinck.[17] The clarifications in square brackets are by Liang, Spekkens, and Wiseman.

> At the Assyrian School of Prophets in Arba'ilu in the time of King Asarhaddon [(681–669 BCE)], there taught a seer from Nineva. He was a distinguished representative of his faculty (eclipses of the sun and moon) and aside from the heavenly bodies, his interest was almost exclusively in his daughter. His teaching success was limited; the subject proved to be dry and required a previous knowledge of mathematics which was scarcely available. If he did not find the student interest which he desired in class, he did find it elsewhere in overwhelming measure. His daughter had hardly reached a marriageable age when he was flooded with requests for her hand from students and young graduates. And though he did not believe that he would always have her by his side, she was in any case still too young and her suitors in no way worthy. In order that the suitors might convince themselves of their unworthiness, he promised them that she would be wed to the one who could solve a prediction task that was posed to them.
>
> Each suitor was taken before a table on which three little boxes stood in a row [each of which might or might not contain a gem], and was asked to predict which of the boxes contained a gem and which did not. But no matter how many times they tried, it seemed impossible to succeed in this task. After each suitor had made his prediction, he was ordered by the father to open any two boxes which he had predicted to be both empty or any two boxes which he had predicted to be both full [in accordance with whether he had predicted there to be at most one gem among the three boxes, or at least two gems, respectively]. But it always turned out that one contained a gem and the other one did not, and furthermore the stone was sometimes in the first and sometimes in the second of the boxes that were opened. But how can it be possible, given three boxes, to neither be able to pick out two as empty nor two as full?
>
> The daughter would have remained unmarried until the father's death, if not for the fact that, after the prediction of the son of a prophet [whom she fancied], she quickly opened two boxes herself, one of which had been indicated to be full and the other empty, and the suitors prediction [for these two boxes] was found, in this case, to be correct. Following the weak protest of her father that he had wanted two other boxes opened, she tried to open the third. But this proved impossible whereupon the father grudgingly admitted that the prediction, being unfalsified, was valid. [The daughter and the suitor were married and lived happily ever after.]

As you might expect, the story has a counterpart in Bananaworld. There are banana trees in Bananaworld—Specker banana trees—that have bunches with just three bananas. The probability of a peeled banana tasting ordinary or intense is 1/2. If you pick and peel any two bananas on a bunch (either from the top end or the stem end, it makes no difference), they always turn out to have different tastes: one tastes ordinary, and the other intense. Once you peel two bananas, the remaining banana is inedible.

Suppose the seer has one of these Specker banana trees, and a suitor is asked to choose a bunch of bananas and predict the tastes of all three bananas before they are peeled. For any prediction, there must be two bananas that are predicted to both taste 0, or two bananas that are predicted to both taste 1. If the seer selects such a pair according to the suitor's prediction and the bananas are peeled and tasted, the prediction will be wrong, because if only two bananas from the same bunch on a Specker banana tree are peeled, they always turn out to have different tastes.

Evidently, a Specker banana can't have a definite taste before it's peeled, independent of the taste of the banana that is peeled with it. If each banana on a Specker bunch did have a definite taste before being peeled, a "being-thus" in Einstein's sense, then either at least two bananas taste ordinary (if one banana tastes intense, or no bananas taste intense), or at least two bananas taste intense (if one banana tastes ordinary, or no bananas taste ordinary). In that case, a suitor would correctly predict that two bananas have the same taste in about one third of the trials, but the facts in the Bananaworld story are that this never happens. So no noncontextual assignment of tastes to the bananas before they are peeled can fit the phenomena. The suitors' predictions always fail because they correspond to such an assignment.

The title of Specker's paper makes clear that the paper is about logic, or propositions, but the title is followed by a quotation from the Swiss philosopher Ferdinand Gonseth: "La logique est d'abord une science naturelle" ("Logic is first of all a natural science"). So Specker evidently had in mind a principle that applies to events in the physical world.[18] The moral of the story for Specker was that you could have a set of propositions such that every pair of propositions is "decidable," in the sense that, for any selected pair of propositions, both propositions have a definite truth value, either true or false, but the whole set is not simultaneously decidable. You can't, in principle, assign definite truth values to all the propositions in the set simultaneously, so you can't check the truth values of all the propositions simultaneously (just as you can't open all the boxes simultaneously, but you can open any selected pair).

Subsequently (anticipated in the 1960 paper), Kochen and Specker proved that you can have a finite set of quantum propositions—corresponding to observables with two possible values, 0 (for "false") and 1 (for "true")—that can't all be assigned truth values simultaneously, so they aren't simultaneously decidable. The observables in such Kochen–Specker sets don't all commute with each other, so only subsets of commuting observables are decidable or "compatible" and can be measured simultaneously.

It's possible to have three quantum observables, $A, B, C$, such that $A$ commutes with $B$, $B$ commutes with $C$, but $A$ doesn't commute with $C$, so $A$ and $B$ can be measured together, and $B$ and $C$ can be measured together, but all three observables can't be measured together because $A$ and $C$ can't be measured together. (See Figure 6.3 in Section 6.3, "The Kochen–Specker Theorem and Klyachko Bananas," in Chapter 6, "Quantum Magic." There, $O$ commutes with $P$, $P$ commutes with $O'$, but $O$ doesn't commute with $O'$.)

If every pair of quantum observables in a set is compatible (so the observables all commute pairwise, unlike the set $A, B, C$ in which $A$ fails to commute with $C$), then the whole set is simultaneously compatible, and every observable in the set can be assigned a value in a simultaneous measurement. The seer's magic boxes, or bunches of Specker bananas, don't have this property. The point of Specker's parable is to show that a broad class of correlations is possible for physical systems if all propositions aren't simultaneously decidable, including quantum correlations as a special case. Specker's principle separates quantum mechanics from other theories in which propositions aren't simultaneously decidable. In Cabello's recorded discussion,[19] Specker puts the principle this way: "If you have, say, ... three questions, and you can answer any two of them, then you can also answer all three of them. And this seems to me very fundamental."

The correlations of Specker bananas are superquantum. There are three possible pairs of bananas in a three-banana bunch, 12, 23, and 31, if the bananas are labeled 1, 2, 3. The sum of the probabilities that two peeled bananas have different tastes, over all three pairs, is 3. If the bananas have definite noncontextual tastes before they are peeled, then the sum of the probabilities is at most 2, and this is also the case for any quantum simulation of the Specker banana correlations.

The simplest case where there is a quantum advantage over any noncontextual assignment of tastes to the bananas before they are peeled is the five-banana bunch of Klyachko bananas. As I showed in Section 6.3 in Chapter 6, "Quantum Magic," the sum of the probabilities that two peeled bananas on a pentagram edge have different tastes, over the four pentagram edges, is at most 4, for any noncontextual assignment of tastes to the bananas before they are peeled. If probabilities are assigned by a certain quantum state, the sum of probabilities is $2\sqrt{5} \approx 4.47$. For Klyachko bananas, the sum of the probabilities is 5.

The exclusivity principle shows that the *maximum* quantum value for a five-banana bunch of Klyachko bananas is in fact $2\sqrt{5}$, and that there is no quantum advantage for a three-banana bunch of Specker bananas, and a lot more has been shown to follow from the principle. Applying the principle, even to these simple cases, is a nontrivial exercise in graph theory that I won't go into, but here's how to understand the motivation behind the principle.

The exclusivity principle refers to events, and the relevant events are measurements yielding outcomes, or sets of simultaneous (compatible) measurements yielding sets of outcomes. Two events are *mutually exclusive* if they can't both happen at the same time. For example, the following events $e_1$ and $e_2$ are mutually exclusive:

$e_1$: the observables $O_1, O_2$ are measured (perhaps both by Alice if the observables commute, or by Alice and Bob in different places if they don't) to have outcomes $0, 1$ respectively;

$e_2$: the observables $O_1, O_2'$ are measured to have outcomes $1, 1$ respectively.

The events $e_1$ and $e_2$ are mutually exclusive because the value 0 for $O_1$ in $e_1$ excludes the value 1 for $O_1$ in $e_2$: the two pairs of measurement outcomes can't both occur at the same time.

Events are *pairwise mutually exclusive* if no two of them can occur at the same time—either neither event occurs, or just one of the events occurs. Events are *jointly exclusive* if the occurrence of an event excludes the possibility of any other events occurring. For example, the possible outcomes of a measurement are jointly exclusive: you can only get one outcome in a measurement. If events are jointly exclusive, so that at most one event can occur, the sum of the probabilities of the events is less than or equal to 1. If the events are also *jointly exhaustive* so that they cover all possibilities (like the set of possible outcomes of a measurement), then the sum of the probabilities is exactly 1. If events are not simultaneously decidable (like the events in Specker's parable, or the measurement outcomes of incompatible quantum observables), they can be pairwise mutually exclusive without being jointly exclusive. In that case, the sum of the probabilities of the events needn't be less than or equal to 1. But if events satisfy Specker's principle, the fact that they are pairwise decidable guarantees that they are simultaneously or jointly decidable. In that case, pairwise mutually exclusive events are are also jointly exclusive, and the sum of the probabilities of the events must be less than or equal to 1. This is the exclusivity principle in Cabello's formulation: the sum of the probabilities of mutually exclusive events is less than or equal to 1.

Imposing the exclusivity principle on events that are not simultaneously decidable restricts the set of possible events. The principle is satisfied by classical theories (because all propositions are simultaneously decidable) and by quantum mechanics (because if every pair of observables in a set commutes, then the observables can all be measured simultaneously, so Specker's principle is satisfied). The hope is that this principle will turn out to be the defining principle that separates quantum mechanics from superquantum theories that are "more contextual" or "more nonlocal" than quantum mechanics.[20]

**The bottom line**

- The exclusivity principle, or the principle of local orthogonality, is another principle that has been proposed as an answer to the question "Why the quantum?" In a 1960 paper, Ernst Specker motivates a related principle ("pairwise decidable propositions are simultaneously decidable") with a story about an ancient seer from Nineva, who requires his daughter's suitors to pass a test that involves predicting which of three boxes hold a gem and which do not. The prediction game can't be won because the boxes have a magical property: for any two selected boxes that are opened (and only two boxes can be opened), one contains a gem and one is empty. That's impossible if each of the three boxes either has a gem or not before being opened.

(continued)

**The bottom line** *(continued)*

- The story has a counterpart in Bananaworld. Specker banana trees have bunches with just three bananas. The probability of a peeled Specker banana tasting ordinary or intense is 1/2. If you pick and peel any two bananas on a Specker bunch (either from the top end or the stem end), they always turn out to have different tastes. Once you peel two bananas, the remaining banana is inedible. A Specker banana can't have a definite taste before it's peeled, independent of the taste of the banana that is peeled with it. The taste must be context dependent, where the context is defined by the paired banana.
- The moral of Specker's parable is that you can have a set of propositions, such that every pair of propositions is "decidable," in the sense that both propositions have a definite truth value, either true or false, but the whole set is not simultaneously decidable. You can check the truth values of any selected pair of propositions, but you can't check the truth values of all the propositions simultaneously (just as you can't open all the boxes in Specker's parable simultaneously, but you can open any selected pair).
- Specker's paper anticipates the Kochen–Specker theorem, discussed in Chapter 6: there are finite sets of quantum propositions, corresponding to observables with two possible values, 0 (for "false") and 1 (for "true"), that can't all be assigned truth values simultaneously, so they aren't simultaneously decidable.
- Imposing the exclusivity principle on events that are not simultaneously decidable restricts the set of possible events. Quantum mechanics satisfies the principle, and the hope is that it will turn out to be the defining principle that separates quantum mechanics from superquantum theories that are "more contextual" or "more unlocal."

## 9.3 MORE

The subsections in this "More" section all involve a lot of unavoidable calculation. The calculations are elementary and don't involve quantum states, but they are tedious to go through. The subsection "The Guessing Game" shows how Alice and Bob can use Popescu–Rohrlich bananas to win the guessing game in which a moderator gives Alice a list of four bits, and Bob a designated bit in Alice's list that he has to guess. Alice is allowed to send Bob one bit of information at each round of the game. There's an illuminating comic strip that shows how they can do this in a specific case, and a detailed explanation of this case and the general case. The subsection "The Probability of Guessing Correctly" is a derivation of a result that's used in the subsection "How Information

Causality Works," where I formulate the principle of information causality explicitly and use some numbers to show how the principle works. The final subsection, "The Tsirelson Bound from Information Causality," is included for readers who would like to see how the Tsirelson bound, $1/\sqrt{2}$, follows from information causality.

### 9.3.1 The Guessing Game

Suppose the moderator gives Alice four bits, $x_0, x_1, x_2, x_3$, at each round of the game. Bob has to guess a particular bit chosen by the moderator. The strategy involves two stages, with two shared pairs of Popescu–Rohrlich bananas, labeled 1 and 2, at the first stage, and a third shared pair labeled 3 at the second stage.

At the first stage, Alice peels her banana from the first pair $S$ or $T$ depending on whether the first two bits, $x_0$ and $x_1$, are the same (in which case she peels $S$) or different (in which case she peels $T$). She peels her banana from the second pair $S$ or $T$ depending on whether the second two bits $x_2$ and $x_3$ are the same (in which case she peels $S$) or different (in which case she peels $T$). Call the taste of the banana from the first pair $a_1$ and the taste of the banana from the second pair $a_2$. If the bit Bob has to guess is one of the first two bits, $x_0$ or $x_1$, he peels his banana from the first pair $S$ for $x_0$ and $T$ for $x_1$, and he does nothing to his banana from the second pair. If the bit he has to guess is one of the second two bits, $x_2$ or $x_3$, he peels his banana from the second pair $S$ for $x_2$ and $T$ for $x_3$, and he does nothing to his banana from the first pair. Call the taste of his peeled banana $b$.

At the second stage, Alice peels her banana from the third pair $S$ if there is an even number of 1's (or no 1's) in $x_0, x_2$, and the tastes, $a_1, a_2$, of bananas 1 and 2. She peels her third banana $T$ if there is an odd number of 1's. Call the taste of Alice's third banana $a_3$. Bob peels his banana from the third pair $S$ if the bit he has to guess is one of the first two bits, $x_0$ or $x_1$, and $T$ if the bit he has to guess is one of the second two bits, $x_2$ or $x_3$. Call the taste of Bob's third banana $b_3$. Alice then sends Bob the message bit $m = 0$ if there are an even number of 1's in $x_0$ and the tastes, $a_1, a_3$, of her first and third bananas, and $m = 1$ if there are an odd number of 1's. Bob guesses 0 if there are an even number of 1's (or no 1's) in $m$ and the tastes of his two bananas, $b$ and $b_3$, and he guesses 1 if there are an odd number of 1's.

This is probably hard to follow, so see Figures 9.2 and 9.3 for a specific case.

Here's why the strategy works. The bit Bob has to guess can be specified by two bits $i_0$ and $i_1$. The index $i_1$ is 0 or 1 depending on whether Bob has to guess one of the first two bits, or one of the second two bits. The index $i_0$ is 0 or 1 depending on whether Bob has to guess $x_0$ or $x_1$ in the first two bits, or $x_2$ or $x_3$ in the second two bits.

In the simple game where Alice has two bits $x_0$ and $x_1$, Bob can guess either bit as he chooses based on the message bit $m$ and the taste of his banana, depending on how he peels his banana. In this case, Alice peels her banana $S$ or $T$ depending on whether $x_0$ and $x_1$ are the same or different. This is equivalent to saying that she peels $S$ if $x_0 \oplus x_1 = 0$ and she peels $T$ if $x_0 \oplus x_1 = 1$. In other words, if Alice peels her banana according to $x_0 \oplus x_1$,

**Figure 9.2** The guessing game if Alice has four bits and Bob has to guess a designated bit.

**Figure 9.3** The guessing game if Alice has four bits and Bob has to guess a designated bit.

Bob can guess either $x_0$ or $x_1$, as he chooses, depending on how he peels his banana. (You might want to re-read the explanation of why the strategy in the simple game works in Section 9.1, "Information Causality," to follow the argument below, which is a lot more convoluted.)

In the case where Alice has four bits, Alice peels her third banana $S$ or $T$ according to whether there is an even number of 1's (or no 1's) in $x_0, a_1, x_2, a_2$, or an odd number of 1's. That's equivalent to saying that she peels her third banana $S$ or $T$ according to whether $(x_0 \oplus a_1) \oplus (x_2 \oplus a_2)$ is 0 or 1. So Bob could guess either $x_0 \oplus a_1$ or $x_2 \oplus a_2$ as $m \oplus b_3$, depending how he peels his third banana. Since $m$ can be expressed as $m = x_0 \oplus a_1 \oplus a_3$, Bob could guess either $x_0 \oplus a_1$ or $x_2 \oplus a_2$ as $x_0 \oplus a_1 \oplus a_3 \oplus b_3$, depending on how he peels his third banana.

Here $a_3 \oplus b_3$, the Boolean sum of the tastes of the third pair of bananas, is equal to the product of the peelings (taking $S$ as 0 and $T$ as 1). Alice peels her third banana according to whether $x_0 \oplus a_1 \oplus x_2 \oplus a_2$ is 0 or 1, and Bob peels his third banana according to whether $i_1$ is 0 or 1. So the product of the peelings is $(x_0 \oplus a_1 \oplus x_2 \oplus a_2) \cdot i_1$, and according to the Popescu–Rohrlich correlation

$$a_3 \oplus b_3 = (x_0 \oplus a_1 \oplus x_2 \oplus a_2) \cdot i_1.$$

If the index bit $i_1 = 0$, $a_3 \oplus b_3 = 0$, so Bob would guess $x_0 \oplus a_1$ as $m \oplus b_3 = x_0 \oplus a_1 \oplus a_3 \oplus b_3 = x_0 \oplus a_1$. If $i_1 = 1$, $a_3 \oplus b_3 = x_0 \oplus a_1 \oplus x_2 \oplus a_2$, so Bob would guess $x_2 \oplus a_2$ as $m \oplus b_3 = x_0 \oplus a_1 \oplus a_3 \oplus b_3 = x_0 \oplus a_1 \oplus (x_0 \oplus a_1 \oplus x_2 \oplus a_2) = x_2 \oplus a_2$ (because Boolean addition of $x_0 \oplus a_1$ twice is zero).

To sum up this preliminary argument,

- if $i_1 = 0$, then $a_3 \oplus b_3 = 0$ and $m \oplus b_3 = x_0 \oplus a_1$;
- if $i_1 = 1$, then $a_3 \oplus b_3 = x_0 \oplus a_1 \oplus x_2 \oplus a_2$ and $m \oplus b_3 = x_2 \oplus a_2$.

Now, if Bob has to guess the value of $x_0$ (corresponding to $i_0 = 0, i_1 = 0$) or $x_1$ (corresponding to $i_0 = 1, i_1 = 0$), he guesses $m \oplus b_3 \oplus b$, where $b$ is the taste of Bob's banana in the first pair of bananas. For both $x_0$ and $x_1$, Bob's index $i_1$ is 0, so for both these bits $a_3 \oplus b_3 = 0$, as I showed in the preliminary argument. In this case,

$$m \oplus b_3 \oplus b = x_0 \oplus a_1 \oplus a_3 \oplus b_3 \oplus b = x_0 \oplus a_1 \oplus b.$$

Since $a_1 \oplus b = (x_0 \oplus x_1) \cdot i_0$ according to the Popescu–Rohrlich correlation for the first pair of bananas, Bob's guess is $x_0 \oplus ((x_0 \oplus x_1) \cdot i_0)$. If $i_0 = 0$, Bob correctly guesses $x_0$. If $i_0 = 1$, Bob correctly guesses $x_1$, because $x_0 \oplus x_0 \oplus x_1 = x_1$.

If Bob has to guess the value of $x_2$ (corresponding to $i_0 = 0, i_1 = 1$) or $x_3$ (corresponding to $i_0 = 1, i_1 = 1$), he guesses $m \oplus b_3 \oplus b$, where $b$ is the taste of Bob's banana in the second pair of bananas. For both $x_2$ and $x_3$, Bob's index $i_1$ is 1, so for both these bits $a_3 \oplus b_3 = x_0 \oplus a_1 \oplus x_2 \oplus a_2$, as I showed in the preliminary argument. In this case

$$m \oplus b_3 \oplus b = x_0 \oplus a_1 \oplus a_3 \oplus b_3 \oplus b = x_0 \oplus a_1 \oplus (x_0 \oplus a_1 \oplus x_2 \oplus a_2) \oplus b = x_2 \oplus a_2 \oplus b,$$

because Boolean addition of $x_0 \oplus a_1$ twice is zero. Since $a_2 \oplus b = (x_2 \oplus x_3) \cdot i_0$ according to the Popescu–Rohrlich correlation for the second pair of bananas, Bob's guess is $x_2 \oplus ((x_2 \oplus x_3) \cdot i_0)$. If $i_0 = 0$, Bob correctly guesses $x_2$. If $i_0 = 1$, Bob correctly guesses $x_3$, because Boolean addition of $x_2$ twice is 0.

So if Alice's data set consists of two bits or four bits and Alice and Bob have a supply of Popescu–Rohrlich bananas, Bob can correctly guess any bit he chooses in Alice's data set if Alice sends him one bit of information. The strategy for four bits is also a strategy for three bits (there's just one less bit that Bob has to worry about). If Alice's data set consists of eight bits, Bob can guess any bit he chooses using a strategy with three stages, with four pairs of Popescu–Rohrlich bananas at the first stage, two pairs of Popescu–Rohrlich bananas at the second stage, and one pair of Popescu–Rohrlich bananas at the third and final stage. This is also a strategy for a data set with any number of bits between five and eight. A similar strategy applies if Alice's data set consists of 16 bits, or fewer than 16 bits, and so on. So there's a strategy for a data set of any size. If Alice has $N = 2^n$ bits, the strategy involves $n$ stages.

The game can be modified to allow Alice to send $M < N$ classical bits of information to Bob at each round, in which case Bob has to guess the values of any randomly chosen set of $M$ bits in Alice's list of $N$ bits. In this case, Alice and Bob apply the strategy for $N$ with $M$ times the number of stages needed for one bit.

**The bottom line**

- I show how the strategy using Popescu–Rohrlich bananas works for the guessing game in which Alice gets four bits at each round of the game and Bob has to guess a particular bit chosen by the moderator, and Alice can communicate one bit to Bob at each round.
- Check out the comic strip to get the idea of how the strategy works for a particular choice of bits. For an explanation, and to see why this strategy works in the general case, you'll have to go through the unavoidably tedious nuts and bolts of the detailed discussion.

## 9.3.2 The Probability of Guessing Correctly

The probability of perfectly simulating a Popescu–Rohrlich correlation is $1/2(1 + K/4) = 1/2(1 + E)$, where $K$ is the Clauser–Horne–Shimony–Holt correlation and $E = K/4$, as I showed in the "Bananaworld" chapter. This is the probability of simulating the correlation with a general no-signaling box, where $E$ depends on the nature of the box. For a classical box or a local box, $E \leq 1/2$; for a quantum box that exploits nonlocal entanglement, $E \leq 1/\sqrt{2}$; and for a superquantum no-signaling box, $1/\sqrt{2} < E \leq 1$.

Consider the guessing game of the previous section, where Alice has four bits supplied by the moderator and Bob has to guess a particular bit that the moderator chooses.

Suppose Alice and Bob share no-signaling boxes of some sort, not necessarily Popescu-Rohrlich bananas, and Alice is allowed to communicate one bit to Bob. Bob's guess $m \oplus b_1 \oplus b_0$ will be correct if $b_1$ and $b_0$, the outputs of Bob's part of the no-signaling boxes at the two stages, are both correct (in the sense that they are both in accord with no-signaling correlations) or both incorrect (since $b_1 \oplus b_0$ will be the same in either case).

The probability of being correct at both stages is

$$\frac{1}{2}(1+E) \cdot \frac{1}{2}(1+E) = \frac{1}{4}(1+E)^2.$$

The probability of being incorrect at both stages is

$$\left(1-\frac{1}{2}(1+E)\right) \cdot \left(1-\frac{1}{2}(1+E)\right) = \frac{1}{2}(1-E) \cdot \frac{1}{2}(1-E) = \frac{1}{4}(1-E)^2.$$

So the probability, $P$, that Bob guesses correctly is

$$P = \frac{1}{4}(1+E)^2 + \frac{1}{4}(1-E)^2 = \frac{1}{2}(1+E^2).$$

If Alice has eight bits, the strategy involves three stages. The probability of being correct at each of the three stages is

$$\frac{1}{2}(1+E) \cdot \frac{1}{2}(1+E) \cdot \frac{1}{2}(1+E),$$

and the probability of being incorrect at two out of the three stages (i.e., for $b_0, b_1$ or $b_0, b_2$ or $b_1, b_2$) is

$$3 \cdot \frac{1}{2}(1-E) \cdot \frac{1}{2}(1-E) \cdot \frac{1}{2}(1+E).$$

So the probability that Bob guesses correctly is

$$P = \frac{1}{8}(1+E)^3 + \frac{3}{8}(1-E)^2(1+E) = \frac{1}{2}(1+E^3).$$

In the general case where Alice has $N = 2^n$ bits, the strategy involves $n$ stages. Bob guesses correctly if he makes an even number of errors over the $n$ stages. So the probability that Bob guesses correctly is

$$P = \frac{1}{2^n}(1+E)^n + \frac{1}{2^n}\sum_{j=1}^{\lfloor \frac{n}{2} \rfloor} \binom{n}{2j}(1-E)^{2j}(1+E)^{n-2j} = \frac{1}{2}(1+E^n),$$

where the sum in $\sum$ is from $j = 1$ to $j = \lfloor \frac{n}{2} \rfloor$, the integer value of $\frac{n}{2}$.

## The bottom line

- For the guessing game where Alice has any number, $n$, bits, the probability that Bob correctly guesses a particular bit is $\frac{1}{2}(1+E^n)$, where $E = K/4$, and $K$ is the Clauser–Horne–Shimony–Holt quantity (the sum of expectation values).

*(continued)*

 **The bottom line** *(continued)*

- This is a preliminary result that's used in the following section about how information causality works. If you're not interested in the derivation, note the result, skip this subsection, and read on.

### 9.3.3 How Information Causality Works

Pawłowski and colleagues formulate information causality as a limitation on Bob's information gain about an unknown data set of Alice: given all Bob's local resources and $M$ classical bits communicated by Alice, Bob's information gain is at most $M$ bits. The no-signaling principle is just information causality for $M = 0$. If Alice communicates nothing to Bob, then there is no information in the statistics of Bob's outputs about Alice's data set. So information causality is a generalization of no-signaling. Amazingly, the Tsirelson bound, $E \leq \frac{1}{\sqrt{2}}$, follows from this apparently innocuous condition.

The restriction to the communication of classical bits is essential here because entanglement correlations can be exploited to allow Alice to send Bob two classical bits by communicating just one quantum bit or qubit, as I showed in Section 8.3 in the chapter "Quantum Feats."

In the simplest case of the guessing game ($M = 1$), Alice has a data set of $N$ bits and is allowed to send Bob a 1-bit message $m$ in each round of the game. Bob has to guess a particular one of Alice's data bits selected by a moderator. Suppose the probability that Bob correctly guesses this bit is $P$. A convenient measure of how much information Bob gains from Alice's message is the binary entropy of $P$, which is defined as

$$h(P) = -P \log P - (1 - P) \log(1 - P),$$

where $\log P$ is the logarithm of $P$.

In information theory, logarithms are to the base 2. The logarithm of a number $n$ to the base 2 is the power to which 2 has to be raised to equal $n$. For example, if $2^l = n$, then $\log_2 n = l$. So $\log_2 1 = 0$ (which is another way of saying that 2 raised to the power 0 is 1, an arithmetic convention about the power 0), $\log_2 2 = 1$ (which is another way of saying that 2 raised to the power 1 is 2), and $\log_2 2^n = n$ (which is another way of saying that 2 raised to the power $n$ is $2^n$).

If Bob knows the value of the bit he has to guess, $P = 1$ and $h(P) = 0$. (The first term in $h(P)$ is 0 because $\log_2 1 = 0$. The second term is also 0 if $P = 1$.) If Bob has no information about the bit he has to guess, his guess is at chance: $P = 1/2$, and $h(P) = 1$. (The first term in $h(P)$ is $-\frac{1}{2} \log_2 \frac{1}{2}$. The second term is also $-\frac{1}{2} \log_2 \frac{1}{2}$. Since $\log_2 \frac{1}{2} = -1$, because $2^{-1} = \frac{1}{2}$, $h(P) = \log_2 \frac{1}{2} = 1$.) So the probability $P$ that Bob guesses the designated bit correctly is between 1 and $1/2$, which is to say that the binary entropy $h(P)$ is between 0 and 1.

If Alice sends Bob one classical bit of information, information causality requires that Bob's information about the $N$ unknown bits in Alice's data set increases by at most one bit. If the bit that Bob has to guess is chosen randomly, Bob's information about an arbitrary bit in Alice's data set can't increase by more than $1/N$ bits. In other words, the effect of Alice's one-bit communication on Bob's information can be measured by the change in the binary entropy of Bob's guess. Information causality limits this effect and says that the binary entropy $h(P)$ is at most $1/N$ closer to 0 from the chance value 1:

$$h(P) \geq 1 - 1/N.$$

In the "Bananaworld" chapter, I showed that the probability of successfully simulating the Popescu–Rohrlich correlation is $1/2(1 + K/4)$, or $1/2(1 + E)$, where $E = K/4$. In the previous subsection, "The Probability of Guessing Correctly," I showed that if Bob has to guess a randomly selected bit in Alice's data set of $N$ bits, then the probability that Bob guesses correctly is $P = \frac{1}{2}(1 + E^n)$, where $n = \log_2 N$, or $N = 2^n$. So information causality is satisfied when

$$h(P) \geq 1 - \frac{1}{2^n}.$$

Using the expression $P = \frac{1}{2}(1 + E^n)$, this can be expressed as

$$h\left(\frac{1}{2}(1 + E^n)\right) \geq 1 - \frac{1}{2^n}.$$

Putting it differently, information causality is *violated* when

$$h\left(\frac{1}{2}(1 + E^n)\right) < 1 - \frac{1}{2^n}.$$

To get a feel for how the information causality condition works, consider some numbers for $E$ and $n$. The binary entropy is expressed in terms of logarithms to the base 2 and logarithmic tables are usually to the base 10, because our arithmetic uses the base 10. (When you count and reach 10, you "carry 1" to mark the first group of 10 and start the count over for the next group of 10.) To use base 10 logarithmic tables, you need to convert the $\log_2$ expressions in the binary entropy to $\log_{10}$ expressions.

Here's how you do this. Suppose $\log_{10} n = k$ and $\log_2 n = l$. Since $\log_{10} n = k$ is just another way of saying that $10^k = n$, and $\log_2 n = l$ is just another way of saying that $2^l = n$, it follows that $10^k = 2^l$. The logarithm to the base 10 of the left-hand side of this equation is just $k$, which is $\log_{10} n$. The logarithm to the base 10 of the right-hand side is $l \log_{10} 2$, where $l = \log_2 n$. So to express $\log_2 n$ in terms of $\log_{10} n$, you divide $\log_{10} n$ by $\log_{10} 2$, which you can look up in a table as having a value of about .301:

$$\log_2 n = \frac{1}{\log_{10} 2} \log_{10} n \approx \frac{1}{.301} \log_{10} n.$$

First consider a couple of cases where $E = E_Q = \frac{1}{\sqrt{2}}$, the Tsirelson bound. Suppose $n = 1$, so Alice's data set has $2^1 = 2$ bits. In this case, $P = \frac{1}{2}(1 + \frac{1}{\sqrt{2}})$ and $1 - P = \frac{1}{2}(1 - \frac{1}{\sqrt{2}})$, so

$$h(P) \approx -\frac{1}{2}\left(1+\frac{1}{\sqrt{2}}\right)\frac{\log_{10}\frac{1}{2}\left(1+\frac{1}{\sqrt{2}}\right)}{.301}$$

$$-\frac{1}{2}\left(1-\frac{1}{\sqrt{2}}\right)\frac{\log_{10}\frac{1}{2}\left(1-\frac{1}{\sqrt{2}}\right)}{.301}$$

$$\approx .600.$$

There is no violation of information causality, because $.600 > 1 - \frac{1}{2^1} = \frac{1}{2}$.

If $n = 10$, Alice's data set has $2^{10} = 1024$ bits:

$$h(P) \approx -\frac{1}{2}\left(1+\frac{1}{\sqrt{2}^{10}}\right)\frac{\log_{10}\frac{1}{2}\left(1+\frac{1}{\sqrt{2}^{10}}\right)}{.301}$$

$$-\frac{1}{2}\left(1-\frac{1}{\sqrt{2}^{10}}\right)\frac{\log_{10}\frac{1}{2}\left(1-\frac{1}{\sqrt{2}^{10}}\right)}{.301}$$

$$\approx .99939.$$

There is still no violation of information causality, because $.99939 > 1 - \frac{1}{2^{10}} = 1 - \frac{1}{1024} \approx .9990$.

Now consider $E > E_Q$. The Tsirelson bound $E_Q$ is about .707. Take $E = .725$, a little larger than the Tsirelson bound, and $n = 6$, so Alice has $2^6 = 64$ bits. To work out $h(P)$ in this case, replace the Tsirelson bound $1/\sqrt{2}$ with .725 and the exponent 10 with $n = 6$ in the above expression. The calculation gives $h(P) \approx .9848$. There is no violation of information causality because $.9848 > 1 - \frac{1}{64} \approx .9844$. But for $n = 7$, when Alice has $2^7 = 128$ bits, $h(P) \approx .99208$ and there *is* a violation of information causality because $.99208 < 1 - \frac{1}{128} \approx .99218$.

The closer $E$ is to the Tsirelson bound, the greater $n$ must be for a violation of information causality. For $E = .708$ and $n = 10$, Alice's data set has $2^{10} = 1024$ bits and $h(P) \approx .99938$. There is no violation of information causality because $.99938 > 1 - \frac{1}{1024} \approx .9990$. In fact, for $E$ this close to the Tsirelson bound, you need a much greater value of $n$ for a violation of information causality—many more bits than 1024 in Alice's data set.

To find a value of $n$ for which information causality would be violated for $E = .708$, it's convenient to use an inequality from the paper by Pawłowski and colleagues:

$$h\left(\frac{1}{2}(1+y)\right) \leq 1 - \frac{y^2}{2\ln 2},$$

where $\ln 2 \approx .693$ is the natural logarithm of 2 (the logarithm to the base $e$, which is approximately 2.718). Since the condition for a violation of information causality is $h(\frac{1}{2}(1+E^n)) < 1 - \frac{1}{2^n}$, and $h(\frac{1}{2}(1+E^n)) < 1 - \frac{E^{2n}}{2\ln 2}$ if you substitute $E^n$ for $y$ in the inequality, you find that information causality is violated if

$$1 - \frac{E^{2n}}{2\ln 2} < 1 - \frac{1}{2^n}.$$

This expression simplifies to $(2E^2)^n > 2\ln 2 \approx 1.386$, or $n \approx 130$ if $E = .708$. So Alice's data set would have to contain about $2^{130}$ bits for a violation of information causality, which is more than 1 followed by 39 zeros!

The Tsirelson bound is $E = 1/\sqrt{2}$, which can be expressed as $E^2 = 1/2$, or $2E^2 = 1$. If $E > 1/\sqrt{2}$, then $2E^2 = 1 + a$, for some $a$, however small. In this case, information causality is violated.

Using the inequality in the paper by Pawłowski and colleagues, information causality is violated if $(2E^2)^n > 1.386$, so if $(1 + a)^n > 1.386$. Taking the logarithm to base 10 of both sides of this inequality, the condition for a violation of information causality becomes $n \log(1 + a) > \log(1.386)$. The logarithm of 1.386 is about .1418, so the condition for a violation is $n > .1418/\log(1 + a)$. As $a$ gets smaller and smaller, $\log(1 + a)$ gets closer and closer to $\log 1$, which is 0. So the right-hand side of the inequality gets larger and larger as $a$ gets smaller and smaller. But there is always a value of $n$ which is larger than this, however close $a$ is to 0. In other words, for any value of $a$, however small, there is a value of $n$, a size for Alice's data set, for which information causality is violated.

In the next section I'll show that information causality is satisfied when $E$ is exactly the Tsirelson bound, no matter how many bits there are in Alice's data set.

Here's another way to look at all this. If $E = E_Q = \frac{1}{\sqrt{2}}$, $P = \frac{1}{2}(1 + E^n)$ tends to $1/2$ and $h(P)$ tends to 1 as $n$ tends to infinity. So, if Alice has a very large data set and sends Bob one bit of information, Bob's ability to correctly guess an arbitrary bit in Alice's list is essentially at chance, if Alice and Bob use a strategy based on measurements on entangled quantum states instead of Popescu–Rohrlich bananas, where the correlations are bounded by the Tsirelson bound $1/\sqrt{2}$. For Popescu–Rohrlich bananas, $E = 1, P = 1$ and $h(P) = 0$. So Bob can correctly guess any arbitrary bit in Alice's list.

## The bottom line

- Pawłowski and colleagues formulate information causality as a limitation on Bob's information gain about an unknown data set of Alice: given all Bob's local resources and $M$ classical bits communicated by Alice, Bob's information gain is at most $M$ bits. The no-signaling principle is just information causality for $M = 0$: if Alice communicates nothing to Bob, then there is no information in the statistics of Bob's outputs about Alice's data set.
- Consider the guessing game where Alice has a list of $N = 2^n$ bits, and Bob has to guess a particular bit in the list. Alice can communicate one bit to Bob in each round of the game (so $M = 1$). I show that information causality is violated when $h(\frac{1}{2}(1 + E^n)) < 1 - \frac{1}{2^n}$, where $h$ is the binary entropy (explained in the

*(continued)*

 **The bottom line** *(continued)*

text), which is a convenient measure of how much information Bob gains from Alice's message.

- I take some specific numbers for $E$ and $n$ to show how the condition works. The closer $E$ is to the Tsirelson bound, the greater $n$ must be for a violation of information causality.
- In the following section I show that there is no violation of information causality at the Tsirelson bound.

### 9.3.4 The Tsirelson Bound from Information Causality

Pawłowski and colleagues prove that information causality is satisfied for any no-signaling theory satisfying some very general information-theoretic constraints. Since these constraints are satisfied for quantum theory, it follows that information causality is satisfied at and below the Tsirelson bound. Here's a direct proof that information causality is satisfied if $E = \frac{1}{\sqrt{2}}$.

The condition that information causality is satisfied is

$$h\left(\frac{1}{2}(1+E^n)\right) \geq 1 - \frac{1}{2^n}.$$

Explicitly from the definition of the binary entropy $h$:

$$-\frac{1}{2}(1+E^n)\log\left(\frac{1}{2}(1+E^n)\right) - \frac{1}{2}(1-E^n)\log\left(\frac{1}{2}(1-E^n)\right) \geq 1 - \frac{1}{2^n}.$$

This simplifies to

$$\log(1-E^{2n}) + E^n \log\frac{1+E^n}{1-E^n} \leq \frac{1}{2^{n-1}},$$

where the logarithms are to the base 2.

The proof involves expanding the left-hand side of this inequality as an infinite series. In books on mathematical tables and on websites like http://mathworld.wolfram.com, you can find expressions for various functions as infinite series. Here are two such expressions for logarithmic functions:

$$\log_e(1-x^2) = -x^2 - \frac{1}{2}x^4 - \frac{1}{3}x^6 - \frac{1}{4}x^8 + \cdots,$$

$$x\log_e\frac{1+x}{1-x} = 2x^2 + \frac{2}{3}x^4 + \frac{2}{5}x^6 + \cdots.$$

The equalities hold if $-1 \leq x \leq 1$. Putting these expressions together,

$$\log_e(1-x^2) + x\log_e\frac{1+x}{1-x} = x^2 + \frac{1}{6}x^4 + \frac{1}{15}x^6 + \cdots.$$

The logarithms are to the base $e$. To convert to base 2, divide the expression by $\log_e 2$ (or multiply by $\log_2 e$). Substituting $E^n$ for $x$ (which satisfies the condition $-1 \leq x \leq 1$), the condition for information causality to be satisfied can be expressed as

$$\frac{1}{\log_e 2}\left(E^{2n} + \frac{1}{6}E^{4n} + \frac{1}{15}E^{6n} + \cdots\right) \leq \frac{1}{2^{n-1}}$$

or

$$2^{n-1}\left(E^{2n} + \frac{1}{6}E^{4n} + \frac{1}{15}E^{6n} + \cdots\right) \leq \log_e 2,$$

which can be written as

$$\frac{1}{2}(2E^2)^n + \frac{1}{12}\frac{(2E^2)^{2n}}{2^n} + \frac{1}{30}\frac{(2E^2)^{3n}}{2^{2n}} + \cdots \leq \log_e 2.$$

From the sources for infinite series for $\log_e(1 - x^2)$ and $x \log_e \frac{1+x}{1-x}$, you'll find an expression for $\log_e 2$ as

$$\log_e 2 = \frac{1}{2} + \frac{1}{3} - \frac{1}{4} + \cdots.$$

Since $\frac{1}{3} - \frac{1}{4} = \frac{1}{12}$, $\frac{1}{5} - \frac{1}{6} = \frac{1}{30}$, and so on, this can be expressed as

$$\log_e 2 = \frac{1}{2} + \frac{1}{12} + \frac{1}{30} + \cdots.$$

So information causality is satisfied if

$$\frac{1}{2}(2E^2)^n + \frac{1}{12}\frac{(2E^2)^{2n}}{2^n} + \frac{1}{30}\frac{(2E^2)^{3n}}{2^{2n}} + \cdots \leq \frac{1}{2} + \frac{1}{12} + \frac{1}{30} + \cdots.$$

If $E = 1/\sqrt{2}$, so $2E^2 = 1$, this becomes

$$\frac{1}{2} + \frac{1}{12} \cdot \frac{1}{2^n} + \frac{1}{30} \cdot \frac{1}{2^{2n}} + \cdots \leq \frac{1}{2} + \frac{1}{12} + \frac{1}{30} + \cdots.$$

Now you can see that information causality is satisfied if $E = 1/\sqrt{2}$. For $n = 0$, the left-hand side is equal to the right-hand side. For $n > 0$, the left-hand side is smaller than the right-hand side, so the inequality is satisfied for any $n$.

## The bottom line

- Pawłowski and colleagues prove that information causality is satisfied for any no-signaling theory satisfying some very general information-theoretic constraints. Since these constraints are satisfied for quantum theory, it follows that information causality is satisfied at and below the Tsirelson bound.
- This section contains a direct proof that information causality is satisfied if $E = \frac{1}{\sqrt{2}}$. The proof involves expanding some expressions as infinite series. It's straightforward, but tedious to work through.

# Notes

1. Wheeler's "Really Big Questions," see *Science and Ultimate Reality: Quantum Theory, Cosmology, and Complexity*, John D. Barrow, Paul C. W. Davies, Charles L. Harper, Jr. (eds.) (Cambridge University Press, Cambridge, 2004).
2. Marcin Pawłowski, Tomasz Patarek, Dagomir Kaszlikowski, Valerio Scarani, Andreas Winter, Marek Zukowski, "A new physical principle: information causality," *Nature* 461, 1101–1104 (2009). The quoted remark is from pp. 1103–1104.
3. Applying information causality to recover part of the boundary of quantum correlations in the space of no-signaling correlations: Jonathan Allcock, Nicolas Brunner, Marcin Pawłowski, Valerio Scarani, "Recovering part of the quantum boundary from information causality," *Physical Review A* 80, 040103 (2009).
4. No bipartite principle can rule out all superquantum no-signaling tripartite correlations: Tsyh Haur Yang, Daniel Cavalcanti, Mafald L. Almeida, Colin Teo, and Valerio Scarani, "Information causality and extremal tripartite correlations," *New Journal of Physics* 14, 013061 (2012).
5. Wim van Dam's thesis: *Nonlocality and Communication Complexity*, Oxford, 2000. See also "Implausible consequences of superstrong nonlocality," arXiv: quant-ph/0501159v1 (2005).
6. Gilles Brassard, Harry Buhrman, Noah Linden, André Allen Méthot, Alain Tapp, Falk Unger, "Limit on nonlocality in any world in which communication complexity is not trivial," *Physical Review Letters* 96, 250401–250404 (2008).
7. Miguel Navascués and Harald Wunderlich, "A glance beyond the quantum model," *Proceedings of the Royal Society* 466, 881–890 (2010).
8. Daniel Cavalcanti, Alejo Salles, and Valerio Scarani, "Macroscopically local correlations can violate information causality," *Nature Communications* 1, 136 (2010).
9. Paul Skrzypczyk, Nicolas Brunner, and Sandu Popescu, "Emergence of quantum correlations from nonlocality swapping," *Physical Review Letters* 102, 110402 (2009).
10. Adán Cabello, "Specker's fundamental principle of quantum mechanics," arXiv: quant-ph/1212.1756 (2012).
11. Tobias Fritz, Ana Belén Sainz, Remigiusz Augusiak, Jonatan Bohr Brask, Rafael Chaves, Anthony Leverrier, and Antonio Acín, "Local orthogonality as a multipartite principle for quantum correlations," *Nature Communications* 4, 2263 (2013); Ana Belén Sainz, Tobias Fritz, Remigiusz Augusiak, Jonatan Bohr Brask, Rafael Chaves, Anthony Leverrier, and Antonio Acín, "Exploring the local orthogonality principle," *Physical Review A* 89, 032117 (2014).
12. The information causality paper: Marcin Pawłowski, Tomasz Patarek, Dagomir Kaszlikowski, Valerio Scarani, Andreas Winter, Marek Zukowski, "A new physical principle: information causality," *Nature* 461, 1101–1104 (2009). The quoted remark from the conclusion of their paper is from pp. 1103–1104. A caution for readers who might want to look at the Pawłowski et al. paper: the authors use $a, b$ for the two inputs and $A, B$ for the two outputs. The convention I've been following throughout this book is to use capital letters for inputs and lower case letters for outputs. Worth reading: Pawłowski and Valerio Scarani, "Information causality," arXiv: quant-ph/1112. 1142v1 (2011).
13. A useful paper on oblivious transfer: Nicolas Gisin, Sandu Popescu, Valerio Scarani, Stephan Wolf, and Jürg Wullschleger, "Oblivious transfer and quantum non-locality," in *International Symposium on Information Theory* (IEEE Proceedings, ISIT 2005), pp. 1745–1748; arXiv: quant-ph/0502030v1 (2005).

14. Ernst Specker, "Die logik nicht gleichzeitig entscheidbarer aussagen," *Dialectica* 14, 239–246 (1960).
15. Adán Cabello has authored or co-authored several papers on what he calls the "exclusivity principle" or "Specker's principle": "Specker's fundamental principle of quantum mechanics," arXiv: quant-ph/1212.1756 (2012); "A simple explanation of the quantum violation of a fundamental inequality," *Physical Review Letters* 110, 060402 (2013). Also, Barbara Amaral, Marcelo Terra Cunha, and Adán Cabello, "The exclusivity principle forbids sets of correlations larger than the quantum set," *Physical Review A* 89, 030101 (2014); Adán Cabello, Simone Severini, and Andreas Winter, "Graph-theoretic approach to quantum correlations," *Physical Review Letters* 112, 040401 (2014). A video by Cabello of Specker talking about the exclusivity principle is available at https://vimeo.com/52923835.
16. Yeong-Cherng Liang, Robert W. Spekkens, and Howard M. Wiseman, "Specker's parable of the over-protective seer: a road to contextuality, nonlocality and complementarity," *Physics Reports* 506, 1–39 (2011).
17. Allen Stairs, "The logic of propositions which are not simultaneously decidable," in C. A. Hooker (ed.), *The Logico-Algebraic Approach to Quantum Mechanics. Volume 1. Historical Evolution* (Reidel, Dordrecht, 1975), pp. 135–140. Michiel Seevinck, "The logic of non-simultaneously decidable propositions," arXiv: quant-ph/1103.4537v1.
18. The point is made by Renato Renner and Stefan Wolf in their interesting article "Ernst Specker and the hidden variables," *Elemente der Mathematik* 67, 122–133 (2012).
19. Adán Cabello, "Ernst Specker and the fundamental theorem of quantum mechanics" [video], *Vimeo*, 17 June, 2009, http://vimeo.com/52923835.
20. Adán Cabello, "Exclusivity principle and the quantum bound of the Bell inequality," *Physical Review A* 90, 062125 (2014). In this paper, Cabello shows how the Tsirelson bound follows from the exclusivity principle.

# 10

# Making Sense of It All

## 10.1 SCHRÖDINGER'S CAT

Readers of this book have very likely heard of Schrödinger's cat. After interacting with a radioactive atom, the cat is in an entangled quantum state: a superposition of a state in which the radioactive atom has not decayed and the cat is alive, and a state in which the radioactive atom has decayed and triggered a device that kills the cat. So, the story goes, the cat is neither alive nor dead.

Here's how Schrödinger described the "diabolical device" (see Figure 10.1):[1]

> A cat is penned up in a steel chamber, along with the following diabolical device (which must be secure against direct interference by the cat): in a Geiger counter there is a tiny amount of radioactive substance, *so* small, that *perhaps* in the course of one hour one of the atoms decays, but also, with equal probability, perhaps none; if it happens, the counter tube discharges and through a relay releases a hammer which shatters a small flask of hydrocyanic acid. If one has left this entire system to itself for an hour, one would say that the cat still lives *if* meanwhile no atom has decayed. The first atomic decay would have poisoned it. The $\psi$-function of the entire system would express this by having in it the living and the dead cat (pardon the expression) mixed or smeared out in equal parts.

As Schrödinger remarks in further comments, the point of the cat is to show that "an indeterminacy originally restricted to the atomic domain becomes transformed into macroscopic indeterminacy, which can then be *resolved* by direct observation." So, he adds, you can't interpret the quantum description of the microworld as a description of some sort of amorphous blurred or fuzzy reality, like a cloud or a bank of fog. The fact that you can resolve the indeterminacy or indefiniteness at the macrolevel by opening the chamber and looking, says Schrödinger, shows that the quantum description is like the sort of description provided by "a shaky or out-of-focus photograph" where some information is lost, in other words an *incomplete* description of something quite *definite*.

In a supportive letter to Schrödinger dated December 22, 1950, Einstein commented (after inexplicably modifying the device to blow up the cat instead of poisoning it):[2]

> You are the only contemporary physicist, besides Laue, who sees that one cannot get around the assumption of reality—if only one is honest. Most of them simply do not see what sort of risky game they are playing with reality—reality as something independent of what is

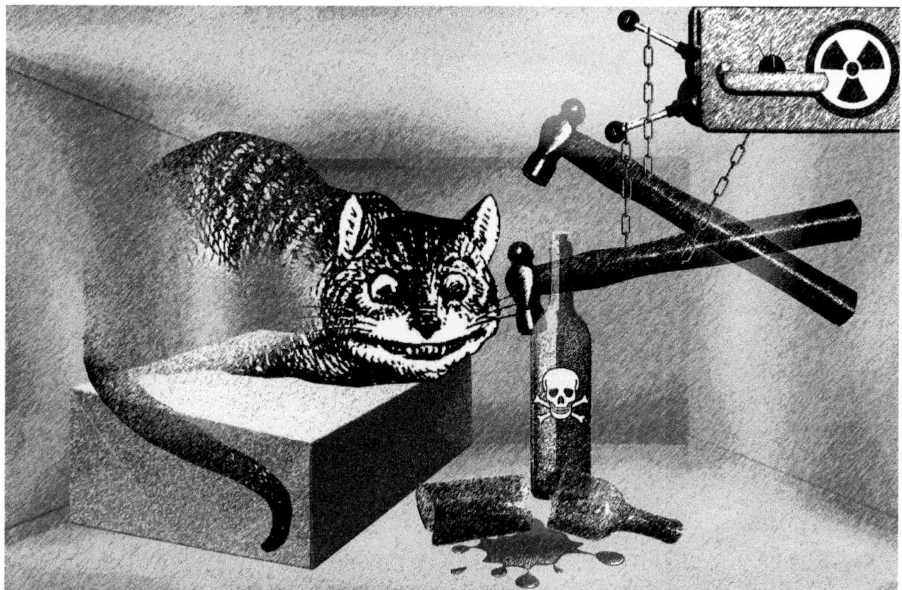

**Figure 10.1** Schrödinger's cat.

> experimentally established. They somehow believe that the quantum theory provides a description of reality, and even a *complete* description; this interpretation is, however, refuted, most elegantly by your system of radioactive atom + Geiger counter + amplifier + charge of gunpowder + cat in a box, in which the $\psi$-function of the system contains the cat both alive and blown to bits. Is the state of the cat to be created only when a physicist investigates the situation at some definite time? Nobody really doubts that the presence or absence of the cat is something independent of the act of observation. But then the description by means of the $\psi$-function is certainly incomplete, and there must be a more complete description. If one wants to consider the quantum theory as final (in principle), then one must believe that a more complete description would be useless because there would be no laws for it. If that were so then physics could only claim the interest of shopkeepers and engineers; the whole thing would be a wretched bungle.

For Einstein and Schrödinger, indeterminacy or indefiniteness at the quantum level, and the intrinsic randomness I talked about in Chapter 4, are only apparent. The suggestion is that something should be added to the theory that resolves the indefiniteness at the microlevel. It's the indefiniteness that's the problem here, not the indeterminism of quantum mechanics per se. Here's what I mean.

Einstein had an extensive correspondence with Max Born. Wolfgang Pauli thought that Born persistently misunderstood Einstein's position on quantum mechanics. At one point, he intervened in the correspondence to set Born straight:[3]

> Also, Einstein gave me your manuscript to read; he was *not at all* annoyed with you, but only said you were a person who will not listen. This agrees with the impression I have formed myself insofar as I was unable to recognise Einstein whenever you talked about him in either

your letter or your manuscript. It seemed to me as if you had erected some dummy Einstein for yourself, which you then knocked down with great pomp. In particular, Einstein does not consider the concept of "determinism" to be as fundamental as it is frequently held to be (as he told me emphatically many times), and he denied energetically that he had ever put up a postulate such as (your letter, para. 3): "the sequence of such conditions must also be objective and real, that is, automatic, machine-like, deterministic." In the same way, he *disputes* that he uses as criterion for the admissibility of a theory the question: "Is it rigorously deterministic?"

Einstein's point of departure is "realistic" rather than "deterministic," which means that his philosophical prejudice is a different one.

Readers might be surprised at Pauli's clarification of Einstein's position, in the light of Einstein's often-quoted comment about God not playing dice. In a 1926 letter to Born, Einstein wrote:[4]

Quantum mechanics is certainly imposing. But an inner voice tells me that it is not yet the real thing. The theory says a lot, but does not bring us any closer to the secret of the "old one." I, at any rate, am convinced that *He* is not playing dice.

Einstein didn't believe that quantum mechanics was "the real thing." But his critique of the theory, as in the Einstein–Podolsky–Rosen argument, doesn't depend on assuming determinism, but rather on what Pauli identifies as a "realistic prejudice." In Bananaworld terms, Einstein's "realist prejudice" is the assumption that a Popescu–Rohrlich banana has a definite taste before being peeled, or that there are some banana variables with definite values that define an "instruction set" for a banana to taste ordinary or intense when peeled a certain way, the "being-thus" of the banana, as Einstein would put it.

In his "hidden variable" theory,[5] David Bohm follows Schrödinger in treating quantum mechanics as a wave theory, but Bohm addresses the realism issue by introducing particles with definite positions (and momenta) at all times. The wave represented by the quantum state acts as a guiding wave for the positions of particles, which are additional "hidden variables" in the theory. Bohm's theory is also deterministic: the wave tells the particles where to go through a deterministic equation of motion for the particle positions that involves the quantum state. There's a cost to introducing definiteness in this way. You could, in principle, exploit the positions of the Bohmian particles to signal instantaneously or violate the free choice principle. Instantaneous signaling is blocked if the particles start out distributed in space as specified by the quantum state according to the Born rule for the probabilities of positions. The dynamics in Bohm's theory then ensures that the Born distribution is preserved over time, which means that you can never get more information about the particle positions in a quantum measurement than is provided by the quantum state according to the Born rule: the definite positions remain hidden. So the decay of the radioactive atom in Schrödinger's thought experiment only seems to be random because of a limitation on what you can measure.

Here's how Bohm's theory would go for a pair of Popescu–Rohrlich bananas. As I showed in Section 4.2, "Really Random Bananas," in Chapter 4, the correlation can be summed up neatly as $a \oplus b = A \cdot B$, where the taste $a$ of Alice's banana and the taste $b$ of Bob's banana take values 0 for ordinary and 1 for intense, and Alice's peeling $A$ and Bob's peeling $B$ take values 0 for peeling from the stem end and 1 for peeling from the top end.

Bohm's theory is nonrelativistic. So assume there's a fact of the matter about who peels first, independent of the reference frames of inertial observers. Suppose the taste of a peeled banana depends on the values of hidden variables, $\lambda_A$ for Alice's banana and $\lambda_B$ for Bob's banana, and a field—call it a $\psi$-field—that pervades the space between the two bananas. Suppose that

- initially the $\psi$-field and the distribution of hidden variable values is such that, for the first peeled banana, the probability of either taste is $1/2$, no matter how the banana is peeled;
- peeling a banana changes the $\psi$-field so that the taste of the second peeled banana is $A \cdot B$ if the first peeled banana tastes ordinary (0), and $(A \cdot B) \oplus 1$ if the first peeled banana tastes intense (1), for all values of the $\lambda$ variable of the second peeled banana.

This produces the Popescu–Rohrlich correlation. Suppose Alice peels her banana first and the taste is ordinary. Then $a = 0$ and $b = A \cdot B$, so $a \oplus b = A \cdot B$. If the taste is intense, then $a = 1$ and $b = (A \cdot B) \oplus 1$, so $a \oplus b = 1 \oplus (A \cdot B) \oplus 1 = A \cdot B$ (because $1 \oplus 1 = 0$).

Given $\lambda_A, \lambda_B$, the no-signaling principle is violated. If Alice peels first, whether she peels $S$ or $T$ is instantaneously revealed in the taste of Bob's banana. If $\lambda_A$ is such that Alice's banana tastes ordinary and Bob peels $T$ (so $B = 1$), then Bob's banana tastes ordinary if Alice peels $S$ ($A = 0$), because $b = A \cdot B = 0 \cdot 1 = 0$, but Bob's banana tastes intense if Alice peels $T$ ($A = 1$), because then $b = A \cdot B = 1 \cdot 1 = 1$. Similarly, if $\lambda_A$ is such that Alice's banana tastes intense and Bob peels $T$, then Bob's banana tastes intense if Alice peels $S$, because $b = (A \cdot B) \oplus 1 = (0 \cdot 1) \oplus 1 = 1$, but Bob's banana tastes ordinary if Alice peels $T$, because then $b = (A \cdot B) \oplus 1 = (1 \cdot 1) \oplus 1 = 0$. So Alice and Bob could signal with a supply of Popescu–Rohrlich bananas if they could control the hidden variables. Adding something to the story to ensure that the hidden variables remain hidden, as in Bohm's theory for quantum mechanics, would prevent this possibility.

Bohm's theory is empirically equivalent to quantum mechanics if the positions of the Bohmian particles are distributed according to the Born probability distribution at any time in the past (because, once the particles are distributed according to the Born distribution, Bohm's equation of motion for the particles ensures that they remain distributed in this way), but conceptually it's a different theory. There's an explicit causal mechanism in the theory for instantaneous action at a distance, rather similar to the action at a distance in Newton's theory of gravitation. In Newton's theory, if you move a rock on the moon, the effect is instantaneously transmitted to the earth through the gravitational field between the moon and the earth (it's infinitesimal, but nonetheless real, according to Newton's theory). As Bell pointed out, "the Einstein–Podolsky–Rosen paradox is resolved in the way which Einstein would have liked least."[6]

In Bohm's theory, the values of hidden variables provide missing information that resolves the indefiniteness at the microlevel and, together with the dynamics, explains how measurement outcomes come about, and the correlations between measurement outcomes at the macrolevel. At least, that's the general idea, but the picture is not as straightforward as it might seem at first sight. There are cases where events are registered as occurring when the particle trajectories in Bohm's theory, which are supposed to explain the occurrence of these events, are located somewhere else. See, for example, the critique by Reinhard Werner that Bohm's theory doesn't really do what it's claimed to do, and the references to "surrealistic Bohmian trajectories"[7] in the "Notes" at the end of the chapter.[8]

Hugh Everett's "many worlds" interpretation[9] is a much more radical proposal that also treats indefiniteness at the quantum level, and the intrinsic randomness of quantum events, as only apparent. Like Bohm, Everett follows Schrödinger in treating quantum mechanics as a wave theory, but the quantum wave is the only fundamental reality. The wave function of the universe is a superposition of components evolving linearly in time according to Schrödinger's equation. This means that a sum of components evolves to a sum of where each component evolves separately. The basic idea of the Everett interpretation is to interpret the quantum theory as telling us that the universal quantum wave function is all there is, and that each component or branch of the wave function evolves separately and represents a complete world as we experience it. Everettians see this as really just standard quantum mechanics, not one interpretation among rivals, but there are some problems and loose ends that complicate the Everettian picture.

The representation of a wave function as a specific superposition of components or branches depends on the basis in the quantum state space, the "coordinate system" with respect to which the wave is represented as superposition. Something physical should determine the basis, since the worlds associated with the branches of the universal wave function depend on the basis. The modern version of the Everett interpretation proposes that the basis is selected by "decoherence," a feature of the interaction between a macrosystem and its environment.

Here's the idea. A macrosystem like a bowling ball is bombarded with a lot of junk in the environment: air molecules, photons, cosmic rays. The scattering interaction between a photon and a bowling ball has a negligible influence on the bowling ball but changes the position of the photon quite a lot. If the bowling ball is initially in a superposition of macroscopically distinguishable states, say $|\text{here}\rangle + |\text{there}\rangle$, the combined state of the bowling ball and photon after the interaction is an entangled state of the form:

$$|\text{here}\rangle|\text{photon scattered here}\rangle + |\text{there}\rangle|\text{photon scattered there}\rangle,$$

where the two photon states represent different positions of the scattered photon. As time passes and the photon continues to move away from the billiard ball, it interacts with other particles in the environment, which leads to further entanglement. In a very short time, the state takes the form

$$|\text{here}\rangle\,|\text{photon scattered here}\rangle\,|\ldots\rangle_h\,|\ldots\rangle_h\ldots +$$
$$|\text{there}\rangle\,|\text{photon scattered there}\rangle\,|\ldots\rangle_t\,|\ldots\rangle_t\ldots,$$

where the states $|\ldots\rangle_h|\ldots\rangle_h\ldots$ and $|\ldots\rangle_t|\ldots\rangle_t\ldots$ represent states of particles in the environment whose positions have become correlated with the two positions of the photon ("*h*" for "here" and "*t*" for "there"). If "here" and "there" are two macroscopically distinct positions, the two photon states, $|\text{photon scattered here}\rangle$ and $|\text{photon scattered there}\rangle$, will be very close to orthogonal in the two-dimensional photon state space. The same goes for the states $|\ldots\rangle_h$ and $|\ldots\rangle_t$ of each of the entangled particles in the environment. As more and more environmental particles become entangled with the states $|\text{here}\rangle$ and $|\text{there}\rangle$, the angle between the product of all the environmental "here" states and the product of all the environmental "there" states become closer and closer to a right angle. For a bowling ball on earth, the approach to orthogonality is virtually instantaneous.

It follows that the quantum state of the bowling ball alone—the "reduced state" of the bowling ball—will be a mixed state representing a probability distribution over the states $|\text{here}\rangle$ and $|\text{there}\rangle$. (See the section "Mixed States" in the Supplement "Some Mathematical Machinery" at the end of the book for the reduced state of a subsystem of a composite system.) In fact, of course, the bowling ball is only in a mixed state if you ignore or average over the environment (that's what's involved in taking the reduced state). The bowling ball is really in an entangled pure state with the environment. But the difference between these two quantum descriptions would only show up in measurements of composite observables of the bowling ball and all the scattered particles in the environment, which is clearly going to be impossible in any remotely practical sense. The probabilities of measurement outcomes of observables of the bowling ball alone are determined by the reduced state of the bowling ball. So for all practical purposes, the bowling ball can be considered to be in a mixed state, and you can't go wrong, with respect to any remotely possible experience concerning the bowling ball, if you consider the bowling ball to be like a classical random bit, either in the state $|\text{here}\rangle$ or in the state $|\text{there}\rangle$, with equal probability (if the coefficients of the states $|\text{here}\rangle$ and $|\text{there}\rangle$ in the initial state of the bowling ball are the same). In this sense the "coherence" between the different parts of the superposition $|\text{here}\rangle + |\text{there}\rangle$ that makes a superposition different from a mixture is destroyed by the interaction with the environment (hence the term "decoherence"). If just two photons hit a bowling ball that is in a superposition of macroscopically distinguishable positions, decoherence is virtually instantaneous (and this will also be the case for an object as small as a dust particle bombarded by air molecules).[10]

Here's how decoherence solves the basis problem for the Everettian. The interaction with the environment in which particles in the environment of a macrosystem are scattered picks out a preferred basis or coordinate system in the state space of the universe associated with position as a preferred observable. The picture is that a multiplicity of worlds emerges as a feature of the universal wave function through decoherence as a dynamical process, which occurs spontaneously with respect to a position basis. In this "multiverse," we ourselves are split into different versions with different experiences associated with the worlds on the different branches of the universal wave function. In particular, the experiences of different outcomes of a quantum measurement occur in worlds on different branches.

The full story is a lot more complicated than this idealized picture suggests. What you get from decoherence is a basis of approximately localized positions, and the worlds that emerge in this dynamical process are only defined in an approximate sense. You can't, for example, specify how many worlds there are of a certain sort, even in principle, or precisely when worlds are formed. Everettians argue that, properly understood, this is not a problem, but the devil is in the details of what turns out to be a complex argument. There's also a problem in making sense of probability in a multiverse, where everything that can happen does happen. One way of dealing with this is to argue that there's a sense in which a rational agent in a world can be uncertain about which branch the agent will end up in after the universal wave function undergoes further branching as the result of a measurement. Such an agent can have rational degrees of belief that satisfy the axioms of probability theory about the outcomes of quantum measurements, even though all possible outcomes occur on different branches and the agent is split between the branches. The claim is that the agent's degrees of belief will agree with the Born probabilities, given certain assumptions about what it means to be rational in the face of uncertainty. The decoherence solution to the basis problem, and the analysis of probability and the derivation of the Born rule, are both problematic and the subject of a continuing debate between Everettians and their opponents. See, in particular, the critique by Adrian Kent.[11]

## The bottom line

- Schrödinger introduced his cat to pose a problem for the quantum description of a microsystem interacting with a macrosystem, and to argue that a solution requires treating quantum mechanics as an incomplete theory.
- Bohm's hidden variable theory replaces quantum mechanics with a deterministic theory. I show how Bohm's theory would go for a pair of Popescu–Rohrlich bananas.
- The basic idea of Everett's "many worlds" interpretation is to interpret the quantum theory as telling us that the wave function of the universe is all there is, and that each component or branch of the wave function evolves separately and represents a complete world as we experience it.
- The full story for Bohm's theory and the Everett interpretation is rather more complicated than the simple picture suggests, and both interpretations involve conceptual problems that are the subject of continuing debate.

## 10.2 THE MEASUREMENT PROBLEM

The interaction between Schrödinger's cat and a radioactive atom is the sort of thing that happens in a quantum measurement. The cat can be regarded as a macroscopic measuring instrument with two "pointer" states, "alive" and "dead," that become correlated with two microstates of the radioactive atom, $|\text{decay in one hour}\rangle$ and $|\text{no decay in one hour}\rangle$. In this sense, the cat ending up alive or dead measures a "yes–no" observable of the atom associated with the atom decaying or not decaying in an hour. The "measurement problem" of quantum mechanics is to explain how the cat can be considered to end up either definitely alive or definitely dead if the final state is an entangled state of the radioactive atom and the cat.

The measurement problem is usually posed as the problem of explaining the "collapse" of the entangled state at the end of a measurement interaction to the component of the entangled state corresponding to the measurement outcome, with the appropriate probability. Von Neumann added a "projection postulate" to the standard theory to cover this transition—the collapse amounts to projecting the state vector of the entangled state onto one of the measurement outcome vectors. Then there are two ways a quantum state can change. There's the deterministic, reversible, linear evolution of the quantum state described by the quantum dynamics for interactions that aren't measurement interactions (see the section "Quantum Dynamics" in the Supplement, "Some Mathematical Machinery" at the end of the book), and there's the stochastic, irreversible, nonlinear transition of the quantum state that occurs when a measurement is performed on a quantum system. The problem is then to explain how a quantum system "knows" that an interaction is supposed to count as a measurement. What's special about a measurement interaction other than the involvement of an observer and the observer's measuring instrument, which are fundamentally just big quantum systems? But then the interaction should be described by the quantum dynamics and produce an entangled state.

It might seem obvious that the entangled state at the end of a measurement is inconsistent with a definite outcome occurring, but there's actually an assumption involved here, an interpretative principle sometimes referred to as the "eigenvalue–eigenstate link." Dirac states the principle as follows:[12]

> The expression that an observable "has a particular value" for a particular state is permissible in quantum mechanics in the special case when a measurement of the observable is certain to lead to the particular value, so that the state is an eigenstate of the observable. . . . In the general case we cannot speak of an observable having a value for a particular state, but we can speak of its having an average value for the state. We can go further and speak of the probability of its having any specified value for the state, meaning the probability of this specified value being obtained when one makes a measurement of the observable.

So a photon in a state like $|0\rangle$, representing horizontal polarization in the $z$ direction, has a definite value for polarization in the $z$ direction, but a photon in a superposition like $|0\rangle + |1\rangle$ does not have a definite value for polarization in the $z$ direction. Similarly, if a measurement isn't a special kind of dynamical process, the entangled state at the end of a measurement, which is a linear superposition of product states associated with different measurement outcomes, is inconsistent with a definite outcome occurring.

So far, the examples of quantum states I've talked about have all been about the polarization of photons or the spin of electrons, not about position or momentum or energy. But you measure the linear polarization of a photon by passing the photon through a beamsplitter and seeing which way it goes, so by detecting its position. Formally in quantum mechanics the state of a photon is represented as a product state of a polarization state and a state representing the photon's position: $|\text{polarization}\rangle\,|\text{position}\rangle$, where $|\text{polarization}\rangle$ could be a superposition of polarization states and $|\text{position}\rangle$ could be a superposition of position states. What happens in the beamsplitter is that the polarization gets correlated with the position in an entangled state of polarization and position. So by detecting a particular position after amplification in a detector, you infer that the photon is polarized horizontally or vertically.

In a polarization measurement, the amplified position of the photon in a photon detector is the instrument "pointer" that measures the polarization, in the sense that you end up with an entangled state in which the position is correlated with horizontal and vertical polarization. After the interaction with the environment, you end up with an entangled state of the photon polarization, the photon position, photon detector variables, and the positions of particles in the environment. As more and more particles in the environment become involved in the interaction, decoherence rapidly occurs as the environmental states in the components of the entangled state evolve to orthogonal states.

Since it's impractical to keep track of all the particles in the environment, if what you're interested in are the probabilities of the two polarization states, you could ignore or average over the environment and consider the reduced state of the photon-plus-beamsplitter-plus-detector alone. If the environmental states are close to orthogonal, the reduced state of the photon-plus-beamsplitter-plus-detector is close to a mixed state of the photon being horizontally or vertically polarized and the detector indicating horizontal or vertical polarization. For all practical purposes, the photon-plus-beamsplitter-plus-detector can be considered to be in a definite state, with the appropriate probability defined by the initial quantum state of the photon. Either the photon is horizontally polarized, moves in the horizontal beam in the beamsplitter, and is detected as horizontally polarized by the detector, or the photon is vertically polarized, moves in the vertical beam in the beamsplitter, and is detected as vertically polarized by the detector. To see any difference between this mixed state (the probabilistic mixture of the "horizontal" and "vertical" states) and the photon polarization states superposed as part of an entangled state with the beamsplitter, the detector, and the environment, you would have to measure some observable of all the systems involved in the interaction, including all the environmental particles, some of which will be on their way to remote regions of the galaxy at the speed of light.

The temptation is to conclude that this explains how the photon ends up in one or other of the polarization states measured by the beamsplitter with the right probabilities, but strictly speaking, the full quantum description of the photon-plus-beamsplitter-plus-detector-plus-environment is an entangled state. The standard critique of this proposed decoherence solution to the measurement problem is that it's only a "FAPP" (for all practical purposes) solution, to use Bell's acronym.[13] The fact that something definite actually happens in a quantum measurement, and happens randomly with the relevant probability, remains unexplained by the dynamics of decoherence.

According to the Everett interpretation, the different outcomes of a quantum measurement occur in different worlds that emerge in the process of environmental decoherence, so every possible outcome actually happens, just in a different world. In Bohm's theory, what happens in a measurement process, followed by the almost instantaneous interaction with the environment, is that you end up with an entangled state as a guiding wave. The wave function—in principle, the wave function of the universe—evolves in a multidimensional "configuration space," not real physical space (see Schrödinger's remarks at the beginning of Section 1.1, "Quantum Blues," in Chapter 1). Each triple of dimensions of this configuration space is associated with the $x, y, z$ position coordinates of one of the particles, so there are three times as many dimensions in configuration space as there are particles in the universe. A point in configuration space defines the positions of all the particles. The guiding wave after a measurement and interaction with the environment is a superposition of components separated into nonoverlapping sharp peaks associated with the different outcomes of the measurement. The point in configuration space representing the positions of the particles is in one of the peaks. So just one of the possible outcomes occurs in a measurement, the outcome corresponding to the peak in the guiding wave containing the point in configuration space.

An Everettian would say that the different tastes of a peeled banana both occur, but in different worlds on different branches of the universal wave function. The dual occurrence emerges as a feature of the decoherence dynamics of the wave function, through the interaction between a banana and its environment. A Bohmian would say that the hidden variables, together with the nonlocal quantum wave function as a guiding field, determine how a banana will taste when peeled from the stem end or the top end as the outcome of a dynamical process. In both these approaches, the intrinsic randomness of events in a quantum universe is only apparent. In Bohm's theory, randomness is only apparent because of a limitation in what we can measure: we can't, in principle, measure position more precisely than allowed by the Born distribution. In the Everett interpretation, randomness is only apparent because of our limited perspective: we live in a multiverse and only have access to our own world in the multiverse.[14]

In the final section of this chapter, I'll clarify what the information-theoretic interpretation has to say about the measurement problem. First I'll consider an ingenious argument by Matthew F. Pusey, Jonathan Barrett, and Terry Rudolph in a 2012 *Nature Physics* paper that seems to exclude an information-theoretic interpretation of quantum mechanics.

**The bottom line**

- The interaction between Schrödinger's cat and a radioactive atom is the sort of thing that happens in a quantum measurement. The cat ending up alive or dead indicates the value of a "yes–no" observable of the atom associated with the atom decaying or not decaying in an hour. The "measurement problem" of quantum mechanics is to explain how the cat can be considered to end up either definitely alive or definitely dead if the final state is an entangled state of the radioactive atom and the cat.
- Interaction with the particles in the environment (air molecules, photons, cosmic rays) is sometimes invoked as a solution to the measurement problem. In a polarization measurement with a beamsplitter and detector, after the interaction with the environment you end up with an entangled state of the photon polarization, the photon position, photon detector variables, and the positions of particles in the environment.
- For all practical purposes, the photon-plus-beamsplitter-plus-detector alone can be considered to be in a definite state, with the appropriate probability defined by the initial quantum state of the photon.
- The standard critique of this proposed decoherence solution to the measurement problem is that it's only a "FAPP" (for all practical purposes) solution, to use Bell's acronym. The fact that something definite actually happens in a quantum measurement, and happens randomly with the relevant probability, remains unexplained.

## 10.3 PUSEY–BARRETT–RUDOLPH BANANAS

Quantum states specify probabilities for the outcomes of measurements, and probabilities, at least in classical theories, are usually associated with ignorance or uncertainty. The Pusey–Barrett–Rudolph argument aims to show that quantum states are "ontic," in the sense that they represent states of physical reality, rather than "epistemic," in which case they merely represent states of incomplete knowledge or uncertainty about the real physical or ontic state.[15]

Suppose quantum states are epistemic, so that a quantum state represents incomplete knowledge about whether the ontic state is $\lambda$ or some other ontic state, with a distribution of probabilities for $\lambda$ and the other possible ontic states. Then you could have two quantum states that assign the same ontic state $\lambda$ different nonzero probabilities. These quantum states would both be consistent with the ontic state actually being $\lambda$. (They would be inconsistent with the quantum state being $\lambda$ only if one quantum state assigned $\lambda$ a nonzero probability and the other quantum state assigned $\lambda$ zero probability.) So if quantum states are epistemic, the probability distributions that two consistent quantum states assign to the ontic states will overlap in a nonempty subset of the space of ontic states. The overlap region would be the set of ontic states to which the consistent quantum states both assign nonzero probabilities—they would be the possible ontic states for these quantum states.

The idea of the proof is to show that two distinct quantum pure states can't define probability distributions that overlap in the ontic state space, so quantum pure states can't be interpreted epistemically as states of information.

In the simple version of the argument, Pusey, Barrett, and Rudolph consider preparing a qubit in one of two quantum states: $|0\rangle$ or $|+\rangle = |0\rangle + |1\rangle$. If the qubit is a photon, $|0\rangle$ represents a state of horizontal linear polarization in the $z$ direction, and $|+\rangle$ represents a state of horizontal linear polarization in a diagonal direction 45° to $z$. Suppose these quantum states are merely epistemic, and that the real physical state of a photon is specified by some ontic state. If the quantum states $|0\rangle$ and $|+\rangle$ overlap in the ontic state space, there's a certain nonzero probability $p$ that if you prepare the system in either of these quantum states, you will actually prepare an ontic state $\lambda$ in the overlap region. So if you prepare two independent noninteracting photons in either of these quantum states, the probability that the photons are in the ontic states $\lambda_1, \lambda_2$, both in the overlap region, is $p^2$.

Call the ontic state of the composite two-photon system $\lambda$. Because the systems are independent and noninteracting, Pusey, Barrett, and Rudolph assume that $\lambda$ is specified by the pair of local ontic states: $\lambda = (\lambda_1, \lambda_2)$. It follows that $\lambda$ is consistent with the four quantum states $|0\rangle|0\rangle$, $|0\rangle|+\rangle$, $|+\rangle|0\rangle$, $|+\rangle|+\rangle$.

Now consider measuring an observable $A$ on the composite two-photon system with four possible values associated with the four entangled eigenstates:

$$|0\rangle|1\rangle + |1\rangle|0\rangle,$$
$$|0\rangle|-\rangle + |1\rangle|+\rangle,$$
$$|+\rangle|1\rangle + |-\rangle|0\rangle,$$
$$|+\rangle|-\rangle + |-\rangle|+\rangle.$$

You can check that these states are orthogonal by calculating the scalar products. (See the Section "Dirac's Ingenious Idea" in the Supplement "Some Mathematical Machinery" at the end of the book.) Take the first two states. The state $|-\rangle$ is $|0\rangle - |1\rangle$, where $|1\rangle$ is orthogonal to $|0\rangle$. So the second state in the list $|0\rangle|-\rangle + |1\rangle|+\rangle$ can be expressed as $|0\rangle(|0\rangle-|1\rangle)+|1\rangle(|0\rangle+|1\rangle) = |0\rangle|0\rangle-|0\rangle|1\rangle+|1\rangle|0\rangle+|1\rangle|1\rangle$. The component $|0\rangle|0\rangle$ has a zero scalar product with $|0\rangle|1\rangle$ and with $|1\rangle|0\rangle$. The component $-|0\rangle|1\rangle$ has a scalar product of $-1$ with $|0\rangle|1\rangle$ and a zero scalar product with $|1\rangle|0\rangle$. The component $|1\rangle|0\rangle$ has a zero scalar product with $|0\rangle|1\rangle$ and a scalar product of 1 with $|1\rangle|0\rangle$. Finally, the component $|1\rangle|1\rangle$ has a zero scalar product with $|0\rangle|1\rangle$ and with $|1\rangle|0\rangle$. So the total scalar product is $-1 + 1 = 0$.

The outcome of a measurement should depend *only* on the ontic state, because the ontic state is the whole story: given the ontic state, an epistemic state provides no further information relevant to an event occurring or not. The ontic state is consistent with the four quantum states $|0\rangle|0\rangle$, $|0\rangle|+\rangle$, $|+\rangle|0\rangle$, $|+\rangle|+\rangle$. As far as the ontic state is concerned, the quantum state could be any of these states. Suppose the initial quantum state is $|0\rangle|0\rangle$. According to quantum mechanics, if you measure the observable $A$ in this quantum state, the outcome corresponding to $|0\rangle|1\rangle + |1\rangle|0\rangle$ is impossible. This outcome has zero probability in the state $|0\rangle|0\rangle$ because $|0\rangle|0\rangle$ is orthogonal to $|0\rangle|1\rangle + |1\rangle|0\rangle$. Similarly:

- $|0\rangle|+\rangle$ is orthogonal to $|0\rangle|-\rangle + |1\rangle|+\rangle$;
- $|+\rangle|0\rangle$ is orthogonal to $|+\rangle|1\rangle + |-\rangle|0\rangle$;
- $|+\rangle|+\rangle$ is orthogonal to $|+\rangle|-\rangle + |-\rangle|+\rangle$.

Since each possible outcome is blocked by a quantum state consistent with the ontic state, and the outcome of a measurement depends only on the ontic state, none of the outcomes can occur! This is a contradiction: if you measure an observable, you must get one of the possible outcomes. Putting it differently, some measurement outcomes must have non-zero probability, because the probabilities of the four possible outcomes must sum to 1.

The conclusion of this part of the argument is that the probability distributions for $|0\rangle$ and $|+\rangle$ can't overlap in the ontic state space: these two quantum states, at least, correspond to different ontic states. The next step is to show the same thing for any pair of quantum states $|\psi\rangle$ and $|\phi\rangle$.

I won't go through this part of the proof, because it's rather more complicated technically. Pusey, Barrett, and Rudolph consider $n$ copies of a qubit, where $n$ depends on the angle between the quantum states $|\psi\rangle$ and $|\phi\rangle$—the smaller the angle, the larger the number $n$ of copies they need. The result they prove is different from Bell's theorem, because they don't assume Bell's locality condition (that the quantum probabilities of joint events are conditionally statistically independent given the ontic state). Instead, they assume that each of the $n$ qubits can be prepared in a pure quantum state independently of the others, and that this also applies to the ontic states. So the ontic state of the composite $n$-qubit system is specified by the set of independent ontic states of all the qubits, which means that there are no "supercorrelations" at the ontic level.

Not surprisingly, the argument has a counterpart in Bananaworld. Pure state bananas grow on trees with large bunches of bananas, and there are four types: $S0$, $S1$, $T0$, $T1$. The pure state bananas $S0$ and $S1$ correspond to the qubit states $|0\rangle$ and $|1\rangle$, the eigenstates of $Z$, and the pure state bananas $T0$ and $T1$ correspond to the qubit states $|x_+\rangle$ and $|x_-\rangle$, the eigenstates of $X$ (represented as $|+\rangle$ and $|-\rangle$ above). So for all the pure state bananas on an $S0$ bunch, if you peel from the stem (which corresponds to measuring $Z$), the banana tastes ordinary (0), but if you peel from the top (which corresponds to measuring $X$), the banana tastes ordinary or intense with probability $1/2$. For all the bananas on an

$S1$ bunch, if you peel from the stem, the banana tastes intense (1), but if you peel from the top, the banana tastes ordinary or intense with probability 1/2. Similarly for $T0$ and $T1$ bananas, where peeling from the top end leads to the same taste, 0 or 1, for all the bananas on the bunch, and peeling from the stem end leads to random tastes.

In the simple version of the argument, Pusey, Barrett, and Rudolph consider preparing two independent noninteracting qubits in either of two pure quantum states corresponding to the banana pure states $S0$ and $T0$, and performing a certain measurement on the combined system. The Pusey–Barrett–Rudolph observable $A$ represents anticorrelation with respect to $Z$ and $Z$, $Z$ and $X$, $X$ and $Z$, and $X$ and $X$ for the two qubits. Measuring $A$ is something you do to the two qubits that changes the state of the composite two-qubit system to one of four possible entangled states, each corresponding to a possible outcome of the measurement. So consider taking two pure state bananas, each in one of the states $S0$ or $T0$, and doing something to the bananas that changes them in a way corresponding to a measurement of the observable $A$. Suppose that if you put the stems of the two bananas into a cup of coffee, they grow together into a pair that is anticorrelated for one of the four possible peelings $SS$, $ST$, $TS$, $TT$, except that:

- if the pair is initially $S0S0$, they never end up anticorrelated for $SS$;
- if the pair is initially $S0T0$, they never end up anticorrelated for $ST$;
- if the pair is initially $T0S0$, they never end up anticorrelated for $TS$;
- if the pair is initially $T0T0$, they never end up anticorrelated for $TT$.

This corresponds to the correlations in the Pusey–Barrett–Rudolph example. If the initial state is $|0\rangle|0\rangle$ the outcome corresponding to $|0\rangle|1\rangle + |1\rangle|0\rangle$ (anticorrelation with respect to $Z$ and $Z$ for the two qubits) never occurs, and similarly for the other three cases.

In the banana pure state $S0$, if you peel $T$, the banana tastes ordinary or intense with probability 1/2. So suppose $S0$ is interpreted as an epistemic state, representing ignorance about some ontic state $\lambda$. Then what happens if you put the stems of two pure state bananas into a cup of coffee depends *only* on their ontic states, $\lambda_1, \lambda_2$.

Suppose two epistemic states like $S0$, $T0$ are consistent with the same ontic state. Then two bananas in the ontic states $\lambda_1, \lambda_2$, each consistent with the epistemic states $S0$, $T0$, would be consistent with the four pairs of epistemic states: $S0S0$, $S0T0$, $T0S0$, $T0T0$. But if that were the case then, contrary to what you find in Bananaworld, the two bananas could never become a pair that is anticorrelated for one of the four possible peelings $SS, ST, TS$, or $TT$, because the initial state $S0S0$ excludes eventual anticorrelation for $SS$, the initial state $S0T0$ excludes eventual anticorrelation for $ST$, the initial state $T0S0$ excludes eventual anticorrelation for $TS$, and the initial state $T0T0$ excludes eventual anticorrelation for $TT$. As in the quantum version of the argument, since each possible outcome is blocked by a banana state consistent with the ontic state, and the taste of a banana depends only on the ontic state, none of the anticorrelated outcomes can occur.

Pusey, Barrett, and Rudolph conclude that "models in which the quantum state is interpreted as mere information about an objective physical state of a system cannot reproduce the predictions of quantum theory."

It would seem that a quantum state must be at least part of the ontic state of a quantum system, the state of physical reality, because different quantum states would have to correspond to different ontic states. Then the only open possibilities, if quantum mechanics is strictly true, are Bohm's theory (in which the full ontic state is the quantum state together with the hidden variables, the position in configuration space of all the particles) or something like it, and the Everett interpretation (in which the ontic state is just the quantum state).

In fact, the Pusey–Barrett–Rudolph argument is a "no go" result for any extension of quantum theory as an explanation of quantum correlations, in which a quantum state represents partial knowledge of an underlying reality. So it addresses the completeness issue at the heart of the debate between Bohr and Einstein—the question whether quantum mechanics, as a probabilistic theory, is the whole story, or whether quantum probabilities should be understood as representing ignorance about the "real" physical state.

On the information-theoretic interpretation outlined in the following section, the quantum state is a complete description of a quantum system. Quantum probabilities and probabilistic correlations are, as von Neumann put it,[16] "sui generis," and reflect the intrinsic randomness of quantum events in a world in which you can simulate Popescu-Rohrlich correlations with a probability of success greater than 3 / 4, the Bell bound—they don't quantify partial knowledge about some ontic state or underlying reality.

### The bottom line

- An argument by Pusey, Barrett, and Rudolph aims to show that quantum states are "ontic," in the sense that they represent states of physical reality, rather than "epistemic," in which case they merely represent states of incomplete knowledge or uncertainty about the real physical or ontic state, where the uncertainty is quantified by the probabilities defined by the state.
- If an epistemic quantum state represents incomplete knowledge about whether the ontic state is $\lambda$ or some other ontic state, with a distribution of probabilities for $\lambda$ and the other possible ontic states, you could have two quantum states that assign the same ontic state $\lambda$ different nonzero probabilities. The idea of the proof is to show that two distinct quantum pure states can't define probability distributions that overlap in the ontic state space—different quantum states would have to correspond to different ontic states. So quantum pure states can't be interpreted epistemically as states of information. The argument has a counterpart in Bananaworld.
- The Pusey–Barrett–Rudolph argument is a "no go" result for any extension of quantum theory as an explanation of quantum correlations in which a quantum state represents partial knowledge of an underlying reality.
- On the information-theoretic interpretation outlined in the following section, the quantum state is a complete description of a quantum system. Quantum

(continued)

 **The bottom line** *(continued)*
probabilities and probabilistic correlations reflect the intrinsic randomness of quantum events in a world in which you can simulate Popescu-Rohrlich correlations with a probability of success greater than 3 / 4, the Bell bound—they don't quantify partial knowledge about some ontic state or underlying reality.

## 10.4 THE INFORMATION-THEORETIC INTERPRETATION

I began this book with Heisenberg's "reinterpretation" of physical quantities as noncommutative, or non-Boolean (see the subsection "Boolean Algebras" in the "More" section of the "Bananaworld" chapter). Heisenberg's insight amounts to the proposal that certain phenomena in our Boolean macroworld that defy a classical physical explanation can be explained probabilistically as a manifestation of collective behavior at a non-Boolean microlevel. The Boolean algebra of physical properties of classical mechanics is replaced by a family of "intertwined" Boolean algebras, one for each set of commuting observables, to use Gleason's term.[17] The intertwinement precludes the possibility of embedding the whole collection into one inclusive Boolean algebra, which means that you can't assign truth values consistently to the propositions about observable values in all these Boolean algebras. So quantum probabilities can't be understood in the classical or Boolean sense as quantifying ignorance about the pre-measurement value of an observable, but cash out in terms of what you'll find if you "measure," which involves considering the outcome, at the macrolevel, of manipulating a quantum system in a certain way.

A quantum "measurement" is a bit of a misnomer and not really the same sort of thing as the measurement of a property of a classical system. It involves putting a microystem, like a photon, in a situation, say a beamsplitter or an analyzing filter, where the photon is forced to make an intrinsically random transition recorded as one of two macroscopically distinct alternatives in a device like a photon detector, just as a banana in Bananaworld makes an intrinsically random transition to one of two alternative tastes when it's peeled.

The registration of the measurement outcome at the Boolean macrolevel is crucial, because it is only with respect to a suitable structure of alternative possibilities that it makes sense to talk about an event as definitely occurring or not occurring, and this structure is a Boolean algebra. George Boole, who came up with the idea in the mid-1800's, introduced Boolean constraints on probability as "conditions of possible experience."[18] Bohr did not refer to Boolean algebras, but the concept is simply a precise way of codifying what Bohr might have had in mind by insisting (his emphasis) that *"however far the phenomena transcend the scope of classical physical explanation, the account of all evidence must be expressed in classical terms,"* by which he meant "unambiguous language with suitable application of the terminology of classical physics"—for the simple reason, as he put it, that we need to be able to "tell others what we have done and what we have learned."[19] Formally speaking, the significance of "classical" here as being able to "tell others what we have done and what we have learned" is that the events in question should fit together as a Boolean algebra.

There are various ways to see how a Boolean macroworld could emerge from the intertwined Boolean algebras at the microlevel, but they are all approximate and to some extent conceptually fuzzy. We are familiar in physics with collective phenomena at the macrolevel that can be quite different from the behavior of individual systems at the microlevel. For example, the air has a temperature, but individual air molecules don't. We understand how the temperature of the air is a collective property of large collections of air molecules, but how to explain the emergence of a commutative or Boolean macrolevel from an underlying noncommutative or non-Boolean microlevel is trickier, and to some extent an ongoing research problem.[20]

I contrasted Heisenberg's understanding of the quantum revolution with Schrödinger's wave-mechanical version of the theory. It's tempting to interpret the wave function representing the quantum state *representationally*, as the analogue of the classical state in stipulating what's true and what's false, rather than *probabilistically*.[21] Since the unitary dynamics produces an entangled wave function after a quantum measurement, a superposition with several components, each associated with a different measurement outcome, a representational interpretation seems to require that the wave function "collapses" dynamically to one component. The measurement problem is then to explain how a measurement, as a quantum interaction described by the deterministic unitary dynamics of the theory, can produce such a collapse, which is stochastic and non-unitary.

The Everettian response is that we should take the wave function of the universe quite literally as a representation of multiple realities. By contrast, the information-theoretic interpretation in the sense I have in mind takes the quantum state as probabilistic, not representational: the quantum state is a bookkeeping device for keeping track of probabilities and probabilistic correlations in a universe that is non-Boolean at the microlevel. A quantum measurement sets up a probabilistic correlation between a micro-observable and a macro-observable. You could replace the wave function as a way of calculating probabilities with correlation arrays, which would express more directly the character of quantum mechanics as a stochastic theory about the irreducibly probabilistic relation between events at the Boolean macrolevel and the underlying non-Boolean microlevel.

I use the term "information-theoretic" to emphasize that the quantum revolution is about new sorts of probabilistic correlations in nature, analogous to the sense in which the relativistic revolution is about new sorts of spatio-temporal relations. A theory of information is fundamentally a theory of probabilistic correlations, and that's what Hilbert space gives you. Probabilities and probabilistic correlations arise as a feature of the non-Boolean structure. They are "uniquely given from the start," to quote von Neumann,[22] not measures over states as they are in a classical or Boolean theory. The intertwinement of commuting and noncommuting observables in Hilbert space imposes objective pre-dynamic probabilistic constraints on correlations between events, analogous to the way in which Minkowski space-time imposes kinematic constraints on events. The probabilistic constraints encoded in the geometry of Hilbert space provide the framework for the physics of a genuinely indeterministic universe. They characterize the way probabilities fit together in a world in which there are nonlocal probabilistic correlations that violate Bell's inequality up to the Tsirelson bound, and these correlations can only occur between intrinsically random events.

Now, a "one-world" interpretation like this is supposed to be excluded by an argument due to Daniela Frauchiger and Renato Renner,[23] a novel re-formulation of the "Wigner's friend" argument[24] exploiting a result known as the Hardy paradox.[25]

Suppose Alice "tosses a biased quantum coin" with probabilities $1/3$ for "heads" and $2/3$ for "tails." More precisely, Alice measures an observable $A$ with eigenstates $|h\rangle_A$, $|t\rangle_A$ on a system in the state $\frac{1}{\sqrt{3}}|h\rangle_A + \frac{\sqrt{2}}{\sqrt{3}}|t\rangle_A$. She prepares a qubit in the state $|0\rangle_B$ if the outcome is $h$, or in the state $\frac{1}{\sqrt{2}}(|0\rangle_B + |1\rangle_B)$ if the outcome is $t$, and sends the qubit to Bob, who measures the observable $B$ with eigenstates $|0\rangle_B$, $|1\rangle_B$. From the perspective of Alice and Bob after obtaining definite outcomes for their measurements, the quantum state of the combined quantum coin and qubit system is $|h\rangle_A|0\rangle_B$ or $|t\rangle_A|0\rangle_B$ or $|t\rangle_A|1\rangle_B$, with equal probability.

The quantum coin and the qubit, as well as Alice and Bob, their measuring instruments and all the systems in their laboratories that become entangled with the measuring instruments in registering and recording the outcomes of the quantum coin toss and the qubit measurement, including the entangled environment, are just two big many-particle quantum systems $S_A$ and $S_B$, which are assumed to be completely isolated from each other. The symbols $A$ and $B$ could represent "super-observables" of the composite systems $S_A$ and $S_B$ that end up with definite values corresponding to the outcomes of Alice's and Bob's measurements on the quantum coin and the qubit.

Consider a super-observer, with vast technological abilities, who contemplates measuring a super-observable $X$ with eigenstates $|\text{fail}\rangle_A = \frac{1}{\sqrt{2}}(|h\rangle_A + |t\rangle_A)$, $|\text{ok}\rangle_A = \frac{1}{\sqrt{2}}(|h\rangle_A - |t\rangle_A)$ on $S_A$, and a super-observable $Y$ with eigenstates $|\text{fail}\rangle_B = \frac{1}{\sqrt{2}}(|0\rangle_B + |1\rangle_B)$, $|\text{ok}\rangle_B = \frac{1}{\sqrt{2}}(|0\rangle_B - |1\rangle_B)$ on $S_B$, where $|h\rangle_A$, $|t\rangle_A$ and $|0\rangle_B$, $|1\rangle_B$ now represent eigenstates of the super-observables $A$ and $B$, and $\{h, t\}$ and $\{0, 1\}$ represent the corresponding eigenvalues. Such a measurement would be extraordinarily difficult to carry out in practice on the whole composite system, including all the systems in the environment, but nothing in quantum mechanics precludes this possibility. From the perspective of such a super-observer, Alice and Bob are just quantum systems, and the system $S_A + S_B$ is a composite entangled quantum system that has evolved unitarily to an entangled state:

$$|\psi\rangle = \frac{1}{\sqrt{3}}(|h\rangle_A|0\rangle_B + |t\rangle_A|0\rangle_B + |t\rangle_A|1\rangle_B)$$

The $A$ and $B$ values simply don't appear anywhere in the super-observer's description of events, so the super-observer sees no reason to conditionalize the state to one of the product states $|h\rangle_A|0\rangle_B$ or $|t\rangle_A|0\rangle_B$ or $|t\rangle_1|1\rangle_B$. For the super-observer, this would seem to require a suspension of unitarity in favor of an unexplained collapse of the quantum state.

But now we have a contradiction. The state $|\psi\rangle$ can also be expressed as:

$$\begin{aligned} |\psi\rangle &= \frac{1}{\sqrt{12}}|\text{ok}\rangle_A|\text{ok}\rangle_B - \frac{1}{\sqrt{12}}|\text{ok}\rangle_A|\text{fail}\rangle_B + \frac{1}{\sqrt{12}}|\text{fail}\rangle_A|\text{ok}\rangle_B + \sqrt{\frac{3}{4}}|\text{fail}\rangle_A|\text{fail}\rangle_B \\ &= \sqrt{\frac{2}{3}}|\text{fail}\rangle_A|0\rangle_B + \frac{1}{\sqrt{3}}|t\rangle_A|1\rangle_B \\ &= \frac{1}{\sqrt{3}}|h\rangle_A|0\rangle_B + \sqrt{\frac{2}{3}}|t\rangle_A|\text{fail}\rangle_B \end{aligned}$$

From the first expression for $|\psi\rangle$, there is a probability $1/12$ that the super-observer finds the pair of outcomes {ok, ok} in a joint measurement of $X$ and $Y$ on the two systems. But this outcome is inconsistent with any pair of outcomes for Alice's and Bob's measurements. From the second expression, the pair $\{\text{ok}, 0\}$ has zero probability, so $\{\text{ok}, 1\}$ is the only possible pair of values for the super-observables $X, B$ if $X$ has the value ok. From the third expression, the pair $\{t, \text{ok}\}$ has zero probability, so $\{h, \text{ok}\}$ is the only possible pair of values for the super-observables $A, Y$ if $Y$ has the value ok. But the pair of values $\{h, 1\}$ for the super-observables $A$ and $B$ has zero probability in the state $|\psi\rangle$, so it does not correspond to a possible pair of measurement outcomes for Alice and Bob.[26]

As Frauchiger and Renner put it, the argument depends only on (1) the one-world assumption, that a measurement has a single outcome, (2) the assumption that quantum mechanics applies to systems of any complexity, including observers, and (3) self-consistency, in particular agreement between an observer and a super-observer. The argument appears to show that there is no consistent story that includes an observer (Alice and Bob here) and a superobserver: a possible outcome, according to quantum mechanics, of a super-observer's measurement of a super-observable of the whole composite Alice-Bob system is inconsistent with the observer obtaining an actual outcome for any of the possible values of the measured observable. It won't do to argue that, as far as we know, there are no super-observers. The actuality of a measurement outcome can't depend on whether or not a super-observer turns up at some point. It's the theoretical possibility of a super-observer that shows the inconsistency of the theory. So what are the options?

There is the Everettian interpretation, which rejects assumption (1), the one-world assumption. This is an ingenious proposal to save quantum mechanics as a representational theory, but looks considerably less plausible when one considers the complexity introduced by invoking environmental decoherence to select a preferred coarse-grained basis to define the worlds, and the decision-theoretic account of the probabilistic correlations that we already know uniquely characterize Hilbert space, according to Gleason's theorem. There is QBism, the quantum Bayesianism of Christopher Fuchs and Ruediger Schack, which is also in some respects "information-theoretic."[27] The QBist rejects assumption (3), the self-consistency assumption. On this view, all probabilities, including quantum probabilities, are understood in the subjective sense as the personal judgements of an agent, based on how the external world responds to actions by the agent. For QBists, the Born rule "is a normative statement ... about the decision-making behavior any individual agent should strive for... not a "law of nature" in the usual sense," and "measurement outcomes *just are* personal experiences for the agent gambling upon them."[28] So there is no requirement that the perspective of an observer and a super-observer should be consistent.

Then there is the information-theoretic interpretation. This has more in common with what I think Bohr and some of the early proponents of the Copenhagen interpretation had in mind than the subjectivism of QBism.[29] Quantum mechanics is quite unlike any theory we have dealt with before in the history of physics, because Heisenberg's proposal to "re-interpret" classical quantities like position and momentum as noncommutative is unprecedented in the implication that events at the Boolean macrolevel reflect collective behavior of elementary systems at a non-Boolean microlevel. Bohr's primary insight was to see that explanation in such a post-classical theory is not the sort of representational explanation we are familiar with in a theory that is commutative or Boolean at the fundamental level. Quantum probabilities are probabilities of what you'll find if you measure, where the reference is to *one ultimate observer* as the end-point of a quantum mechanical analysis, as Bohr repeatedly emphasized:[30]

> The only significant point is that in each case some ultimate measuring instruments ... must always be described entirely on classical [I would say Boolean] lines, and consequently be kept outside the system subject to quantum mechanical treatment.

The outcome of a measurement is an intrinsically random event at the macrolevel and so outside the theory. To quote Pauli:[31]

> Observation thereby takes on the character of *irrational, unique actuality* with unpredictable outcome. ... Contrasted with this *irrational aspect* of concrete phenomena which are determined in their *actuality*, there stands the *rational aspect* of an abstract ordering of the *possibilities* of statements by means of the mathematical concept of probability and the $\psi$-function [I would say "by means of the geometry of Hilbert space"].

On the information-theoretic interpretation of Hilbert space, just one observer perspective is legitimate in the application of quantum mechanics: the perspective of the observer for whom an actual measurement outcome occurs at the macrolevel. If Alice and Bob represent a composite ultimate observer for whom there are definite events at the macrolevel, the observer perspective is legitimate, in which case the final state of the combined quantum coin and qubit system is $|h\rangle_A|0\rangle_B$ or $|t\rangle_A|0\rangle_B$ or $|t\rangle_A|1\rangle_B$. If a super-observer subsequently measures the super-observables $X, Y$ on the whole composite Alice-Bob system, the probability of obtaining the pair of outcomes $\{\text{ok}, \text{ok}\}$ is $1/4$ for any of the product states $|h\rangle_A|0\rangle_B$ or $|t\rangle_A|0\rangle_B$ or $|t\rangle_A|1\rangle_B$ (here interpreted as representing eigenstates of the super-observables $A, B$). After the measurement, the super-observables $A, B$ are indefinite, and so are the corresponding quantum coin and qubit observables. There is no contradiction because the argument from the alternative expressions for the entangled state no longer applies. If the super-observer perspective is legitimate, there are no definite Alice and Bob events at the macrolevel and the state is the entangled state. The probability of a super-observer finding the pair of outcomes {ok, ok} is 1/12, but there is no contradiction because there are no measurement outcome events for Alice and Bob.

In effect, the information-theoretic interpretation rejects assumption (2), the assumption that quantum mechanics is universal—not by restricting the universality of the unitary dynamics or any part of quantum mechanics, but by placing a constraint on how the theory is applied. It's not that unitarity is suppressed at a certain level of complexity, where non-Booleanity becomes Booleanity and quantum becomes classical. Rather, there is a macrolevel, which is Boolean (Bohr would say "classical"), and there are actual events at the macrolevel. But any system, of any complexity, is fundamentally a quantum system and can be treated as such, in principle. A unitary dynamical analysis of a measurement process goes as far as you would like it to go, to whatever level of precision is convenient. The collapse, as a conditionalization of the quantum state, is something you put in by hand after observing the actual outcome. The physics doesn't give it to you.

Special relativity, as a theory about the structure of space-time, provides an explanation for length contraction and time dilation through the geometry of Minkowski space-time, but that's as far as it goes. This explanation didn't satisfy Lorentz, who wanted a dynamical explanation in terms of forces acting on physical systems used as rods and clocks.[32] Quantum mechanics, as a theory about randomness and nonlocality, provides an explanation for probabilistic constraints on events through the geometry of Hilbert space, but that's as far as it goes. This explanation doesn't satisfy Everettians, who insist on a representational story about how nature pulls off the trick of producing intrinsically random events at the macrolevel, with nonlocal probabilistic correlations constrained by the Tsirelson bound. Do we really want to give up the concept of measurement as a procedure that provides information about the actual value of an observable to preserve the ideal of representational explanation in physics? It seems far more rational to accept that if current physical theory has it right, the nature of reality, the way things are, limits the sort of explanation that a physical theory can provide.

### The bottom line

- An argument by Frauchiger and Renner shows that there are circumstances in which the "inner" perspective of an observer is inconsistent with the "outer" perspective of a super-observer with vast, but not impossible, technological abilities who measures the observer and systems entangled with the observer. The argument depends only on (1) the one-world assumption (one definite

(continued)

 **The bottom line** *(continued)*

outcome to a measurement), (2) the assumption that quantum mechanics applies to systems of any complexity, including observers, and (3) self-consistency, in particular agreement between an observer and a super-observer.

- The Everettian interpretation rejects (1). QBism, which interprets all probabilities, including quantum probabilities, as the personal judgements of an agent based on how the external world responds to actions by the agent, rejects (3).
- Quantum mechanics is quite unlike any theory we have dealt with before in the history of physics, because Heisenberg's proposal to "re-interpret" classical quantities like position and momentum as noncommutative is unprecedented in the implication that events at the Boolean macrolevel reflect collective behavior of elementary systems at a non-Boolean microlevel. Explanation in such a post-classical theory is not the sort of representational explanation we are familiar with in a theory that is commutative or Boolean at the fundamental level.
- The information-theoretic interpretation rejects (2)—not by restricting the universality of the unitary dynamics or any part of quantum mechanics, but by placing a constraint on how the theory is applied. Quantum probabilities are probabilities of what you'll find if you measure, where the reference is to *one ultimate observer* as the end-point of a quantum mechanical analysis, as Bohr repeatedly emphasized. So just one observer perspective is legitimate in the application of quantum mechanics: the perspective of the observer for whom an actual measurement outcome occurs at the macrolevel.
- It's not that unitarity is suppressed at a certain level of complexity, where non-Booleanity becomes Booleanity and quantum becomes classical. Any system, of any complexity, is fundamentally a quantum system and can be treated as such. The collapse, as a conditionalization of the quantum state, is something you put in by hand after observing the actual outcome. The physics doesn't give it to you.

## Notes

1. Erwin Schrödinger's famous cat was introduced on p. 812 of his three-part paper "Die Gegenwärtige Situation in der Quantenmechanik," *Die Naturwissenschaften* 23, 807–812, 823–828, 844–849 (1935). An English translation by John D. Trimmer was published in *Proceedings of the American Philosophical Society* 124, 323–338 (1980) and reprinted in J. A. Wheeler and W. H. Zurek (eds.), *Quantum Theory and Measurement* (Princeton University Press, Princeton, 1983). The quotation is from the Trimmer translation in the Wheeler and Zurek anthology, p. 157.

2. Einstein's comment on Schrödinger's cat in a letter to Schrödinger dated December 22, 1950 is in K. Przibram (ed.), *Letters on Wave Mechanics* (Philosophical Library, New York, 1967), p. 39.
3. Wolfgang Pauli's letter to Max Born about Einstein (dated March 31, 1954) is in M. Born (ed.), *The Born–Einstein Correspondence* (Walker and Company, New York, 1971), p. 221.
4. Einstein's comment that God doesn't play dice is from a letter to Max Born dated 4 December, 1926, in M. Born (ed.), *The Born–Einstein Correspondence* (Walker and Company, New York, 1971), p. 91.
5. David Bohm's hidden variable theory was published as "A suggested interpretation of the quantum theory in terms of "hidden" variables, I and II," *Physical Review* 85, 166–179, 180–193 (1952). See also the article "Bohmian mechanics," by Sheldon Goldstein in Edward N. Zalta (ed.), *The Stanford Encyclopedia of Philosophy* (Spring 2013 edition).
6. John Bell's remark is from his paper "On the problem of hidden variables in quantum mechanics," *Reviews of Modern Physics* 38, 447–452 (1966); p. 452.
7. On "surrealistic Bohmian trajectories," see Berthold-Georg Englert, Marlin O. Scully, G. Süssmann, and H. Walther, "Surrealistic Bohmian trajectories," *Zeitschrift für Naturforschung* A47, 1175 (1992), and Marlan O. Scully, "Do Bohm trajectories always provide a trustworthy physical picture of particle motion?" *Physica Scripta* 76, 41–46 (1998). For a Bohmian rebuttal, see Detlef Dürr, Walter Fusseder, Sheldon Goldstein, and Nino Zanghi, "Comment on 'Surrealistic Bohm Trajectories" *Zeitschrift für Naturforschung* 48a, 1261–1262 (1993).
8. Reinhardt Werner has pointed out difficulties with Bohm's theory. See his guest post and the following discussion on Tobias J. Osborne's research notes at https://tjoresearchnotes.wordpress.com/2013/05/13/guest-post-on-bohmian-mechanics-by-reinhard-f-werner/.
9. Hugh Everett's original formulation of the interpretation now called the Everett or "many worlds" interpretation was published as "Relative state formulation of quantum mechanics," *Reviews of Modern Physics* 29, 454–462 (1957), and "The theory of the universal wave function," in B. De Witt and N. Graham (eds.), *The Many-Worlds Interpretation of Quantum Mechanics* (Princeton University Press, Princeton, 1973). For the modern version of the Everett interpretation, see David Wallace, *The Emergent Multiverse: Quantum Theory according to the Everett Interpretation* (Oxford University Press, Oxford, 2012) and Simon Saunders, Jonathan Barrett, Adrian Kent, and David Wallace (eds.) *Many Worlds? Everett, Quantum Theory, and Reality* (Oxford University Press, Oxford, 2010).
10. The reference to decoherence and a bowling ball is from Maximilian A. Schlosshauer, *Decoherence and the Quantum-to-Classical Transition* (Springer, Berlin, 2008), p. 94: "For example, photons scattering off a bowling ball will hardly affect the ball's motion in any way, while they will lead to virtually instantaneous decoherence of a superposition state involving macroscopically distinguishable positions of the ball."
11. Adrian Kent has been a persistent critic of the Everett interpretation. For a readable account of the measurement problem, with critical comments on Everett (and on the Copenhagen interpretation, Bohm's theory, and the Ghirardi–Rimini–Weber theory), see his article "Our quantum problem" in the online magazine *Aeon*, http://aeon.co/magazine/science/our-quantum-reality-problem/. For more on this, see his article "One world versus many: the inadequacy of Everettian accounts of evolution, probability, and scientific confirmation," in Simon Saunders, Jonathan Barrett, Adrian Kent, and David Wallace (eds.), *Many Worlds? Everett, Quantum Theory, and Reality* (Oxford University Press, Oxford, 1912), pp. 307–354.
12. The quotation from Dirac is from P. A. M. Dirac, *Quantum Mechanics*, 4th edition (Clarendon Press, Oxford, 1958), p. 47.

13. Bell's acronym FAPP is from his article "Against measurement," *Physics World* 3, 33–40 (1990), reprinted in M. Bell, K. Gottfried, and M. Veltman (eds.), *John S. Bell on the Foundations of Quantum Mechanics* (World Scientific, Singapore, 2001), pp. 208–215.
14. The Ghirardi–Rimini–Weber theory, with later variations by Philip Pearle and Roderich Tumulka, has also been proposed as a solution to the measurement problem. See GianCarlo C. Ghirardi, Alberto Rimini, and T. Weber, "Unified dynamics for microscopic and macroscopic systems," *Physical Review* D34, 470–491 (1986) for the original version of the theory, and GianCarlo Ghirardi, "Collapse theories," in Edward N. Zalta (ed.), *The Stanford Encyclopedia of Philosophy* (Winter 2011 edition) for more recent versions and extensions of the original theory, with references to work by Pearle and Tumulka. Roughly, the idea is to modify the quantum dynamics by adding a stochastic term that has a negligible effect for a microsystem like a radioactive atom, but adds up for a macrosystem like a cat, which is composed of an enormous number of microsystems. In the interaction involving the radioactive atom and the cat, the result is that the entangled state of the atom-plus-diabolical device-plus-cat very rapidly undergoes a dynamical "collapse" to one of the components of the entangled state, in which the cat is either definitely alive, or definitely dead. I mentioned the Ghirardi–Rimini–Weber theory in Chapter 3. It's really a rival theory to standard quantum mechanics that makes different predictions in certain situations. These predictions are hard to check, but at some point with the advance of quantum technologies there's likely to be an experimental test.
15. The Pusey-Barrett-Rudolph argument: Matthew Pusey, Jonathan Barrett, and Terry Rudolph, "On the reality of the quantum state," *Nature Physics* 8, 475–478 (2012). Their conclusion that "models in which the quantum state is interpreted as mere information about an objective physical state of a system cannot reproduce the predictions of quantum theory" is on p. 477.
16. The quote from John von Neumann about quantum probabilities being "sui generis" is from "Quantum Logics: Strict- and Probability-Logics," a 1937 unfinished manuscript published in A. H. Taub (ed.), *Collected Works of John von Neumann*, Volume 4 (Pergamon Press, Oxford and New York, 1961), pp. 195–197.
17. The term 'intertwined' was used by A.N. Gleason in a seminal paper, "Measures on the Closed Subspaces of Hilbert Space," *Journal of Mathematics and Mechanics* 6, 885–893 (1957), to refer to the way in which observables in quantum mechanics are related. See p. 886.
18. See Itamar Pitowsky, "George Boole's "conditions of possible exprience" and the quantum puzzle," *British Journal for the Philosophy of Science* 45, 95–125 (1994).
19. Niels Bohr, "Discussions with Einstein on epistemological problems in modern physics," in P. A. Schilpp (ed.), *Albert Einstein: Philosopher-Scientist*, The Library of Living Philosophers, Volume 7 (Open Court, Evanston, 1949), pp. 201–241. The quotation is from p. 209.
20. For the treatment of macrosystems in quantum mechanics, see N. P. Landsman, "Between classical and quantum," in Jeremy Butterfield and John Earman (eds.), *Philosophy of Physics*, Part A, pp. 417–553, a volume in Dov M. Gabbay, Paul Thagard, and John Woods (eds.), *Handbook of the Philosophy of Science* (North-Holland, Amsterdam, 2007). See also Itamar Pitowsky, "Macroscopic objects in quantum mechanics: a combinatorial approach," *Physical Review* A 70, 022103 (2004).
21. The "representational" versus "probabilistic" distinction is from David Wallace's paper, "What is orthodox quantum mechanics?" arXiv eprint quant- ph/1604.05973.
22. John von Neumann, "Unsolved problems in mathematics," an address to the International Mathematical Congress, Amsterdam, September 2, 1954. In Miklós Rédei and Michael Stöltzner (eds.), *John von Neumann and the Foundations of Quantum Physics*, pp. 231–245 (Kluwer Academic Publishers, Dordrecht, 2001). The quotation is from p. 245. In other places,

as pointed out in the previous section, von Neumann refers to quantum probabilities as "sui generis." Gleason's theorem, cited above, is usually referenced as proving the uniqueness of the Born rule for quantum probabilities from structural features of Hilbert space. Von Neumann proved a similar result in 1927 in "Wahrscheinlichkeitstheoretischer Aufbau der Quantenmechanik," *Nachrichten von der Gesellschaft der Wissenschaften zu Göttingen. Mathematisch-Physikalische Klasse 1927*, 245–272 (1927). See section 6 in Anthony Duncan and Michel Janssen, "(Never) Mind your ps and q's: von Neumann versus Jordan on the foundations of quantum theory," *The European Physical Journal* H 38, 175–259 (2013).

23. Daniela Frauchiger and Renato Renner, "Single-World interpretations of quantum mechanics cannot be self-consistent," arXiv eprint quant-ph/1604.07422.
24. Eugene Wigner, 'Remarks on the mind-body question," in I.J. Good (ed.), *The Scientist Speculates*, (Heinemann, London, 1961). In the argument below, Alice and Bob correspond to Wigner's friend and the super-observer corresponds to Wigner in Wigner's original version.
25. Lucien Hardy "Quantum mechanics, local realistic theories, and Lorentz-invariant physical theories," *Physical Review Letters* 68, 2981–2984 (1992). See also Paul G. Kwiat and Lucien Hardy, "The mystery of the quantum cakes," *American Journal of Physics* 68, 33–36 (2000)
26. There could be two super-observers, one for $S_A$ and one for $S_B$. Since the composite systems $S_A$ and $S_B$ are assumed to be completely isolated from each other, the no-signaling principle ensures that the outcome of a measurement on one of the systems can't depend on whether or not a measurement is performed on the other system.
27. Christopher A. Fuchs, N. David Mermin, and Ruediger Schack, "An introduction to QBism with an application to the locality of quantum mechanics," *American Journal of Physics* 82, 749–754 (2014). Also Christopher Fuchs and Ruediger Schack, "Quantum measurement and the Paulian idea," in H. Atmanspacher and C.A. Fuchs (eds.), *The Pauli–Jung Conjecture and its Impact Today* (Imprint Academic, Exeter, 2014).
28. Christopher A. Fuchs, "Notwithstanding Bohr, the reasons for QBism," arXiv eprint quant-ph/1705.03483.
29. My thinking on quantum mechanics has been influenced a great deal, over the years, by discussions with William Demopoulos. See his paper "Effects and propositions," *Foundations of Physics* 40, 368–389 (2010) for his analysis of the transition from classical to quantum mechanics, and the sequel "Generalized probability measures and the framework of effects," in Yemima Ben-Menahem and Meir Hemmo (eds.), *Probability in Physics* (Springer, New York, 2012), pp. 201–217.
30. Niels Bohr, "The causality problem in atomic physics," in *New Theories in Physics* (International Institute of Intellectual Cooperation, Warsaw,1939), p. 24.
31. Wolfgang Pauli, "Probability and physics," in Charles P. Enz and Karl von Meyenn (eds.), *Wolfgang Pauli: Writings on Physics and Philosophy* (Springer, Berlin, 1994), p. 46. The article was first published in *Dialectica* 8, 112–124 (1954).
32. See Michel Janssen, "Drawing the line between kinematics and dynamics in special relativity," *Studies in History and Philosophy of Modern Physics* 40, 26–52 (2009).

# SUPPLEMENT: SOME MATHEMATICAL MACHINERY

In this Supplement, I'll show how quantum states of single systems and complex multipartite systems can be represented in a now standard notation invented by Dirac, how observables can be represented by noncommuting operators, and how multipartite systems can be in entangled states.

The discussion is not as daunting as you might expect, but if you are encountering the joys of the Dirac notation for the first time, you might need a pen and paper, and perhaps a glass of wine, to play with the formalism as you read it through. A formal expression is opaque if you are not already familiar with the notation and you read it symbol by symbol instead of seeing it as a concise way of saying something that would be clumsy to say in words. So the temptation is to read through the text and skip the equations and other bits of displayed formalism. I'll talk through the formal constructions in some detail rather than leaving it to you to figure them out. Think of it as getting a grip on the software running the universe. The glass of wine will help.

## DIRAC'S INGENIOUS IDEA

In the section "Quantum States" of Chapter 2, "Qubits," I introduced Dirac's notation for representing a quantum state as a vector, denoted by the symbol $|\cdots\rangle$, where $\cdots$ is simply a label for the state.

For a general qubit, two states labeled $|0\rangle$ and $|1\rangle$ associated with the possible values of some two-valued observable are chosen as the standard or "computational" states—the reference states for computations. To remind you: the state $|0\rangle$ is just another qubit state like $|1\rangle$ and should not be confused with the number 0. For a photon, these states are usually the horizontal and vertical states of linear polarization in some standard direction, $|H\rangle$ and $|V\rangle$. For an electron, $|0\rangle$ and $|1\rangle$ usually represent the "up" and "down" states of spin in the direction $z$, for some standard direction chosen as the $z$ direction.

The qubit states $|0\rangle$ and $|1\rangle$ are represented by orthogonal vectors of unit length in a two-dimensional state space. In the infinite-dimensional case, the state space is technically a Hilbert space (don't worry—this doesn't play any explicit role in the following). The term is also used for a finite-dimensional space like the space of qubit states. The states $|0\rangle$ and $|1\rangle$ define a coordinate system or "basis" in the state space called the standard or computational basis, just as the unit vectors $\vec{x}$ and $\vec{y}$ in the directions of the $x$ and $y$ coordinates of a Cartesian coordinate system define the two orthogonal directions of a basis in the Cartesian plane. Any qubit state $|\psi\rangle$ represented by a unit vector at an angle $\theta$ to $|0\rangle$ can be represented as a linear combination or sum of its projections along $|0\rangle$ and $|1\rangle$ as $|\psi\rangle = \cos\theta\,|0\rangle + \sin\theta\,|1\rangle$, where the coefficients $\cos\theta$ and $\sin\theta$ represent the lengths of the projections of the unit state vector onto the directions defined by the unit length basis vectors $|0\rangle$ and $|1\rangle$. A linear combination or sum of states like $|\psi\rangle = \cos\theta\,|0\rangle + \sin\theta\,|1\rangle$ is referred to as a "superposition" of the states $|0\rangle$ and $|1\rangle$. Any state can be expressed as a superposition of other states in an infinite number of ways, just as any vector can be expressed as a linear combination of other vectors.

There is a difference between the quantum state space and the space of vectors in the Cartesian plane. To accommodate the probability relations between different polarization states, you need to consider linear combinations of unit vectors with coefficients that can be complex numbers.

If you are unfamiliar with complex numbers, here's a quick tutorial. A complex number is a number of the form $c = a + ib$, where $i = \sqrt{-1}$, so $i^2 = -1$. In effect, a new number $i$ is defined as the square root of –1: if you square $i$, you get –1. This gives you a whole new set of numbers to play with, the complex numbers, that are sums of "real" numbers and multiples of real numbers by $i$. The number $a$ is referred to as the real part of $c$, denoted by $\mathrm{Re}(c)$, and $b$ as the imaginary part of $c$, denoted by $\mathrm{Im}(c)$. The real numbers then become a subset of the complex numbers—the complex numbers with a zero imaginary part.

The "complex conjugate" of $c$ is defined as $c^* = a - ib$. So the real part of $c$ is $\mathrm{Re}(c) = \frac{1}{2}(c + c^*)$, because you get $2a$ when you add $a + ib$ and $a - ib$. The imaginary part of $c$ is $\mathrm{Im}(c) = \frac{i}{2}(c^* - c)$, because subtracting $a + ib$ from $a - ib$ gives you $-2ib$, which becomes 2b when you multiply by $i$, since $i \cdot i = -1$. The product of $c$ and $c^*$ is a real number:

$$cc^* = (a + ib)(a - ib) = a^2 + b^2$$

(the product of $ib$ and $-ib$ is $-i \cdot i \cdot b^2 = b^2$). This is taken as the square of the absolute value of the complex number $c$ and denoted by $|c|^2$.

The only other feature of complex numbers relevant to the representation of quantum states here is the Euler relation:

$$e^{i\theta} = \cos\theta + i\sin\theta,$$

where $e$ is the base of the natural logarithm. If you have a hard time with the idea of an imaginary exponent, take the left-hand side of the Euler equation as a shorthand expression for the right-hand side, which is just a complex number. The relation follows from the expansion of $e^{i\theta}$, $\cos\theta$, and $\sin\theta$ as infinite series. The complex number $e^{i\theta}$ has absolute value 1. The complex conjugate of $e^{i\theta}$ is $e^{-i\theta}$, and $e^{i\theta}e^{-i\theta} = 1$. Alternatively, this follows because $(\cos\theta + i\sin\theta)(\cos\theta - i\sin\theta) = \cos^2\theta + \sin^2\theta = 1$.

In quantum mechanics, observables or physical quantities, like the polarization of a photon or the spin of an electron, are represented by operators that transform states represented by vectors to new states.

The neat thing about the Dirac notation is the clever way of representing operators. The projection operator $P_{|0\rangle}$ onto the basis vector $|0\rangle$ takes a general qubit state represented by a unit vector $|\psi\rangle$ as input and produces as output the vector that is the projection of $|\psi\rangle$ onto $|0\rangle$. So $|\psi\rangle$ can be expressed in terms of its projections onto the basis states $|0\rangle$ and $|1\rangle$ as a superposition, $|\psi\rangle = P_{|0\rangle}|\psi\rangle + P_{|1\rangle}|\psi\rangle$. This is just like expressing a vector $\vec{v}$ in terms of its projections $P_x\vec{v}$ and $P_y\vec{v}$ onto the $x$ and $y$ axes of a Cartesian coordinate system as a linear combination $\vec{v} = P_x\vec{v}+P_y\vec{v}$. (Here, $P_{|0\rangle}|\psi\rangle = \cos\theta\,|0\rangle$ and $P_{|1\rangle}|\psi\rangle = \sin\theta\,|1\rangle$ because $|\psi\rangle$ is a unit vector.) In the Dirac notation, $P_{|0\rangle}$ is written as $|0\rangle\langle 0|$ and $P_{|1\rangle}$ as $|1\rangle\langle 1|$. Dirac calls a vector like $|0\rangle$ a "ket" and a vector like $\langle 0|$ a "bra." The bra vector of $c|\psi\rangle$, for some complex number $c$, is $\langle\psi|c^*$, where $c^*$ is the complex conjugate of $c$.

Here's how the notation works: to obtain the output of the projection of a state vector $|\psi\rangle$ onto another state vector $|\phi\rangle$, apply the projection operator $|\phi\rangle\langle\phi|$ to $|\psi\rangle$. The output $|\phi\rangle\langle\phi||\psi\rangle$ is read as $|\phi\rangle\langle\phi|\psi\rangle$: the two vertical middle lines together are treated as a single vertical line. Dirac calls the expression $\langle\phi|\psi\rangle$ a "bra(c)ket," hence the "bra," "ket" terminology. It represents a number: the "length" of the projection of $|\psi\rangle$ onto $|\phi\rangle$, or technically the "scalar product" of $|\phi\rangle$ and $|\psi\rangle$. The term "scalar" refers to a number rather than a vector, so a scalar product is a type of product of vectors that produces a number. The term "length" is in quotes here because the number could be a complex number.

Since $\langle\phi|\psi\rangle$ is just a number, it can be moved to a position in front of the state $|\phi\rangle$, so:

$$\underbrace{|\phi\rangle\langle\phi|}\,|\psi\rangle = |\phi\rangle\,\underbrace{\langle\phi|\psi\rangle} = \underbrace{\langle\phi|\psi\rangle}\,|\phi\rangle.$$

Dirac's neat idea is that regrouping the expressions according to the underbraces allows you to see $|\phi\rangle\langle\phi|$ operating on $|\psi\rangle$ as a multiple of the unit vector $|\phi\rangle$, which is what you would expect after projecting $|\psi\rangle$ onto $|\phi\rangle$.

To sum up:

- The projection operator $|\phi\rangle\langle\phi|$ transforms a unit vector $|\psi\rangle$ to the vector $|\phi\rangle\langle\phi|\psi\rangle = \langle\phi|\psi\rangle|\phi\rangle$, which is the projection of $|\psi\rangle$ onto $|\phi\rangle$. In other words, a projection operator is a function that takes a vector as input and produces another vector as output, the projection of the original vector along a particular direction in the state space.
- The expression $|\phi\rangle\langle\phi|$ represents an operator, the projection operator onto $|\phi\rangle$. In general, an expression like $|\phi\rangle\langle\psi|$—note the $\langle\psi|$ here—represents an operator that transforms a vector $|\chi\rangle$ to a multiple of $|\phi\rangle$: $\underbrace{|\phi\rangle\langle\psi|}\,|\chi\rangle = |\phi\rangle\,\underbrace{\langle\psi|\chi\rangle} = \underbrace{\langle\psi|\chi\rangle}\,|\phi\rangle$.
- The expression $\langle\phi|\psi\rangle$ represents a number, the scalar product of $|\phi\rangle$ and $|\psi\rangle$, or the "length" of the projection of $|\psi\rangle$ onto $|\phi\rangle$. In particular, $\langle 0|1\rangle = \langle 1|0\rangle = 0$. The length of the projection of $|1\rangle$ onto $|0\rangle$ or $|0\rangle$ onto $|1\rangle$ is 0, because these are orthogonal vectors. Also, $\langle 0|0\rangle = \langle 1|1\rangle = 1$. The length of the projection of $|0\rangle$

onto itself, or $|1\rangle$ onto itself, is 1, because these are unit length vectors. For a non-unit vector like $|\psi\rangle = c|0\rangle$, for some complex number $c$, $\langle\psi|\psi\rangle = \langle 0|c^*c|0\rangle = |c|^2\langle 0|0\rangle = |c|^2 = |\psi|^2$, where $|\psi|^2$ denotes the square of the absolute value of the "length" of $|\psi\rangle$.

- In the Dirac notation, a state $|\psi\rangle$ is represented as the sum of its projections onto the computational basis states as $|0\rangle\langle 0|\psi\rangle + |1\rangle\langle 1|\psi\rangle = \langle 0|\psi\rangle|0\rangle + \langle 1|\psi\rangle|1\rangle$, where $\langle 0|\psi\rangle$ is the "length" of the projection of $|\psi\rangle$ onto $|0\rangle$ and $\langle 1|\psi\rangle$ is the "length" of the projection of $|\psi\rangle$ onto $|1\rangle$.

You might want to read this section a couple of times to get the hang of the notation, especially the difference between $\langle\phi|\psi\rangle$, which is a number or scalar, and $|\phi\rangle\langle\psi|$, which is an operator.

 **The bottom line**

- This section shows how quantum states are represented as vectors in a state space, in a commonly used notation invented by Dirac. The standard states for a qubit, also called the "computational" states (the reference states for computations), are represented as $|0\rangle$ and $|1\rangle$ in the Dirac notation.
- Observables or physical quantities, like the polarization of a photon or the spin of an electron, are represented by operators that act on quantum states to transform them to new states.
- There's also a brief tutorial on complex numbers in this section.

## THE PAULI OPERATORS

You can add operators like $|\cdots\rangle\langle\cdots|$ with complex numbers as coefficients to form new operators, and you can multiply operators by real or complex numbers.

So take the operator $X = |0\rangle\langle 1| + |1\rangle\langle 0|$ and the two unit vectors, $|x_+\rangle = \frac{1}{\sqrt{2}}|0\rangle + \frac{1}{\sqrt{2}}|1\rangle$ and $|x_-\rangle = \frac{1}{\sqrt{2}}|0\rangle - \frac{1}{\sqrt{2}}|1\rangle$. These quantum state vectors are often denoted as $|+\rangle$ and $|-\rangle$. Here I want to consider operators $X$ and $Y$ and distinguish the state vectors associated with $X$ from the state vectors associated with $Y$, so I'll write $|x_+\rangle$ and $|x_-\rangle$ rather than $|+\rangle$ and $|-\rangle$. The coefficients $\frac{1}{\sqrt{2}}$ and $-\frac{1}{\sqrt{2}}$ in $|x_+\rangle$ and $|x_-\rangle$ are just "normalizing" factors that ensure that the vectors $|x_+\rangle$ and $|x_-\rangle$ are unit vectors, with length or "norm" 1.

To find the length of a vector, you take the squares of the lengths of the orthogonal projections, add them, and take the square root. This is an application of Pythagoras's theorem: the square of the length of the diagonal of a triangle is the sum of the squares of the lengths of the two orthogonal sides. The length of the diagonal here is the length of the vector, and the orthogonal sides of the triangle are the projections of the vector onto $|0\rangle$ and $|1\rangle$. The length of $|x_+\rangle$, then, is the square root of $(\frac{1}{\sqrt{2}})^2+(\frac{1}{\sqrt{2}})^2 = \frac{1}{2}+\frac{1}{2} = 1$, and the length of $|x_-\rangle$ is the square root of $(\frac{1}{\sqrt{2}})^2 + (-\frac{1}{\sqrt{2}})^2 = \frac{1}{2} + \frac{1}{2} = 1$.

Alternatively, you can find the length of $|x_+\rangle$ by calculating $\langle x_+|x_+\rangle$, the length of the projection of the unit vector $|x_+\rangle$ onto itself, and the length of $|x_-\rangle$ by calculating $\langle x_-|x_-\rangle$.

It's easier to read a quantum state represented by a superposition of unit vectors with equal coefficients without the coefficients: $|0\rangle + |1\rangle$ instead of $\frac{1}{\sqrt{2}}|0\rangle + \frac{1}{\sqrt{2}}|1\rangle$. In that case, the state is said to be "unnormalized," which is a technical way of saying that the state vector doesn't have unit length (if the two coefficients are both 1, the length of the state vector is $\sqrt{1+1} = \sqrt{2}$, by Pythagoras's theorem). I'll routinely leave out the coefficients when they are all equal and it's clear how to put them back in to "normalize" the state to unit length. (So they are left out in the expressions for the quantum states in the following paragraph.) If there are two terms in the superposition, the two equal coefficients are $\frac{1}{\sqrt{2}}$. If there are three terms in the superposition, the three equal coefficients are $\frac{1}{\sqrt{3}}$, and so on.

The unit vectors $|x_+\rangle = |0\rangle + |1\rangle$ and $|x_-\rangle = |0\rangle - |1\rangle$ are uniquely associated with $X$, in the sense that $X|x_+\rangle = |x_+\rangle$ and $X|x_-\rangle = -|x_-\rangle$. The result of operating on $|x_+\rangle = |0\rangle + |1\rangle$ with the operator $X = |0\rangle\langle 1| + |1\rangle\langle 0|$ is the sum of the results of operating on each of the components of this vector in turn.

To work this out, take the operation on the components of $|x_+\rangle$ in order. The result of the first term $|0\rangle\langle 1|$ in the operator $X$ operating on the *first* component $|0\rangle$ of $|x_+\rangle$ is $|0\rangle\langle 1|0\rangle = 0$, because $\langle 1|0\rangle = 0$. The result of the second term $|1\rangle\langle 0|$ in $X$ operating on the component $|0\rangle$ is $|1\rangle\langle 0|0\rangle = |1\rangle$, because $\langle 0|0\rangle = 1$. So $X$ operating on the first component $|0\rangle$ of $|x_+\rangle$ produces the vector $|1\rangle$.

The first term $|0\rangle\langle 1|$ in $X$ operating on the *second* component $|1\rangle$ of $|x_+\rangle$ produces $|0\rangle$, and the second term $|1\rangle\langle 0|$ in $X$ operating on the component $|1\rangle$ produces 0. So $X$ operating on the second component $|1\rangle$ of $|x_+\rangle$ produces $|0\rangle$.

The net effect of $X$ operating on $|x_+\rangle = |0\rangle + |1\rangle$, then, is to produce $|1\rangle + |0\rangle = |0\rangle + |1\rangle$, so $X$ operating on $|x_+\rangle$ reproduces $|x_+\rangle$. Applied to $|x_-\rangle = |0\rangle - |1\rangle$, the operator $X$ produces $|1\rangle - |0\rangle$, which is $-|x_-\rangle = -(|0\rangle - |1\rangle)$.

The operator $X$ operating on $|x_+\rangle$ or $|x_-\rangle$ leaves these vectors unchanged, except for multiplication by a factor of 1 or –1, respectively. This is not going to be the case for any other vectors in the qubit state space, because they are all represented by linear superpositions of these basis vectors, with non-zero coefficients. The vectors $|x_+\rangle$ and $|x_-\rangle$ are referred to as the "eigenvectors" of the operator $X$, and 1 and –1 as the "eigenvalues" of $X$ (from the German word "eigen," which in this sense means "characteristic of").

A minor caveat is necessary here, because the previous comment is not quite correct. If $X$ operates on $-|x_+\rangle$ or $-|x_-\rangle$, the effect is also to leave these vectors unchanged, except for multiplication by a factor 1 or –1. In fact, the vectors $|\psi\rangle$ and $e^{i\theta}|\psi\rangle$, for any $\theta$, are regarded as representing the same qubit state. This covers the case $-|\psi\rangle$, because $e^{i\pi} = \cos\pi + i\sin\pi = -1$ (the cosine of $\pi$ is –1, and the sine of $\pi$ is 0), and the case $i|\psi\rangle$, because $e^{i\frac{\pi}{2}} = \cos\pi/2 + i\sin\pi/2 = i$ (the cosine of $\pi/2$ is 0 and the sine of $\pi/2$

is 1). The set of vectors $\{e^{i\theta}\,|\,\psi\rangle\}$, for all $\theta$, defines a one-dimensional subspace or "ray." Any unit vector in the ray can be taken as representing the ray. The factor $e^{i\theta}$, which has absolute value 1, is referred to as a "phase," and identifies different vectors in the ray. So, to be precise, a qubit state is represented by a ray or one-dimensional subspace in the state space, not a vector, but for simplicity you can take a unit vector in the subspace to specify the state. Alternatively, take the projection operator onto the one-dimensional subspace to represent the state.

While the global phase makes no difference to the probabilities defined by a quantum state vector—$e^{i\theta}\,|\,\psi\rangle$ assigns the same probabilities to the possible values of observables as the state vector $|\,\psi\rangle$—relative phase does make a difference. In the section "Relative Phase" I show that the state vectors $|x_+\rangle = |0\rangle + |1\rangle$ and $|x_\theta\rangle = |0\rangle + e^{i\theta}\,|1\rangle$ are different qubit states and assign different probabilities to the possible outcomes of an $X$-measurement.

The eigenvectors of $X$ are orthogonal: their scalar product $\langle x_+|x_-\rangle$ or $\langle x_-|x_+\rangle$ is 0. This is always the case for the different eigenvectors of an operator. To compute an expression like $\langle x_+|x_-\rangle$ when $|x_+\rangle$ and $|x_-\rangle$ are linear superpositions, express $\langle x_+|$ explicitly as $\langle 0| + \langle 1|$ and take the scalar products of the terms in this expression with the terms in the expression $|0\rangle - |1\rangle$ for $|x_-\rangle$, in order, adding or subtracting the results as appropriate depending on whether the coefficients are 1 or –1. The scalar products of $\langle 0| + \langle 1|$ and $|0\rangle - |1\rangle$ term by term are $\langle 0|0\rangle + \langle 1|0\rangle - \langle 0|1\rangle - \langle 1|1\rangle$, so $\langle x_+|x_-\rangle = \frac{1}{2}(\langle 0|0\rangle + \langle 1|0\rangle - \langle 0|1\rangle - \langle 1|1\rangle) = \frac{1}{2}(1 + 0 - 0 - 1) = 0$.

Similarly, you can check that the operator $Y = -i|0\rangle\langle 1| + i|1\rangle\langle 0|$ has the eigenvectors $|y_+\rangle = |0\rangle + i|1\rangle$ and $|y_-\rangle = |0\rangle - i|1\rangle$ with eigenvalues $\pm 1$, and the computational basis states $|0\rangle$ and $|1\rangle$ are the eigenvectors of the operator $Z = |0\rangle\langle 0| - |1\rangle\langle 1|$, so $Z|0\rangle = |0\rangle$ and $Z|1\rangle = -|1\rangle$.

The three operators $X, Y, Z$ are called the Pauli operators. They represent three qubit observables, each with possible values $\pm 1$, the eigenvalues of the observables. (I'll use the same symbol to represent the observable and the corresponding operator.) The two eigenvectors of each observable define a basis associated with the observable. As I mentioned before, the convention is to take the eigenvectors of $Z$ as the standard or computational basis, and to represent the eigenvector $|z_+\rangle$ as $|0\rangle$ and the eigenvector $|z_-\rangle$ as $|1\rangle$. Then the eigenvector $|0\rangle$ corresponds to the eigenvalue 1 and the eigenvector $|1\rangle$ corresponds to the eigenvalue –1, which might be a little confusing at first.

To emphasize again: the eigenvalues are just convenient labels for the two possible values of the observables. It's convenient for computational and information-theoretic purposes to represent the eigenvalues of a conventionally chosen "computational basis" observable of a qubit as the bits, 0 or 1, but for other purposes it's convenient to take the eigenvalues of spin observables or polarization observables as $\pm 1$. What's physically relevant is the relation between different observables, expressed by the geometrical relation between their associated eigenvectors, the different basis states corresponding to the observables.

Here's a summary:

- $X = |0\rangle\langle 1| + |1\rangle\langle 0|$. Eigenvectors: $|x_+\rangle = |0\rangle + |1\rangle$, $|x_-\rangle = |0\rangle - |1\rangle$;
- $Y = -i|0\rangle\langle 1| + i|1\rangle\langle 0|$. Eigenvectors: $|y_+\rangle = |0\rangle + i|1\rangle$, $|y_-\rangle = |0\rangle - i|1\rangle$;
- $Z = |0\rangle\langle 0| - |1\rangle\langle 1|$. Eigenvectors: $|z_+\rangle = |0\rangle$, $|z_-\rangle = |1\rangle$.

Finally, the identity operator, $I$, can be expressed as $I = |0\rangle\langle 0| + |1\rangle\langle 1|$. The identity is like the number 1, which leaves any number the same when you multiply by 1. So, for example, $I \cdot Z = (|0\rangle\langle 0| + |1\rangle\langle 1|) \cdot (|0\rangle\langle 0| - |1\rangle\langle 1|) = |0\rangle\langle 0| - |1\rangle\langle 1| = Z$, and $I|\psi\rangle = |\psi\rangle$ for any state $|\psi\rangle$. You can check that $X^2 = Y^2 = Z^2 = I$ (see the section "Quantum Dynamics" for an explicit calculation).

The identity operator is the sum of projection operators onto two orthogonal directions in the two-dimensional qubit space, so it could equally well be expressed as $I = |x_+\rangle\langle x_+| + |x_-\rangle\langle x_-|$ or as $I = |y_+\rangle\langle y_+| + |y_-\rangle\langle y_-|$, or in terms of any other basis in the state space.

The notation $X, Y, Z$ for these operators suggests an orthogonal coordinate system, with $x, y, z$ axes. In fact, for electrons, $X, Y, Z$ represent spin observables for three orthogonal directions of spin: spin in the direction $x$, spin in the direction $y$, and spin in the direction $z$. For photons, the eigenstates of $Z$ are the horizontal and vertical linear polarization states for some standard direction $\vec{z}$: $|H\rangle = |0\rangle$, $|V\rangle = |1\rangle$; the eigenstates of $X$ are the two orthogonal states of diagonal polarization at an angle $45^o$ to $\vec{z}$: $|\nearrow\rangle = |x_+\rangle$, $|\searrow\rangle = |x_-\rangle$; and the eigenstates of $Y$ are the two states of right and left circular polarization: $|\circlearrowright\rangle = |y_+\rangle$, $|\circlearrowleft\rangle = |y_-\rangle$.

You might wonder why the eigenstates of $Y$ involve complex numbers while the eigenstates of $X$ and $Z$ do not. The short answer is that you can't fit the representation of spins on the three-dimensional Bloch sphere or polarizations on the three-dimensional Poincaré sphere onto a two-dimensional qubit space without using complex numbers. So some observables and eigenstates will need to involve complex numbers in their expressions, and the choice ultimately depends on getting the experimental probabilities right.

The representation $Z = |z_+\rangle\langle z_+| - |z_-\rangle\langle z_-|$, in which the operator is represented as a sum of the projection operators onto the eigenstates of the operator, with the eigenvalues as coefficients, is called the "spectral representation" of the operator. The "spectrum" of an operator or the corresponding observable is the set of eigenvalues, the possible values of the observable, in this case labeled 1 for the eigenvalue that is the coefficient of $|z_+\rangle\langle z_+|$ and $-1$ for the eigenvalue that is the coefficient of $|z_-\rangle\langle z_-|$. (Alternatively, you could write $Z = |0\rangle\langle 0| - |1\rangle\langle 1|$ to make explicit that $|z_+\rangle$ and $|z_-\rangle$ are the computational basis eigenvectors $|0\rangle$ and $|1\rangle$.)

You can check that the spectral representations of $X$ and $Y$ are $X = |x_+\rangle\langle x_+| - |x_-\rangle\langle x_-|$ and $Y = |y_+\rangle\langle y_+| - |y_-\rangle\langle y_-|$ by substituting the expressions for the eigenvectors in the computational basis to obtain the representations of $X$ and $Y$ in the computational basis: $X = |0\rangle\langle 1| + |1\rangle\langle 0|$ and $Y = -i|0\rangle\langle 1| + i|1\rangle\langle 0|$. So, to be clear, $|x_+\rangle\langle x_+| - |x_-\rangle\langle x_-| = |0\rangle\langle 1| + |1\rangle\langle 0|$ and $|y_+\rangle\langle y_+| - |y_-\rangle\langle y_-| = -i|0\rangle\langle 1| + i|1\rangle\langle 0|$.

Not all operators in the state space represent observables. Operators that represent observables and can be associated with eigenvectors and real-valued eigenvalues in this way form a special class of operators in the state space: the "self-adjoint" or "Hermitian" operators. The adjoint of an operator is obtained by replacing $i$ with $-i$ (so replacing complex numbers with their complex conjugates) and transposing bras and kets (so replacing $|\psi\rangle\langle\phi|$ with $|\phi\rangle\langle\psi|$). In the section "States and Operators as Matrices" I'll show that operators can be represented by matrices. Transposing bras and kets of an operator in the Dirac notation is equivalent to transposing rows and columns of the corresponding matrix, which produces the "transpose" of the matrix.

The adjoint of $X = |0\rangle\langle 1| + |1\rangle\langle 0|$ is $X$, because the coefficients are real numbers, and exchanging bras and kets in the two terms changes the first term to the second term, and the second term to the first term. Since the terms are added, the operator is unchanged. Similarly, the adjoint of $Z = |0\rangle\langle 0| - |1\rangle\langle 1|$ is $Z$, because all the coefficients are real numbers, and switching bras and kets leaves the operator unchanged. In the case of $Y$, denoting the adjoint by $Y^*$, you get $Y^* = (-i|0\rangle\langle 1| + i|1\rangle\langle 0|)^* = i|1\rangle\langle 0| - i|0\rangle\langle 1| = Y$. So the three Pauli operators are all self-adjoint.

To sum up: The quantum states of a qubit are represented by unit vectors (more properly by one-dimensional subspaces or rays) on a complex vector space or linear space ("linear" referring to the way vectors are composed by vector addition to produce new vectors; "complex" because the coefficients multiplying the vectors are real or complex numbers, and real numbers are just a special case of complex numbers). Observables are represented by self-adjoint operators on the state space. Each such operator is associated with a basis of orthogonal unit vectors in the state space, with each vector in the basis corresponding to one of the two possible binary values of the observable, representing the possible outcomes of a measurement of the observable.

This is the case for *pure* quantum states. There are also mixed states representing mixtures: probability distributions or convex combinations of pure states. For example, a qubit could be prepared in such a way that it is either in the state $|0\rangle$ with probability $p_0$ or in the state $|1\rangle$ with probability $p_1$, in which case the qubit is said to be in a mixed state. I'll show how to represent mixed states by operators in the section "Mixed States" later in this Supplement.

It turns out that in quantum mechanics, unlike in classical theories, the representation of a mixed state as a particular mixture of pure states is not unique. The maximally mixed state of a qubit represented as an equal weight mixture of the pure states $|0\rangle$ and $|1\rangle$ with $p_1 = p_2 = 1/2$ is equivalent to an equal weight mixture of the pure states $|x_+\rangle$ and $|x_-\rangle$. Measuring $Z$ or $X$ or any other observable on either of these mixtures will produce the same distribution of measurement outcomes. So there's no way to distinguish a qubit that has been prepared in such a way that it is either in the state $|0\rangle$ or in the state $|1\rangle$ with probability $1/2$ from a qubit that has been prepared in such a way that it is either in the state $|x_+\rangle$ or in the state $|x_-\rangle$ with probability $1/2$. In fact, a maximally mixed state is equivalent to *any* equal weight mixture of orthogonal pure states, as well as an infinite set of mixtures of nonorthogonal pure states.

In the "Bananaworld" chapter, peeling $S$ corresponds to measuring the $Z$ observable of a qubit, peeling $T$ corresponds to measuring the $X$ observable, and the tastes 0 and 1 correspond to the two possible values of $Z$ and $X$. In the chapter "Quantum Magic," the correspondence between quantum observables and peelings is different.

You could define pure state bananas in Bananaworld; they play a role in the discussion of Pusey–Barrett–Rudolph bananas in the chapter "Making Sense of It All." Pure state bananas grow on trees with large bunches of bananas, and there are four types: $S0, S1, T0, T1$. The pure state bananas $S0$ and $S1$ correspond to the qubit states $|0\rangle$ and $|1\rangle$, the eigenstates of $Z$, and the pure state bananas $T0$ and $T1$ correspond to the qubit states $|x_+\rangle$ and $|x_-\rangle$, the eigenstates of $X$. So for all the pure state bananas in an $S0$ bunch, if you peel from the stem (which corresponds to measuring $Z$), the banana tastes ordinary (0), but if you peel from the top (which corresponds to measuring $X$), the banana tastes ordinary (0) or intense (1) with probability $1/2$. For all the bananas in an $S1$ bunch, if you peel $S$, the banana tastes intense (1), but if you peel $T$, the banana tastes ordinary or intense with probability $1/2$. Similarly for $T0$ and $T1$ bananas.

## The bottom line

- The quantum states of a qubit are represented by unit vectors, and the observables by operators. A measurement of a qubit observable has two possible outcomes, represented by the two "eigenvalues" of the observable and two corresponding orthogonal state vectors, the "eigenvectors" of the observable. The eigenvectors define a coordinate system or basis in the state space of the qubit.
- The Pauli observables, denoted by $X, Y, Z$, are three qubit observables that play a special role throughout the book. For electrons, $X, Y, Z$ represent spin observables for three orthogonal directions of spin: spin in the direction $x$, spin in the direction $y$, and spin in the direction $z$. For photons, the eigenstates of $Z$ are the horizontal and vertical linear polarization states for some standard direction $z$, the eigenstates of $X$ are the two orthogonal states of diagonal polarization at an angle $45^o$ to $z$, and the eigenstates of $Y$ are the two states of right and left circular polarization. The convention is to take the eigenbasis of $Z$ as the standard or computational basis, and to represent the eigenvectors as $|0\rangle$ and $|1\rangle$.
- There are also mixed states representing mixtures: probability distributions of pure states. For example, a qubit could be prepared in such a way that it is either in the state $|0\rangle$ with probability $p_0$ or in the state $|1\rangle$ with probability $p_1$. I show how to represent mixed states by operators in the section "Mixed States."

(continued)

 **The bottom line** *(continued)*

- In the "Bananaworld" chapter, peeling $S$ corresponds to measuring the $Z$ observable of a qubit, peeling $T$ corresponds to measuring the $X$ observable, and the tastes 0 and 1 correspond to the two possible values of $Z$ and $X$. In the chapter "Quantum Magic," the correspondence between quantum observables and peelings is different.

# THE BORN RULE

The relation of states and observables to what we see in quantum phenomena is given by the Born probability rule. If a qubit in the pure state $|\psi\rangle$ is subjected to a measurement, say an $X$-measurement, the probabilities of the two possible outcomes, $\pm 1$ corresponding to the eigenstates $|x_+\rangle$, $|x_-\rangle$, are given by the squares of the absolute values of the "lengths" of the projections of $|\psi\rangle$ onto the eigenstates of $X$ (the quotes are there because this "length" could be a complex number):

$$\text{prob}_{|\psi\rangle}(1) = |P_{|x+\rangle}|\psi\rangle|^2 = |\langle x_+|\psi\rangle|^2,$$
$$\text{prob}_{|\psi\rangle}(-1) = |P_{|x-\rangle}|\psi\rangle|^2 = |\langle x_-|\psi\rangle|^2.$$

After the measurement, the qubit will either be in the state $|x_+\rangle$ or in the state $|x_-\rangle$, depending on the outcome. For example, a photon subjected to a measurement of linear polarization in a particular direction $z$ by passing it through a polarizing beamsplitter will exit the beamsplitter in one of two states of linear polarization: it will either be linearly polarized in the $z$ direction, or in a direction orthogonal to $z$. Although it's customary, it's a bit odd to refer to an $X$-measurement of a qubit in the state $|\psi\rangle$ as a measurement to check whether the qubit is in the state $|x_+\rangle$ or the state $|x_-\rangle$, since a qubit in any state, if subjected to an $X$-measurement, is forced to "choose" between the two states $|x_+\rangle$ and $|x_-\rangle$. A measurement of a qubit, unlike a measurement in a classical theory, is a process that induces the qubit to make an intrinsically random transition to one of two alternative states, as I showed in Chapter 4, "Really Random."

Another way of expressing the Born rule is in terms of the expectation value of an observable in a quantum state. The expectation value or average value of $X$ in the pure state $|\psi\rangle$, denoted by $\langle X\rangle_{|\psi\rangle}$, is the sum of the possible values $\pm 1$ of $X$ weighted by the probabilities $p_+ = |P_{|x+\rangle}|\psi\rangle|^2$ for 1 and $p_- = |P_{|x-\rangle}|\psi\rangle|^2$ for $-1$ in the state $|\psi\rangle$: $\langle X\rangle_{|\psi\rangle} = p_+ - p_-$.

There is a concise way of expressing the expectation value in the Dirac notation. The projection operator $P_{|x+\rangle}$ is self-adjoint (this is obvious because of its form: $P_{|x+\rangle} = |x_+\rangle\langle x_+|$) and "idempotent": $P_{|x+\rangle} \cdot P_{|x+\rangle} = P_{|x+\rangle}$ (projecting twice is the same as projecting once). So, just as $|\psi|^2 = \langle\psi|\psi\rangle$ is the square of the absolute value of the "length" of the vector $|\psi\rangle$, so $|P_{|x+\rangle}|\psi\rangle|^2$ can be expressed as $\langle\psi|P_{|x+\rangle}\cdot P_{|x+\rangle}|\psi\rangle =$ $\langle\psi|P_{|x+\rangle}|\psi\rangle$. Similarly, $|P_{|x-\rangle}|\psi\rangle|^2 = \langle\psi|P_{|x-\rangle}|\psi\rangle$. So

$$\begin{aligned}\langle X\rangle_{|\psi\rangle} &= \langle\psi|P_{|x+\rangle}\psi\rangle - \langle\psi|P_{|x-\rangle}\psi\rangle \\ &= \langle\psi|P_{|x+\rangle} - P_{|x-\rangle}|\psi\rangle \\ &= \langle\psi|X|\psi\rangle,\end{aligned}$$

because $P_{|x+\rangle} - P_{|x-\rangle}$ is $X$ in the spectral representation. This is a general expression for the expectation value of any observable $O$ in a pure quantum state $|\psi\rangle$: $\langle O\rangle_{|\psi\rangle} = \langle\psi|O|\psi\rangle$.

The Born rule has an extension for mixed states. See the section "Mixed States."

The Pauli operators are "mutually unbiased." The absolute value of the "length" of the projection of any eigenvector in one basis onto any eigenvector in another basis is the same: $1/\sqrt{2}$. This is clear from the expressions for the $X$-eigenvectors and the $Y$-eigenvectors in terms of the $Z$-eigenvectors $|0\rangle$ and $|1\rangle$, because the coefficients are $1/\sqrt{2}$ for $|0\rangle$ and $\pm 1/\sqrt{2}$ or $\pm i/\sqrt{2}$ for $|1\rangle$, and these all have absolute value $1/\sqrt{2}$. To see that this is so for the projections of $Y$-eigenvectors onto $X$-eigenvectors, or $Z$-eigenvectors onto $Y$-eigenvectors, you could express $Y$-eigenvectors in terms of $X$-eigenvectors, or $Z$-eigenvectors in terms of $Y$-eigenvectors, which would be rather tedious, or simply observe that the relations between $X, Y, Z$ are unchanged if you permute them cyclically. For example, the relations between these observables are unchanged if you relabel $X$ as $Y$ and $Y$ as $Z$ and $Z$ as $X$.

So, by the Born rule, if a qubit in an eigenstate of one Pauli observable is subjected to a measurement of a different Pauli observable, the probability is $1/2$—the square of $\pm 1/\sqrt{2}$ or the absolute value of the square of $\pm i\sqrt{2}$—that the qubit will be detected as having the eigenvalue 1, and $1/2$ that the qubit will be detected as having the eigenvalue $-1$. For example, if a photon in any state represented by a vector in one of the Pauli bases hits a polarizing filter associated with any other Pauli basis, it has an even chance of getting through and an even chance of being blocked.

## The bottom line

- The relation of states and observables to what we see in quantum phenomena is given by the Born probability rule. If a qubit in the pure state $|\psi\rangle$ is subjected to a measurement, say an $X$-measurement, the probabilities of the two possible outcomes, $\pm 1$ corresponding to the eigenstates $|x_+\rangle$, $|x_-\rangle$, are given by the squares of the absolute values of the "lengths" of the projections of $|\psi\rangle$ onto the eigenstates of $X$ (the quotes are there because this "length" could be a complex number).
- Another way of expressing the Born rule is in terms of the expectation value of an observable in a quantum state. There is a concise way of expressing the expectation value in the Dirac notation.

*(continued)*

**The bottom line** *(continued)*

- The Pauli operators are "mutually unbiased." By the Born rule, if a qubit in an eigenstate of one Pauli observable is subjected to a measurement associated with a different Pauli observable, the probability is $1/2$ of the qubit being detected as having the eigenvalue 1 and $1/2$ of the qubit being detected as having the eigenvalue $-1$. For example, if a photon in any state represented by a vector in one of the Pauli bases hits a polarizing filter associated with any other Pauli basis, it has an even chance of getting through and an even chance of being blocked.

## NONCOMMUTATIVITY AND UNCERTAINTY

Operators can be multiplied and the multiplication is associative: $A(B + C) = AB + AC$, for any operators $A, B, C$. As an example, take $A = |0\rangle\langle 0|$, $B = |0\rangle\langle 1|$, and $C = |1\rangle\langle 0|$. Then $|0\rangle\langle 0| \cdot (|0\rangle\langle 1| + |1\rangle\langle 0|) = |0\rangle\langle 0|0\rangle\langle 1| + |0\rangle\langle 0|1\rangle\langle 0| = |0\rangle\langle 1|$, because $\langle 0|0\rangle$ sandwiched between the vectors $|0\rangle$ and $\langle 1|$ is just the number $\langle 0|0\rangle = 1$, and $\langle 0|1\rangle$ sandwiched between the vectors $|0\rangle$ and $\langle 0|$ is the number $\langle 0|1\rangle = 0$.

The operation of multiplying two operators, $A$ and $B$, needn't commute. It could be that $AB$ operating on a vector doesn't produce the same output as $BA$ operating on the vector—the order of applying the operations, first $B$ and then $A$, or first $A$ and then $B$, could make a difference. For example, $XY = (|0\rangle\langle 1| + |1\rangle\langle 0|) \cdot (-i|0\rangle\langle 1| + i|1\rangle\langle 0|) = i|0\rangle\langle 0| - i|1\rangle\langle 1| = iZ$, but $YX = (-i|0\rangle\langle 1| + i|1\rangle\langle 0|) \cdot (|0\rangle\langle 1| + |1\rangle\langle 0|) = -i|0\rangle\langle 0| + i|1\rangle\langle 1| = -iZ$. So $XY - YX = 2iZ$.

The difference $XY - YX$, called the "commutator" of $X$ and $Y$, is denoted by $[X, Y]$, so $[X, Y] = 2iZ$. Similarly, you can check that $YZ = iX, ZY = -iX$, so $[Y, Z] = 2iX$, and $ZX = iY, XZ = -iY$, so $[Z, X] = 2iY$. The commutation relations are the same for the triples $X, Y, Z$ and $Y, Z, X$ and $Z, X, Y$, so for cyclic permutations of $X, Y, Z$.

Operators associated with different bases in the state space don't commute. Operating on a state vector in the state space with two of these operators in succession (represented by multiplying the two operators in a certain order) is generally not the same thing as operating on the state vector with the two operators in the reverse order: the output vector of the combined operation is different in the two cases.

Heisenberg's uncertainty or indeterminacy principle is a consequence of noncommutativity. In Heisenberg's original version, the noncommuting operators represent position and momentum, usually denoted by $\hat{x}$ for position and $\hat{p}$ for momentum. In this case the commutator turns out to be a multiple of the identity operator $I$: $[\hat{x}, \hat{p}] = i\hbar I$, or $[\hat{x}, \hat{p}] = iI$ in units in which $\hbar$, Planck's constant divided by $2\pi$, is equal to 1, which I've implicitly assumed throughout the book. From this relation between the position and momentum operators, you get a reciprocal relation between the extent to which it is possible to fix the range of possible position values and the range of possible momentum values for a system in any quantum state.

There is a similar uncertainty principle for the Pauli observables of a qubit. If a qubit is in a particular eigenstate of $X$, $Y$, or $Z$, there's an even probability of yielding the outcome $\pm 1$ in a measurement of any other Pauli observable. The values of the different Pauli observables can't be specified more precisely than this: if one Pauli observable has a definite value, the values of the other Pauli observables are completely indefinite. It follows from the Colbeck and Renner argument in Chapter 4, "Really Random," that for a qubit in an eigenstate of a Pauli observable, the outcome of a measurement of a different Pauli observable is an intrinsically random event—but this doesn't follow from the uncertainty principle.

### The bottom line

- Operators associated with different bases in the state space don't commute. Operating on a state vector in the state space with two of these operators in succession (represented by multiplying the two operators in a certain order) is generally not the same thing as operating on the state vector with the two operators in the reverse order: the output vector of the combined operation is different in the two cases.
- Heisenberg's uncertainty or indeterminacy principle is a consequence of noncommutativity. In Heisenberg's original version, the noncommuting operators represent position and momentum. There is a similar uncertainty principle for the Pauli observables of a qubit: if a qubit is in a particular eigenstate of $X$, $Y$, or $Z$, there's an even probability of yielding the outcome $\pm 1$ in a measurement of any other Pauli observable. The values of the different Pauli observables can't be specified more precisely than this.

## ENTANGLEMENT

If two qubits, $A$ and $B$, are separately in states $|a\rangle$ and $|b\rangle$, the state of the combined system is represented as $|a\rangle \otimes |b\rangle$, where the symbol $\otimes$ represents a product called a tensor product. Technically, the states $|a\rangle$ and $|b\rangle$ are represented by vectors in the two-dimensional state spaces of the separate systems $A$ and $B$, and the state of the composite system, $|a\rangle \otimes |b\rangle$, is a vector in a new state space, the "tensor product" of the two state spaces, which is a four-dimensional space. The tensor product space is constructed as the state space consisting of all product vectors of the form $|a\rangle \otimes |b\rangle$ (the Cartesian product of the two state spaces, the set of all ordered pairs of states from each of the state spaces), together with superpositions like $c|a_1\rangle \otimes |b_1\rangle + d|a_2\rangle \otimes |b_2\rangle$, where $|a_1\rangle$, $|a_2\rangle$ are two states of $A$ and $|b_1\rangle$, $|b_2\rangle$ are two states of $B$, and $c$ and $d$ are complex numbers.

It's often easier to read such expressions without the tensor product symbol $\otimes$, so unless it's needed for clarity or to avoid ambiguity, it's usual to write $c|a_1\rangle|b_1\rangle + d|a_2\rangle|b_2\rangle$ and even, for short, $c|a_1\rangle b_1 + d|a_2\rangle b_2$.

Some linear superpositions like $|0\rangle|1\rangle + |1\rangle|1\rangle$ are really product states in disguise. This state is just $(|0\rangle + |1\rangle) \otimes |1\rangle = |+\rangle|1\rangle$. So it's a tensor product of the state $|+\rangle$ for qubit $A$ and the state $|1\rangle$ for qubit $B$. The two qubits are said to be separable, and measurements on the qubits are uncorrelated.

Other linear superpositions in the tensor product space, like $|\psi^-\rangle = |0\rangle|1\rangle - |1\rangle|0\rangle$, cannot be expressed as product states. Such two-qubit pure states are referred to as "entangled" pure states. An entangled two-qubit pure state is nonseparable and cannot be expressed as a product state of two one-qubit pure states. Measurement outcomes on the two qubits are correlated. In the case of $|\psi^-\rangle$, measurement outcomes are perfectly anticorrelated (oppositely correlated) if the same observable is measured on both qubits. A mixed state of a bipartite system is entangled if it can't be expressed as a mixture or probability distribution of product states of the two systems. You can have entangled states of any number of qubits or qutrits or any sort of quantum system.

Here's how to see that $|\psi^-\rangle$ can't be expressed as a product state. If $|\psi^-\rangle$ could be expressed as a product state $(a|0\rangle + b|1\rangle) \otimes (c|0\rangle + d|1\rangle) = ac|0\rangle|0\rangle + bd|1\rangle|1\rangle + ad|0\rangle|1\rangle + bc|1\rangle|0\rangle$, the complex numbers $a, b, c, d$ would have to satisfy the conditions $ac = 0, bd = 0, ad = 1, bc = -1$. If $ac = 0$, then at least one of $a$ or $c$ would have to be zero. If $a = 0$ then $ad = 0$, which is inconsistent with the condition $ad = 1$. If $c = 0$, then $bc = 0$, which is inconsistent with the condition $bc = -1$. So the conditions can't be satisfied.

The entangled state $|\psi^-\rangle$ is a rather special state: it takes the same form *for any basis* (possibly up to multiplication by a global phase $e^{i\theta}$). For example: $|0\rangle|1\rangle - |1\rangle|0\rangle = |+\rangle|-\rangle - |-\rangle|+\rangle$, where $|+\rangle, |-\rangle$ are the $X$-eigenstates, and similarly for any other basis. What the entangled state $|\psi^-\rangle$ expresses is perfect anticorrelation. If Alice and Bob share two qubits in this entangled state, and Alice measures any observable of her qubit and Bob measures the same observable of his qubit, their measurement outcomes will be anticorrelated. (Reminder: the missing coefficients are both $1/\sqrt{2}$ here to "normalize" the state to unit length.) If Alice and Bob both measure Z, the probability is $\frac{1}{2}$ (the square of $\frac{1}{\sqrt{2}}$) that Alice will find the state $|0\rangle$ and Bob will find the state $|1\rangle$, and $\frac{1}{2}$ that Alice will find the state $|1\rangle$ and Bob will find the state $|0\rangle$. Similarly, if Alice and Bob measure $X$, there is an even probability that Alice will find the state $|+\rangle$ and Bob the state $|-\rangle$, or that Alice will find the state $|-\rangle$ and Bob the state $|+\rangle$, and similarly for any other qubit observable.

What if Alice measures Z and Bob does nothing? After the measurement, Alice's qubit is in one of the eigenstates of Z, $|0\rangle$ or $|1\rangle$. In the four-dimensional two-qubit state space, each outcome is associated with one of two orthogonal two-dimensional subspaces or planes, defined by the projection operators $|0\rangle\langle 0|_A \otimes I_B$ and $|1\rangle\langle 1|_A \otimes I_B$. Here, $I_B$ is the identity operator on Bob's two-dimensional qubit space: operating on any vector in the space, the identity operator reproduces the vector. (I use subscripts here to distinguish Alice operators from Bob operators.)

According to the Born rule, if the state is $|\psi^-\rangle$, the probability of a particular outcome (if Alice measures $Z$ and Bob does nothing) is the square of the absolute value of the length of the projection of $|\psi^-\rangle$ onto the plane in the two-qubit space associated with the outcome. The projections onto the planes associated with the two possible measurement outcomes are $|0\rangle\langle 0|_A \otimes I_B \cdot |\psi^-\rangle$ and $|1\rangle\langle 1|_A \otimes I_B \cdot |\psi^-\rangle$. (The "·" here is introduced to visually separate the projection operator from the state for clarity.) What are the lengths of these projections?

To work out the length of the projection $|0\rangle\langle 0|_A \otimes I_B \cdot |\psi^-\rangle$, write it out as $|0\rangle\langle 0|_A \otimes I_B \cdot (\frac{1}{\sqrt{2}}|0\rangle_A|1\rangle_B - \frac{1}{\sqrt{2}}|1\rangle_A|0\rangle_B)$, with the coefficients explicit. The $A$-operator acts on vectors in the $A$-space, and the $B$-operator acts on vectors in the $B$-space. Consider the first term in the state $\frac{1}{\sqrt{2}}|0\rangle_A|1\rangle_B - \frac{1}{\sqrt{2}}|1\rangle_A|0\rangle_B$. The identity $I_B$ operating on the vector $|1\rangle_B$ reproduces the vector $|1\rangle_B$, and $|0\rangle\langle 0|_A$ operating on $|0\rangle_A$ produces $|0\rangle_A$ (because $\langle 0|0\rangle_A = 1$). So the operator $|0\rangle\langle 0|_A \otimes I_B$ acting on the first term in the state $\frac{1}{\sqrt{2}}|0\rangle_A|1\rangle_B - \frac{1}{\sqrt{2}}|1\rangle_A|0\rangle_B$ produces the vector $\frac{1}{\sqrt{2}}|0\rangle_A|1\rangle_B$. Since the operator $|0\rangle\langle 0|_A \otimes I_B$ acting on the second term produces 0 (because $\langle 0|1\rangle_A = 0$), the projection is the vector $\frac{1}{\sqrt{2}}|0\rangle_A|1\rangle_B$, which has length $1/\sqrt{2}$ (because $|0\rangle_A$ and $|1\rangle_B$ are both unit vectors). The probability is the square of this length, which is $1/2$. A similar calculation shows that the probability for the projection operator $|1\rangle\langle 1|_A \otimes I_B \cdot |\psi^-\rangle$ is also $1/2$.

There are three other states like $|\psi^-\rangle$ that form a basis with $|\psi^-\rangle$, called the "Bell basis," in the two-qubit state space:

$$|\phi^+\rangle = \frac{1}{\sqrt{2}}|0\rangle|0\rangle + \frac{1}{\sqrt{2}}|1\rangle|1\rangle,$$

$$|\phi^-\rangle = \frac{1}{\sqrt{2}}|0\rangle|0\rangle - \frac{1}{\sqrt{2}}|1\rangle|1\rangle,$$

$$|\psi^+\rangle = \frac{1}{\sqrt{2}}|0\rangle|1\rangle + \frac{1}{\sqrt{2}}|1\rangle|0\rangle,$$

$$|\psi^-\rangle = \frac{1}{\sqrt{2}}|0\rangle|1\rangle - \frac{1}{\sqrt{2}}|1\rangle|0\rangle.$$

As you can check, these states are mutually orthogonal: the scalar product of any two Bell states is zero. The correlations defined by $|\psi^-\rangle$ are the anticorrelations in the Bohm version of the Einstein–Podolsky–Rosen argument. The correlations of the state $|\phi^+\rangle$ are the correlations of Einstein–Podolsky–Rosen bananas. Peeling $S$ corresponds to measuring $Z$, and peeling $T$ corresponds to measuring $X$. The two tastes, 0 or 1, correspond to the eigenvalues 1 and –1. Like the state $|\psi^-\rangle$, the state $|\phi^+\rangle$ takes the same form in the $X$-basis and the $Z$-basis (but, in general, for Bell states other than $|\psi^-\rangle$, transforming between bases takes you from one Bell state to another Bell state). The four mutually orthogonal product states $|0\rangle|0\rangle$, $|0\rangle|1\rangle$, $|1\rangle|0\rangle$, $|1\rangle|1\rangle$ form an alternative computational basis in the two-qubit state space.

Bell's insight was to see that, while the Einstein–Podolsky–Rosen correlations can be simulated with local resources (which is to say that there is a common cause explanation of these correlations), this is not the case for the probabilistic correlations of a Bell state if Alice and Bob measure different observables. For a particular choice of observables, it's possible to reach the Tsirelson bound and produce correlations that are as far as is physically possible from correlations that can be simulated with local resources, and as close as is physically possible to the Popescu–Rohrlich correlations. So in a simulation game where Alice and Bob are required to simulate the correlations of Popescu–Rohrlich bananas, they can win the game with an optimal probability $\frac{1}{2}(1 + \frac{1}{\sqrt{2}}) \approx .85$ if they are allowed unlimited access to shared entangled quantum states. This is as close as you can get with quantum resources to the maximum probability 1 for a perfect simulation, and closer than the optimal probability .75 with local resources. They each measure specifically chosen observables to the prompts in the game and respond with the measurement outcomes. Subsection 3.3, "Simulating Popescu–Rohrlich Bananas," in the "Bananaworld" chapter does the calculation with specific observables for the state $|\phi^+\rangle$.

**The bottom line**

- Some linear superpositions of two-qubit states like $|\psi^-\rangle = |0\rangle|1\rangle - |1\rangle|0\rangle$ are nonseparable and cannot be expressed as a product state of two one-qubit pure states. Such two-qubit pure states are referred to as "entangled" pure states. A mixed state of a bipartite system is entangled if it can't be expressed as a mixture or probability distribution of product states of the two systems.
- Measurement outcomes on two entangled qubits are correlated. In the case of $|\psi^-\rangle$, measurement outcomes are perfectly anticorrelated if the same observable is measured on both qubits.
- The state $|\psi^-\rangle$ is one of four so-called Bell states, $|\psi^-\rangle$, $|\psi^+\rangle$, $|\phi^-\rangle$, $|\phi^-\rangle$, that form a basis in the two-qubit state space. The correlations defined by $|\psi^-\rangle$ are the anticorrelations in the Bohm version of the Einstein–Podolsky–Rosen argument. The correlations of the state $|\phi^+\rangle$ are the correlations of Einstein–Podolsky–Rosen bananas.

## QUANTUM DYNAMICS

Quantum states change and evolve in time as systems interact dynamically, and the dynamics should preserve the Born probability rule for quantum states. The Born rule says that if you measure an observable with eigenvectors that include $|\alpha\rangle$ on a quantum system in the state $|\psi\rangle$, the probability of finding the eigenstate $|\alpha\rangle$ is the square of the absolute value of the "length" of the projection of $|\psi\rangle$ onto $|\alpha\rangle$, which is the square of the absolute value of the scalar product $\langle\alpha|\psi\rangle$. So when states change under a dynamical evolution defined by some operator $U(t)$ that transforms $|\psi\rangle \rightarrow U(t)|\psi\rangle$ and $|\alpha\rangle \rightarrow U(t)|\alpha\rangle$ in time $t$, the scalar product should be preserved, because the "length" of a vector is defined by the scalar product.

If $U(t)$ operates on the state represented by the ket vector $|\alpha\rangle$, the corresponding operator on the bra vector $\langle\alpha|$ is the adjoint operator $U^*(t)$. You get $U^*(t)$ by transposing bras and kets in $U(t)$ and replacing $i$ by $-i$. So the bra vector of $U(t)|\alpha\rangle$ is $\langle\alpha|U^*(t)$ (the operator always operates on the flat side of a bra or ket).

The scalar product of $U(t)|\alpha\rangle$ and $U(t)|\psi\rangle$ is $\langle\alpha|U^*U|\psi\rangle$. The condition that the scalar product should be preserved under the transformation defined by $U(t)$ is then $\langle\alpha|U^*U|\psi\rangle = \langle\alpha|\psi\rangle$. This condition should hold for any states $|\alpha\rangle$ and $|\psi\rangle$, which means that $U^*(t)U(t) = I$, where $I$ is the identity operator. Equivalently, $U^{-1} = U^*$: the inverse of the operator defining the dynamics should be equal to the adjoint of the operator. Operators satisfying this condition are called unitary operators, so dynamical evolutions in quantum mechanics are unitary transformations.

Geometrically, this means that when two states transform under a dynamical evolution, the angle between the vectors remains the same, because it's the angle that's relevant for the probability according to the Born rule (the projection of one unit vector onto another depends only on the angle). Since orthogonality relations are preserved, a unitary transformation takes a set of orthogonal basis vectors to another set of orthogonal basis vectors.

To define a unitary operator, it's sufficient to define the action of $U$ on a set of basis vectors. This fixes the transformation on any vector, because any vector can be represented in terms of its projections onto the basis vectors. For example, the two-qubit transformation

$$
\begin{aligned}
|0\rangle|0\rangle &\rightarrow |0\rangle|0\rangle \\
|0\rangle|1\rangle &\rightarrow |0\rangle|1\rangle \\
|1\rangle|0\rangle &\rightarrow |1\rangle|1\rangle \\
|1\rangle|1\rangle &\rightarrow |1\rangle|0\rangle
\end{aligned}
$$

is unitary, because the only change is that the second states in the product states $|1\rangle|0\rangle$ and $|1\rangle|1\rangle$, the states $|0\rangle$ and $|1\rangle$, are flipped. This is called a "controlled-not" or CNOT transformation. The first qubit is the "control" and the second qubit is the "target." The control qubit doesn't change in the transformation. The target qubit doesn't change if the control state is $|0\rangle$, but if the control state is $|1\rangle$, a "not" operation is applied to the target, in the sense that $|0\rangle \rightarrow |1\rangle$ and $|1\rangle \rightarrow |0\rangle$.

A CNOT transformation can be part of a sequence of transformations in a quantum circuit. There's a standard symbolism for representing various quantum transformations as "quantum gates" in a circuit diagram. In a real circuit, these gates would be specific hardware devices. A CNOT gate is represented as in Figure S.1 by two circles, one solid, joined by a line. The diagrams are to be read from left to right, with input states on the left, and output states on the right. The top input "wire" or channel on the left side of a gate carries the control qubit state, indicated by a subscript "c." The bottom input wire on the left side carries the target qubit state, indicated by a subscript "t." The two output wires or channels on the right side of a gate carry the output states. The way the

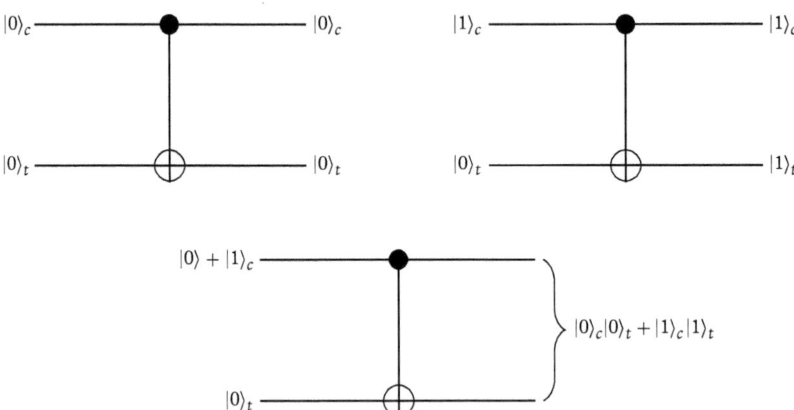


**Figure S.1** Circuit diagrams showing the action of a CNOT gate. If the control is a superposition, as in the bottom diagram, the output is an entangled state.

gate functions is represented in the top two circuit diagrams for the control input states $|0\rangle_c$ and $|1\rangle_c$. The initial state of the target qubit is $|0\rangle$ in both cases here. The bottom circuit diagram shows how the gate functions when the input state of the control qubit is a linear superposition $|0\rangle_c + |1\rangle_c$. In that case, the output state is an entangled state of the control and target qubits on the output wires: $|0\rangle_c|0\rangle_t + |1\rangle_c|1\rangle_t$. That's because the gate implements a unitary transformation, which is linear. So if $|0\rangle_c|0\rangle_t \rightarrow |0\rangle_c|0\rangle_t$ and $|1\rangle_c|0\rangle_t \rightarrow |1\rangle_c|1\rangle_t$, then $(|0\rangle_c + |1\rangle_c)|0\rangle_t \rightarrow |0\rangle_c|0\rangle_t + |1\rangle_c|1\rangle_t$.

The first and third lines of the CNOT transformation,

$$|0\rangle_c|0\rangle_t \stackrel{CNOT}{\longrightarrow} |0\rangle_c|0\rangle_t$$
$$|1\rangle_c|0\rangle_t \stackrel{CNOT}{\longrightarrow} |1\rangle_c|1\rangle_t,$$

provide a model for an idealized measurement interaction between two quantum systems. The measured qubit is the control qubit $c$. The target qubit $t$ represents the instrument "pointer." It starts out in the state $|0\rangle_t$ (the pointer reads 0 or "ready") and ends up mirroring the state of the control qubit (the pointer reads 0 or 1, depending on whether the state of the control is $|0\rangle_c$ or $|1\rangle_c$). Now, because unitary transformations are linear, a CNOT transformation must transform a linear superposition or sum of $|0\rangle_c|0\rangle_t$ and $|1\rangle_c|0\rangle_t$ to the sum of the transformations for $|0\rangle_c|0\rangle_t$ and $|1\rangle_c|0\rangle_t$. So

$$(|0\rangle_c + |1\rangle_c)|0\rangle_t = |0\rangle_c|0\rangle_t + |1\rangle_c|0\rangle_t \stackrel{CNOT}{\longrightarrow} |0\rangle_c|0\rangle_t + |1\rangle_c|1\rangle_t.$$

This means that if the qubit being measured starts out in a linear superposition like $|0\rangle + |1\rangle$, the final state of the combined system, measuring instrument plus measured system, ends up in an entangled state. You might reasonably expect that the final state should reflect what actually happens in a measurement, which is that with probability $1/2$ the combined system ends up in the state $|0\rangle|0\rangle$ and with probability $1/2$ it ends up in the state $|1\rangle|1\rangle$. The difference between what you get from the theory and what you expect is one way of putting the measurement problem of quantum mechanics. I take this up in Section 10.2, "The Measurement Problem," in the chapter "Making Sense of It All."

The (not explicitly written) coefficients of the state of the measured qubit (the control in the CNOT transformation) are both $1/\sqrt{2}$ here. But a qubit in any state with coefficients $c_1, c_2$ could be measured by sending it through a CNOT gate:

$$(c_1|0\rangle + c_2|1\rangle)|0\rangle \overset{CNOT}{\longrightarrow} c_1|0\rangle|0\rangle + c_2|1\rangle|1\rangle.$$

The Pauli operators $X, Y, Z$ in the section "The Pauli Operators" are self-adjoint and represent qubit observables. They are also unitary operators and so can represent dynamical transformations.

The Pauli operators are self-adjoint, $X = X^*, Y = Y^*, Z = Z^*$, and unitarity is the condition $XX^* = I, YY^* = I, ZZ^* = I$. So to show that the Pauli operators are unitary, all you need to prove is that $X^2 = Y^2 = Z^2 = I$.

For $X$, $(|0\rangle\langle 1| + |1\rangle\langle 0|)\cdot(|0\rangle\langle 1| + |1\rangle\langle 0|) = |0\rangle\langle 0| + |1\rangle\langle 1| = I$, because the two terms $|0\rangle\langle 1| \cdot |0\rangle\langle 1|$ and $|1\rangle\langle 0| \cdot |1\rangle\langle 0|$ are both zero, and $|0\rangle\langle 1| \cdot |1\rangle|0\rangle = |0\rangle\langle 0|$ and $|1\rangle\langle 0| \cdot |0\rangle|1\rangle = |1\rangle\langle 1|$. Similar calculations show that $YY = I$ and $ZZ = I$. You can also see this by multiplying the matrix for each of these operators with itself. (See the section "States and Operators as Matrices.")

Considered as a unitary transformation, the operator $X$ flips $|0\rangle$ and $|1\rangle$:

$$|0\rangle \overset{X}{\rightarrow} |1\rangle,\ |1\rangle \overset{X}{\rightarrow} |0\rangle.$$

Explicitly: $(|0\rangle\langle 1| + |1\rangle\langle 0|)\cdot|0\rangle = 0 + |1\rangle = |1\rangle$ and $(|0\rangle\langle 1| + |1\rangle\langle 0|)\cdot|1\rangle = |0\rangle + 0 = |0\rangle$.

The operator $Z$ does nothing to $|0\rangle$ and multiples $|1\rangle$ by $-1$:

$$|0\rangle \overset{Z}{\rightarrow} |0\rangle,\ |1\rangle \overset{Z}{\rightarrow} -|1\rangle.$$

That doesn't change the states $|0\rangle$ and $|1\rangle$ as far as the probabilities defined by the Born rule are concerned, but it changes the relative phase of these two states in a general qubit state: $Z(\alpha|0\rangle + \beta|1\rangle) \rightarrow \alpha|0\rangle - \beta|1\rangle$. (See the section "Relative Phase.")

The operator $Y$ implements the transformation

$$|0\rangle \overset{Y}{\rightarrow} i|1\rangle,\ |1\rangle \overset{Y}{\rightarrow} -i|0\rangle.$$

You can think of the operator $iY$ as representing a transformation $X$ followed by a transformation $Z$, in that order, because $ZX = iY$ ($X$ followed by $Z$ operating on a state $\psi$ is $ZX|\psi\rangle$). To see this, multiply out the operator representatives of $Z$ and $X$ and show that this product is $iY$, or multiply the matrices of $Z$ and $X$ and show that the matrix product is equal to the matrix for $iY$. In fact, the Pauli operators are related by the conditions $XY = iZ$, $YZ = iX$, and $ZX = iY$ and $YX = -iZ$, $ZY = -iX$, and $XZ = -iY$.

For a photon, you can implement the transformations of the Pauli operators by passing the photon through an optical device that alters the polarization of the photon in an appropriate way. For an electron, $X$ rotates the spin of the electron by 180° around the $x$-axis, $Y$ rotates the spin 180° around the $Y$ axis, and $Z$ rotates the spin 180° around the $z$-axis.

Although this plays no role in the book, it's worth mentioning that the dynamics is implemented by a unitary operator that takes the form $U = e^{iHt}$. The operator $H$ is a self-adjoint operator, called the Hamiltonian of the system. It represents the energy observable, with the possible energy values as eigenvalues. The Hamiltonian is unitary because the adjoint, $U^*$, is equal to $e^{-iH^*t}$, which is equal to $e^{-iHt}$, because $H$ is self-adjoint. Since $e^{iHt} \cdot e^{-iHt} = I$, $U^* = U^{-1}$, which is the definition of a unitary operator.

You can understand what this expression with an operator as exponent means via the expansion of $e^x$ as an infinite series: $e^x = 1 + x + \frac{x^2}{2!} + \frac{x^3}{3!} + \cdots$, where $n!$ ("$n$ factorial") is defined as $1 \cdot 2 \cdot 3 \cdots n$. So $U = e^{iHt}$ is equivalent to an infinite sum of powers of operators: $I + iHt + \frac{(iHt)^{2!}}{2!} + \frac{(iHt)^{3!}}{3!} + \cdots$. When applied to a quantum state, this gives you the change of the quantum state in time $t$. The rate of change at each instant is given by the differential equation

$$i\frac{d\,|\,\psi\rangle}{dt} = H\,|\,\psi\rangle.$$

This is Schrödinger's equation.

### The bottom line

- Quantum states change under dynamical evolutions defined by unitary operators. The unitary dynamics preserves the Born probability rule for quantum states.
- The Born rule says that if you measure an observable with eigenvectors that include $|\alpha\rangle$ on a quantum system in the state $|\psi\rangle$, the probability of finding the eigenstate $|\alpha\rangle$ is the square of the absolute value of the "length" of the projection of $|\psi\rangle$ onto $|\alpha\rangle$. Geometrically, a unitary transformation ensures that when two states transform under a dynamical evolution, the angle between the vectors remains the same, because it's the angle that's relevant for the probability according to the Born rule (the projection of one unit vector onto another depends only on the angle). Since orthogonality relations are preserved, a unitary transformation takes a set of orthogonal basis vectors to another set of orthogonal basis vectors.
- An example of a unitary transformation is a controlled-not or CNOT transformation, which is a model for an idealized measurement interaction between two quantum systems.
- The Pauli operators $X, Y, Z$ are unitary operators and so can represent dynamical transformations as well as qubit observables. The operator $X$ flips $|0\rangle$ and $|1\rangle$. The operator $Z$ does nothing to $|0\rangle$ and multiples $|1\rangle$ by $-1$. The

*(continued)*

 **The bottom line** *(continued)*

operator $Y$ implements the transformation $|0\rangle \rightarrow i|1\rangle$ and $|1\rangle \rightarrow -i|0\rangle$. So $iY$ represents a transformation $X$ followed by a transformation $Z$.

I've separated the following five topics—"Relative Phase," "Informationally Complete Observables," "The Biorthogonal Decomposition Theorem," "States and Operators as Matrices," and "Mixed States"—because they rather more challenging than the previous sections. I do refer to them at various places in the book, so you might need to dip into them as needed to follow the argument.

## RELATIVE PHASE

I mentioned in the section "The Pauli Operators" that the *global phase* is irrelevant to the state: $-|\psi\rangle$ represents the same quantum state as $|\psi\rangle$ and generates the same probabilities by the Born rule, and more generally $e^{i\theta}|\psi\rangle$ represents the same state as $|\psi\rangle$. But the *relative phase* does make a difference. The vectors

$$|x+\rangle = |0\rangle + |1\rangle$$

and

$$|\theta\rangle = |0\rangle + e^{i\theta}|1\rangle$$

represent different qubit states. An $X$-measurement on a qubit in the state $|x+\rangle$ will produce the outcome $+1$ with probability 1, but an $X$-measurement on a qubit in the state $|\theta\rangle$ will produce the outcome $+1$ with probability $\cos^2\theta/2$ and $-1$ with probability $\sin^2\theta/2$.

It's worth working through this example to see how probabilities are derived from quantum states in the Dirac notation via the Born rule. First, express $|0\rangle$ and $|1\rangle$ in the $\{|x+\rangle, |x-\rangle\}$ basis explicitly as

$$|0\rangle = \frac{1}{\sqrt{2}}(|x+\rangle + |x-\rangle),$$
$$|1\rangle = \frac{1}{\sqrt{2}}(|x+\rangle - |x-\rangle).$$

Then

$$|\theta\rangle = \frac{1}{2}(|x+\rangle + |x-\rangle) + e^{i\theta}\frac{1}{2}(|x+\rangle - |x-\rangle)$$
$$= \frac{1+e^{i\theta}}{2}|x+\rangle + \frac{1+e^{i\theta}}{2}|x-\rangle.$$

By the Born rule, the probability that a qubit in the state $|\theta\rangle$ subjected to an $X$-measurement will produce the outcome 1 is

$$\left|\frac{1+e^{i\theta}}{2}\right|^2 = \left(\frac{1+e^{i\theta}}{2}\right)\left(\frac{1+e^{-i\theta}}{2}\right) = \frac{2+e^{i\theta}+e^{-i\theta}}{4}.$$

Since $e^{i\theta} + e^{-i\theta} = 2\cos\theta$, this becomes

$$\frac{1+2\cos\theta}{2} = \cos^2\frac{\theta}{2}.$$

The calculation for the probability of the outcome –1 is similar, or just observe that this probability is $1 - \cos^2\frac{\theta}{2} = \sin^2\frac{\theta}{2}$.

**The bottom line**

- The *global phase* is irrelevant to the state: $-|\psi\rangle$ represents the same quantum state as $|\psi\rangle$ and generates the same probabilities by the Born rule, and more generally $e^{i\theta}|\psi\rangle$ represents the same state as $|\psi\rangle$. But the *relative phase* does make a difference: $|\theta\rangle = |0\rangle + e^{i\theta}|1\rangle$ is not the same state as $|x+\rangle = |0\rangle + |1\rangle$.
- An explicit calculation in the Dirac notation shows how different probabilities are derived from the states $|\theta\rangle$ and $|x+\rangle$ via the Born rule.

## INFORMATIONALLY COMPLETE OBSERVABLES

The expectation value or average value of an observable in a quantum state $|\psi\rangle$ is the weighted sum of the eigenvalues, with the probabilities defined by $|\psi\rangle$ as the weights. So the expectation value of Z with eigenvalues $\pm 1$ in the state $|\psi\rangle$, denoted by $\langle Z\rangle_\psi$, is:

$$\begin{aligned}\langle Z\rangle_\psi &= \text{prob}(1) - \text{prob}(-1)\\ &= |\langle 0|\psi\rangle|^2 - |\langle 1|\psi\rangle|^2.\end{aligned}$$

The probabilities are the absolute values of the "lengths" of the projections of $|\psi\rangle$ onto the eigenstates $|0\rangle$ and $|1\rangle$ for the eigenvalues 1 and –1, according to the Born rule.

The expectation value can be derived from the representation of Z as the operator $|0\rangle\langle 0| - |1\rangle\langle 1|$:

$$\begin{aligned}\langle Z\rangle_\psi &= \langle\psi|Z|\psi\rangle\\ &= \langle\psi|(|0\rangle\langle 0| - |1\rangle\langle 1|)|\psi\rangle\\ &= \langle\psi|0\rangle\langle 0|\psi\rangle - \langle\psi|1\rangle\langle 1|\psi\rangle\\ &= |\langle 0|\psi\rangle|^2 - |\langle 1|\psi\rangle|^2.\end{aligned}$$

In the tutorial for complex numbers in the section "Dirac's Ingenious Idea," I pointed out that the square of the absolute value of a complex number $c$ is $cc^*$, where $c^*$ is the complex conjugate of $c$. Here $\langle 0|\psi\rangle = \langle\psi|0\rangle^*$ and $\langle 1|\psi\rangle = \langle\psi|1\rangle^*$, which explains the final step in the above derivation.

To see this, express the vector $|\psi\rangle$ in terms of its components in the $|0\rangle$, $|1\rangle$ basis as

$$|\psi\rangle = \langle 0|\psi\rangle|0\rangle + \langle 1|\psi\rangle|1\rangle,$$

and the corresponding bra vector as

$$\langle\psi| = \langle 0|\psi\rangle^*\langle 0| + \langle 1|\psi\rangle^*\langle 1|.$$

Taking the scalar product of $\langle\psi|$ with $|0\rangle$ on the right, you get

$$\langle\psi|0\rangle = \langle 0|\psi\rangle^*,$$

because $\langle 0|0\rangle = 1$ and $\langle 1|0\rangle = 0$. Similarly, $\langle\psi|1\rangle = \langle 1|\psi\rangle^*$.

With an appropriate choice of the basis states $|0\rangle$ and $|1\rangle$ (so an appropriate choice of the observable $Z$), and ignoring the global phase, a general qubit can be represented as a unit vector:

$$|\psi\rangle = \cos\theta|0\rangle + e^{i\phi}\sin\theta|1\rangle.$$

The projection operator onto this state is

$$\begin{aligned}|\psi\rangle\langle\psi| &= (\cos\theta|0\rangle + e^{i\phi}\sin\theta|1\rangle)(\cos\theta\langle 0| + e^{-i\phi}\sin\theta\langle 1|)\\ &= \cos^2\theta|0\rangle\langle 0| + \sin^2\theta|1\rangle\langle 1| + e^{i\phi}\cos\theta\sin\theta|1\rangle\langle 0| + e^{-i\phi}\cos\theta\sin\theta|0\rangle\langle 1|.\end{aligned}$$

The three expectation values for the Pauli observables in the state $|\psi\rangle$ turn out to be

$$\begin{aligned}\langle X\rangle_\psi &= \cos\theta\sin\theta\cos\phi,\\ \langle Y\rangle_\psi &= \cos\theta\sin\theta\sin\phi,\\ \langle Z\rangle_\psi &= \cos^2\theta - \sin^2\theta,\end{aligned}$$

and $|\psi\rangle\langle\psi|$ can be expressed in terms of these expectation values as

$$\frac{1}{2}\left(I + \langle X\rangle_\psi X + \langle Y\rangle_\psi Y + \langle Z\rangle_\psi Z\right),$$

where $I$ is the identity operator: $I = |0\rangle\langle 0| + |1\rangle\langle 1|$.

You can check this by showing that, in the expression for $|\psi\rangle\langle\psi|$,

$$\cos^2\theta|0\rangle\langle 0| + \sin^2\theta|1\rangle\langle 1| = \frac{1}{2}(I + \langle Z\rangle_\psi Z)$$

and

$$e^{i\phi}\cos\theta\sin\theta|1\rangle\langle 0| + e^{-i\phi}\cos\theta\sin\theta|0\rangle\langle 1| = \frac{1}{2}(\langle X\rangle_\psi X + \langle Y\rangle_\psi Y).$$

So the projection operator onto the state of a qubit is specified by the expectation values of the three Pauli observables, $X, Y, Z$. This means that the quantum state of a qubit can be determined experimentally to arbitrary accuracy if you have a beam of qubits, all in the same quantum state $|\psi\rangle$. Take a sample of a sufficiently large number of qubits in the beam and divide the sample into three sets, and separately measure $X$, $Y$, and $Z$ on the qubits in the sets. Then you can determine the expectation values $\langle X\rangle_\psi, \langle Y\rangle_\psi, \langle Z\rangle_\psi$ from these measurements to an accuracy that depends on the size of the sample. The expectation values are three real numbers that are sufficient to define the quantum state of the qubits in the beam.

It's a general feature of quantum systems, not just qubits, that the state of a system can be specified by the probabilities of the outcomes of measurements on some "informationally complete" or "fiducial" set of observables. The probabilities (which determine the expectation values) can be estimated from the statistics of a sufficiently large number of measurements. Of course, in this process of "state tomography," as it's called, you need a set of quantum systems all in the same quantum state. The number of observables in the informationally complete set will depend on the dimension of the state space. For a qubit, the three Pauli observables are an informationally complete set. For other quantum systems you need more than three observables. A set of informationally complete observables is not unique: *any* observable can be part of some informationally complete set.

 **The bottom line**

- It's a general feature of quantum systems that the state of a system can be specified by the probabilities of the outcomes of measurements on some "informationally complete" or "fiducial" set of observables.
- For a qubit, the three Pauli observables, $X, Y, Z$, are an informationally complete set.

## THE BIORTHOGONAL DECOMPOSITION THEOREM

There's a really useful theorem about entangled states, the biorthogonal decomposition theorem or Schmidt decomposition theorem, that gets applied in the quantum simulation of Aravind–Mermin bananas in Subsection 6.4.2, "Simulating Aravind–Mermin Bananas," in the "More" section of the "Quantum Magic" chapter. The theorem says that for any pure state $|\Phi\rangle$ of a binary quantum system $S_1 + S_2$ (where $S_1$ and $S_2$ could each be composed of subsystems), there is an orthogonal set of unit vectors $|u_1\rangle, |u_2\rangle, \ldots$ associated with a basis in the state space of $S_1$, and a corresponding orthogonal set of unit vectors $|v_1\rangle, |v_2\rangle, \ldots$ for $S_2$ associated with a basis in the state space of $S_2$, such that $|\Phi\rangle$ can be expressed as a linear superposition of product states $|u_i\rangle|v_i\rangle$, for $i = 1, 2, \ldots$, so a "biorthogonal" decomposition. The sum in the superposition could reduce to just one term, in which case $|\Phi\rangle$ is just a product state, but more generally the sum is an entangled state, a superposition of product states.

If the absolute values of the coefficients of the linear superposition are all different, the biorthogonal decomposition is unique. Otherwise, the biorthogonal decomposition is not unique. (If a coefficient is a real number $\pm c$, the absolute value is $c$. If the coefficient is a complex number, the absolute value is $\sqrt{cc*}$.)

For example, the entangled two-qubit Bell state $|\phi^+\rangle = |0\rangle|0\rangle + |1\rangle|1\rangle$ is a biorthogonal decomposition of the state $|\phi^+\rangle$ into products of corresponding computational basis states for the two qubits. In this case, the absolute values of the coefficients are the same, so the biorthogonal decomposition is not unique. The state $|\phi^+\rangle$ could equally well be expressed as a biorthogonal decomposition of $X$-eigenstates of the two qubits $|\phi^+\rangle = |+\rangle|+\rangle + |-\rangle|-\rangle$ or, in fact, as a biorthogonal decomposition with respect to any orthogonal basis states, $|\phi^+\rangle = |s_1\rangle|s_1\rangle + |s_2\rangle|s_2\rangle$, where the basis $|s_1\rangle, |s_2\rangle$ is related to the computational basis by a unitary transformation with real coefficients.

That's the case if you transform from the computational basis to the $X$-basis where the coefficients are real numbers. If you transform to a basis like the $Y$-basis that is related to the computational basis by a unitary transformation with complex coefficients, $|y_+\rangle = |0\rangle + i|1\rangle$, $|y_-\rangle = |0\rangle - i|1\rangle$, then you get $|y_+\rangle|y_-\rangle + |y_-\rangle|y_+\rangle$, which is the Bell state $|\psi^-\rangle$ rather than $|\phi^+\rangle$.

To see why the theorem might be useful, suppose Alice and Bob share two qutrits, and consider two bases, $|1\rangle, |2\rangle, |3\rangle$ and $|1\rangle', |2\rangle', |3\rangle'$, where $|1\rangle = |1\rangle'$ and $|2\rangle', |3\rangle'$ are rotated relative to $|2\rangle, |3\rangle$ in the plane orthogonal to $|1\rangle$. See Figure S.2.

The maximally entangled state

$$|\Phi\rangle = |1\rangle_A|1\rangle_B + |2\rangle_A|2\rangle_B + |3\rangle_A|3\rangle_B$$

has equal coefficients for all the terms in the biorthogonal decomposition. If Alice and Bob both measure the observable $O$ with eigenstates $|1\rangle, |2\rangle, |3\rangle$ of $O$, then they will both get the same outcome, either the outcome corresponding to $|1\rangle$, or the outcome corresponding to $|2\rangle$, or the outcome corresponding to $|3\rangle$, with equal probability $1/3$. (As usual, I've left out the equal coefficient $1/\sqrt{3}$ for each of the terms in the state $|\Phi\rangle$.) Because the coefficients are all equal, the biorthogonal decomposition of $|\Phi\rangle$ is not unique, and $|\Phi\rangle$ could equally well be expressed as

$$|\Phi\rangle = |1\rangle'_A|1\rangle'_B + |2\rangle'_A|2\rangle'_B + |3\rangle'_A|3\rangle'_B = |1\rangle_A|1\rangle_B + |2\rangle'_A|2\rangle'_B + |3\rangle'_A|3\rangle'_B.$$

So if Alice and Bob both measure the observable $O'$ with eigenstates $|1\rangle' = |1\rangle$, $|2\rangle', |3\rangle'$, they will again get the same outcome with equal probability.

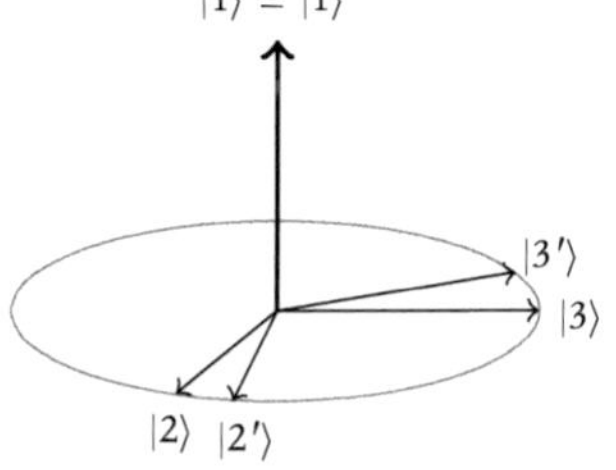


**Figure S.2** Two bases with a common state vector.

What if Alice measures $O$ and Bob measures $O'$? Then Alice will get the outcome corresponding to $|1\rangle$ for $O$ if and only if Bob gets the corresponding outcome for $O'$. If Alice gets one of the other outcomes, then Bob will too, but not necessarily the outcome corresponding to Alice's outcome (so Alice could get the outcome corresponding to the eigenvector $|2\rangle$ and Bob could get the outcome corresponding to $|3\rangle'$, or Alice could get the outcome corresponding to $|3\rangle$ and Bob could get the outcome corresponding to $|2\rangle'$).

To see why, express Bob's states in the biorthogonal decomposition as $O'$-eigenstates. Suppose

$$\begin{aligned}|1\rangle &= |1\rangle,' \\ |2\rangle &= a|2\rangle' + b|3\rangle', \\ |3\rangle &= c|2\rangle' + d|3\rangle',\end{aligned}$$

where all the coefficients $a, b, c, d$ are real numbers. Then

$$\begin{aligned}|\Phi\rangle &= |1\rangle_A|1\rangle_B + |2\rangle_A|2\rangle_B + |3\rangle_A|3\rangle_B \\ &= |1\rangle_A|1\rangle'_B + |2\rangle_A(a|2\rangle' + b|3\rangle') + |3\rangle_A(c|2\rangle' + d|3\rangle') \\ &= |1\rangle|1\rangle' + a|2\rangle|2\rangle' + b|2\rangle|3\rangle' + c|3\rangle|2\rangle' + d|3\rangle|3\rangle'.\end{aligned}$$

Remember that all these terms should be multiplied by the coefficient $1/\sqrt{3}$. According to the Born rule, then, the probability that Alice gets the outcome corresponding to $|1\rangle$ for $O$ and Bob gets the corresponding outcome $|1\rangle' = |1\rangle$ for $O'$ is $1/3$. There's also a probability $|a|^2/3$ that Alice gets the outcome corresponding to $|2\rangle$ and Bob gets the outcome corresponding to $|2\rangle'$, a probability $|b|^2/3$ that Alice gets the outcome corresponding to $|2\rangle$ and Bob gets the outcome corresponding to $|3\rangle'$, a probability $|c|^2/3$ that Alice gets the outcome corresponding to $|3\rangle$ and Bob gets the outcome corresponding to $|2\rangle'$, and a probability $|d|^2/3$ that Alice gets the outcome corresponding to $|3\rangle$ and Bob gets the outcome corresponding to $|3\rangle'$. In other words, if Alice and Bob measure two different observables that have a common eigenvector, then either they will both get the outcome corresponding to the common eigenvector, or they will both get some other outcome. The probability is zero that Alice will get the outcome corresponding to the common eigenvector and Bob won't, or that Bob will and Alice won't.

Now consider an analogous but slightly more complicated case that comes up in the quantum simulation of Aravind–Mermin bananas in the "More" section of Chapter 6, "Quantum Magic." Alice and Bob share three pairs of qubits, each pair in the entangled state $|\phi^+\rangle = |0\rangle|0\rangle + |1\rangle|1\rangle$, so the six-qubit state is

$$\begin{aligned}|\phi^+\rangle \otimes |\phi^+\rangle \otimes |\phi^+\rangle = \frac{1}{\sqrt{8}}(|0\rangle_A|0\rangle_B + |1\rangle_A|1\rangle_B) \otimes (|0\rangle_A|0\rangle_B \\ + |1\rangle_A|1\rangle_B) \otimes (|0\rangle_A|0\rangle_B + |1\rangle_A|1\rangle_B),\end{aligned}$$

where I've indicated the Alice-states and Bob-states by subscripts.

If you multiply out the terms in this expression, you get a linear combination of eight terms like

$$|0\rangle_A|0\rangle_B \otimes |0\rangle_A|0\rangle_B \otimes |0\rangle_A|0\rangle_B$$

(this is the product of the first three terms in the three states $|\phi^+\rangle$). Taking all the Alice-states together and all the Bob states together, these terms can be reordered as products of Alice-states and Bob-states, like

$$|000\rangle_A|000\rangle_B,$$

where $|000\rangle_A$ is short for $|0\rangle_A|0\rangle_A|0\rangle_A$. The whole expression is

$$|\phi^+\rangle|\phi^+\rangle|\phi^+\rangle = |000\rangle_A|000\rangle_B + |001\rangle_A|001\rangle_B + \cdots$$

for all of the eight triples of bits 000, 001, 010, 011, 100, 101, 110, 111.

This is a biorthogonal decomposition for the Alice and Bob systems (which each consists of three qubits) with equal coefficients in the *ZZZ* basis, so the state will take the same form in the *XXX* basis. Now suppose Alice measures her three qubits in the Z-basis, so she measures *ZZZ*, and Bob measures his first qubit in the Z-basis but his other two qubits in the *X*-basis, so he measures *ZXX*. (This is the sort of observable relevant to the simulation of Aravind–Mermin bananas.) As in the previous case, the biorthogonal decomposition ensures that Alice and Bob will get the same outcomes for the Z-measurements on their first qubits.

To work out the probabilities explicitly, you would express Alice's and Bob's three-qubit states in terms of Z-eigenstates and *X*-eigenstates, as appropriate. So

$$|\phi^+\rangle|\phi^+\rangle|\phi^+\rangle = |000\rangle_A|0\rangle_B)(|+\rangle_B + |-\rangle_B)(|+\rangle_B + |-\rangle_B) + \cdots .$$

In the expression after the equal sign, I've replaced each of Bob's second and third qubit states $|0\rangle_B$ in the first term by $|+\rangle_B + |-\rangle_B$. The $\cdots$ refers to the other seven product states. For the next state $|001\rangle_A|001\rangle_B$, Bob's second qubit state $|0\rangle_B$ is replaced by $|+\rangle_B + |-\rangle_B$, and his third qubit state $|1\rangle_B$ is replaced by $|+\rangle_B - |1\rangle_B$, and so on.

Now what's relevant here is that you get four states in the sum of the form $|0\cdots\rangle_A|0\rangle_B(\cdots)_B(\cdots)_B$ and four states of the form $|1\cdots\rangle_A|1\rangle_B(\cdots)_B(\cdots)_B$. Here, $|0\cdots\rangle_A$ are the four states $|000\rangle_A$, $|001\rangle_A$, $|010\rangle_A$, $|011\rangle_A$, and $|1\cdots\rangle_A$ are the four states $|100\rangle_A$, $|101\rangle_A$, $|110\rangle_A$, $|111\rangle_A$. The states of the form $(\cdots)_B$ are either $|+\rangle_B + |-\rangle_B$ or $|+\rangle_B - |-\rangle_B$. So, again, it follows from the Born rule for quantum probabilities that Alice gets the outcome corresponding to the state $|0\rangle$ for the Z-measurement on her first qubit if and only if Bob gets the corresponding outcome for the Z-measurement on his first qubit, and she gets the outcome corresponding to the state $|1\rangle$ on her first qubit if and only if Bob gets the corresponding outcome for the Z-measurement on his first qubit, and each of these possibilities occurs with probability $1/2$.

 **The bottom line**

- The biorthogonal decomposition theorem says that you can always express the state of a composite system $AB$ as a linear superposition of product states for $A$ and $B$, where the $A$-states and the $B$-states are orthogonal states associated with a basis in the $A$ state space and a basis in the $B$ state space, respectively.
- For example, the entangled two-qubit Bell state $|\phi^+\rangle = |0\rangle|0\rangle + |1\rangle|1\rangle$ is a biorthogonal decomposition.
- The theorem has many applications. In particular, it gets applied in the quantum simulation of Aravind–Mermin bananas in Subsection 6.4.2, "Simulating Aravind–Mermin Bananas," in the "More" section of the "Quantum Magic" chapter. I work out an example relevant to the simulation of Aravind–Mermin bananas explicitly.

## STATES AND OPERATORS AS MATRICES

Operators can be represented by matrices. I don't explicitly use matrices in this book, but it's useful to see how Dirac's "bra–ket" notation for an operator is related to the rows and columns of the corresponding matrix. In particular, it's useful to see this for the Pauli operators.

The idea behind matrices is pretty straightforward. A pair of linear equations like

$$x' = ax + by$$
$$y' = cx + dy$$

can be expressed as a matrix equation, in which the single-column matrix $\binom{x}{y}$ is operated on, or multiplied by, a four-element square matrix to produce the new single-column matrix $\binom{x'}{y'}$:

$$\begin{pmatrix} x' \\ y' \end{pmatrix} = \begin{pmatrix} a & b \\ c & d \end{pmatrix} \begin{pmatrix} x \\ y \end{pmatrix}.$$

The matrix multiplication rule simply follows the multiplication and addition operations in the linear equations. To get the first element of the new single-column matrix, $x'$, multiply the two elements in the first row of the four-element square matrix by the two elements in the single-column matrix sequentially and add the products. To get the second element, $y'$, multiply the two elements in the second row of the four-element matrix by the two elements in the single-column matrix sequentially and add the products.

To multiply two four-element square matrices follow the same rule for each element in the resulting product matrix. The element in the *first row* of the *first column* is obtained by multiplying the two elements in the *first row* of the first matrix with the two elements in the *first column* of the second matrix sequentially and adding the products. The element in the *first row* of the *second column* is obtained by multiplying the two elements in the *first row* of the first matrix with the two elements in the *second column* of the second matrix sequentially and adding the products. So,

$$\begin{pmatrix} a & b \\ c & d \end{pmatrix}\begin{pmatrix} e & f \\ g & h \end{pmatrix} = \begin{pmatrix} ae + bg & af + bh \\ ce + dg & cf + dh \end{pmatrix}.$$

In matrix notation, states represented by ket vectors in a qubit state space are represented by single-column matrices (so they are referred to as column vectors), bra vectors by single-row matrices (so row vectors), and operators by four-element square matrices. The computational basis ket vectors $|0\rangle$ and $|1\rangle$ are represented by the column vectors

$$|0\rangle = \begin{pmatrix} 1 \\ 0 \end{pmatrix}, \quad |1\rangle = \begin{pmatrix} 0 \\ 1 \end{pmatrix},$$

and the corresponding bra vectors are represented as the row vectors

$$\langle 0| = \begin{pmatrix} 1 \; 0 \end{pmatrix}, \quad \langle 1| = \begin{pmatrix} 0 \; 1 \end{pmatrix}.$$

The matrix associated with the operator $|0\rangle\langle 1|$ is a four-element square matrix obtained as the product

$$|0\rangle\langle 1| = \begin{pmatrix} 1 \; 0 \end{pmatrix}\begin{pmatrix} 0 \\ 1 \end{pmatrix} = \begin{pmatrix} 0 & 0 \\ 1 & 0 \end{pmatrix}.$$

To get the first column of the four-element matrix, multiply the first element 1 in the row matrix with the column matrix $\binom{0}{1}$. To get the second column of the four-element matrix, multiply the second element 0 in the row matrix with the column matrix $\binom{0}{1}$. Similarly,

$$|1\rangle\langle 0| = \begin{pmatrix} 0 \; 1 \end{pmatrix}\begin{pmatrix} 1 \\ 0 \end{pmatrix} = \begin{pmatrix} 0 & 1 \\ 0 & 0 \end{pmatrix}.$$

So the Pauli operator $X = |0\rangle\langle 1| + |1\rangle\langle 0|$ can be expressed in matrix form as

$$X = |0\rangle\langle 1| + |1\rangle\langle 0| = \begin{pmatrix} 0 & 1 \\ 1 & 0 \end{pmatrix},$$

and the Pauli operators $Y$ and $Z$ can be expressed as

$$Y = -i|0\rangle\langle 1| + i|1\rangle\langle 0| = \begin{pmatrix} 0 & i \\ -i & 0 \end{pmatrix},$$

$$Z = |0\rangle\langle 0| - |1\rangle\langle 1| = \begin{pmatrix} 1 & 0 \\ 0 & -1 \end{pmatrix}.$$

The identity operator $I$ becomes

$$I = |0\rangle\langle 0| + |1\rangle\langle 1| = \begin{pmatrix} 1 & 0 \\ 0 & 1 \end{pmatrix}.$$

Here's another way of thinking about this. The top and bottom row of a four-element matrix can be numbered as the 0 row and the 1 row, and the left and right columns as the 0 column and the 1 column. The two elements in the top row are the 00 element (row 0, column 0) and the 01 element (row 0, column 1). The two elements in the bottom row are the 10 element (row 1, column 0) and the 11 element (row 1, column 1). The matrix elements of an operator depend on the basis chosen. In the standard or computational basis, the matrix elements of a general qubit operator $Q$ on the top row are $\langle 0|Q|0\rangle$ for the 00 element and $\langle 0|Q|1\rangle$ for the 01 element, and the matrix elements on the bottom row are $\langle 1|Q|0\rangle$ for the 10 element and $\langle 1|Q|1\rangle$ for the 11 element. So for the operator $|0\rangle\langle 1|$, the matrix elements in the top row are $\langle 0|0\rangle\langle 1|0\rangle = \langle 1|0\rangle = 0$ and $\langle 0|0\rangle\langle 1|1\rangle = \langle 1|1\rangle = 1$, and the matrix elements in the bottom row are $\langle 1|0\rangle\langle 1|0\rangle = 0$ and $\langle 1|0\rangle\langle 1|1\rangle = 0$.

A tensor product state of two qubits is represented by a four-element column vector:

$$|0\rangle \otimes |1\rangle = \begin{pmatrix} 1 \\ 0 \end{pmatrix} \otimes \begin{pmatrix} 0 \\ 1 \end{pmatrix} = \begin{pmatrix} 0 \\ 1 \\ 0 \\ 0 \end{pmatrix}.$$

You multiply the top element 1 in the first column vector by the elements in the second column vector to get the first two rows in the four-element column vector (the first and second elements, reading down), and you multiply the bottom element 0 in the first column vector by the elements in the second column vector to get the second two rows in the four-element column vector (the third and fourth elements, reading down).

Similarly,

$$|1\rangle \otimes |0\rangle = \begin{pmatrix} 0 \\ 1 \end{pmatrix} \otimes \begin{pmatrix} 1 \\ 0 \end{pmatrix} = \begin{pmatrix} 0 \\ 0 \\ 1 \\ 0 \end{pmatrix},$$

so

$$|\psi^-\rangle = |0\rangle|1\rangle - |1\rangle|0\rangle = \begin{pmatrix} 0 \\ 1 \\ 0 \\ 0 \end{pmatrix} - \begin{pmatrix} 0 \\ 0 \\ 1 \\ 0 \end{pmatrix} = \begin{pmatrix} 0 \\ 1 \\ -1 \\ 0 \end{pmatrix}.$$

### The bottom line

- Operators can be represented by matrices. Although matrices are not used in this book, this section is useful in showing how Dirac's "bra–ket" notation for an operator is related to the rows and columns of the corresponding matrix.
- In particular, it's useful to see this for the Pauli operators, and for the identity operator.

## MIXED STATES

A qubit can be in a pure state that is a linear superposition of other pure states, like the pure state $|+\rangle = |0\rangle + |1\rangle$. It can also be in a mixture of pure states. For example, a qubit could be prepared in such a way that it is either in the pure state $|0\rangle$ with probability $p_0$ or in the pure state $|1\rangle$ with probability $p_1$. You could have a whole collection of qubits prepared in this way, and sampling a qubit from such a mixture of qubit states would produce a qubit in the state $|0\rangle$ with probability $p_0$ or a qubit in the state $|1\rangle$ with probability $p_1$. The probabilities $p_1$ and $p_2$ are referred to as the "weights" of the pure states in the mixture.

The expectation values or average values of an observable $Q$ in the pure states $|0\rangle$ and $|1\rangle$ are $\langle Q\rangle_{|0\rangle} = \langle 0|Q|0\rangle$ and $\langle Q\rangle_{|1\rangle} = \langle 1|Q|1\rangle$. So the expectation value of $Q$ in a mixture of states $|0\rangle$ and $|1\rangle$ with weights $p_1$ and $p_2$ is the weighted sum of the expectation values in the two states: $\langle Q\rangle = p_0\langle 0|Q|0\rangle + p_1\langle 1|Q|1\rangle$.

There's a neat way of expressing the expectation value of an observable $Q$ in a mixture in terms of an operator called the "density operator" of the mixture, and a function called the "trace" function, represented by Tr(. . .):

$$\langle Q\rangle_\rho = \mathrm{Tr}(\rho Q),$$

where $\rho = p_0|0\rangle\langle 0| + p_1|1\rangle\langle 1|$ is the "density operator" representing the mixture. The density operator of a mixture of orthogonal pure states is the weighted sum of the projection operators onto the pure states in the mixture, where the weights are the weights of the pure states in the mixture. (I've been representing operators by capital Roman letters, but it's usual to represent a density operator by the symbol $\rho$, which is also used to represent a probability distribution or mixture of classical pure states.)

Here's how to see this. The trace of an operator $R$ is the sum of the elements along the diagonal of the matrix representing $R$. For a qubit, a matrix representing an operator has four elements, and these elements depend on the basis. The two diagonal elements are $\langle 0|R|0\rangle$ and $\langle 1|R|1\rangle$ in the computational basis. So the trace of $R$ is $\mathrm{Tr}(R) = \langle 0|R|0\rangle + \langle 1|R|1\rangle$.

The trace is an invariant function of an operator: it takes the same value for any basis. For example, with respect to the basis $|+\rangle$, $|-\rangle$, $\mathrm{Tr}(R) = \langle + | R | + \rangle + \langle - | R | - \rangle$. Now, $|+\rangle = \frac{1}{\sqrt{2}}(|0\rangle + |1\rangle)$ and $|-\rangle = \frac{1}{\sqrt{2}}(|0\rangle - |1\rangle)$. So the first term, $\langle + | R | + \rangle$, is the sum $\frac{1}{2}(\langle 0 | R | 0 \rangle + \langle 0 | R | 1 \rangle + \langle 1 | R | 0 \rangle + \langle 1 | R | 1 \rangle)$. The second term, $\langle - | R | - \rangle$, is the same, except that the off-diagonal terms $\langle 0 | R | 1 \rangle$ and $\langle 1 | R | 0 \rangle$ appear with a – sign and so cancel out with the corresponding elements in $\langle + | R | + \rangle$.

The trace is a linear function: $\mathrm{Tr}(cR) = c\mathrm{Tr}(R)$, and $\mathrm{Tr}(R + S) = \mathrm{Tr}(R) + \mathrm{Tr}(S)$. So the trace of $\rho Q = (p_0 |0\rangle\langle 0| + p_1 |1\rangle\langle 1|)Q$ is $p_0\mathrm{Tr}(|0\rangle\langle 0|Q) + p_1\mathrm{Tr}(|1\rangle\langle 1|Q)$. Since the trace is invariant with respect to the choice of basis, you can choose any convenient basis to calculate the trace. Taking the trace with respect to the computational basis, the trace of $|0\rangle\langle 0|Q$ is $\langle 0|0\rangle\langle 0|Q|0\rangle + \langle 1|0\rangle\langle 0|Q|1\rangle = \langle 0|Q|0\rangle$ because $\langle 0|0\rangle = 1$ and $\langle 1|0\rangle = 0$. Similarly, the trace of $|1\rangle\langle 1|Q$ is $\langle 1|Q|1\rangle$. So $\mathrm{Tr}(\rho Q) = p_0\langle 0|Q|0\rangle + p_1\langle 1|Q|1\rangle$, which is the expectation value of $Q$ in the mixture.

From the density operator for a mixture, you can also calculate the probabilities of the eigenvalues or possible values of an observable $Q$. Suppose $Q$ is a qubit observable with two possible values, $q_1$ and $q_2$. Then $Q$ can be expressed in the spectral representation as $Q = q_1 P_{|q1\rangle} + q_2 P_{|q2\rangle}$, where the operators $P_{|q1\rangle}$ and $P_{|q2\rangle}$ are the projection operators onto the eigenvectors corresponding to the eigenvalues $q_1$ and $q_2$. (See the section "The Pauli Operators.") In terms of the density operator $\rho$, you can express the expectation value of $Q$ as $\mathrm{Tr}(\rho Q) = \mathrm{Tr}(\rho(q_1 P_{|q1\rangle} + q_2 P_{|q2\rangle})) = q_1\mathrm{Tr}(\rho P_{|q1\rangle}) + q_2\mathrm{Tr}(\rho P_{|q2\rangle})$. This is the weighted sum of the eigenvalues of $Q$ with the weights $\mathrm{Tr}(\rho P_{|q1\rangle})$ and $\mathrm{Tr}(\rho P_{|q2\rangle})$, so these weights are the probabilities of the eigenvalues.

Since you can derive the probabilities and expectation values of any observable in a mixture from the density operator, this operator plays the same role for mixtures as the state vector does for pure states. A quantum system in a mixture is said to be in a "mixed state" represented by the density operator of the mixture, and the Born rule for a mixed state $\rho$ is

$$\mathrm{prob}_\rho(q) = \mathrm{Tr}(\rho P_q),$$
$$\langle Q\rangle_\rho = \mathrm{Tr}(\rho Q).$$

The Born rule in this form is quite general and applies to pure and mixed states. For a pure state $|\psi\rangle$, the density operator $\rho$ representing the state is the projection operator $P_\psi = |\psi\rangle\langle\psi|$ onto the state. Then $\mathrm{prob}_\psi(q) = \mathrm{Tr}(|\psi\rangle\langle\psi|P_q) = \mathrm{Tr}(|\psi\rangle\langle\psi|q\rangle\langle q|)$. To calculate the trace, use the basis defined by the eigenvectors of $Q$. Then $\mathrm{Tr}(|\psi\rangle\langle\psi|q\rangle\langle q|) = \langle q|\psi\rangle\langle\psi|q\rangle\langle q|q\rangle$ + terms that are zero $= \langle q|\psi\rangle\langle\psi|q\rangle = |\langle q|\psi\rangle|^2$. This is the probability of $q$ in the state $|\psi\rangle$ according to the Born rule: the square of the absolute value of the projection of $|\psi\rangle$ onto $|q\rangle$. The terms that are zero are terms like $\langle q'|\psi\rangle\langle\psi|q\rangle\langle q|q'\rangle$ for eigenvectors orthogonal to $|q\rangle$, which are zero because $\langle q|q'\rangle = 0$.

For the expectation value $\langle Q\rangle_\psi = \mathrm{Tr}(|\psi\rangle\langle\psi|Q\rangle)$, take the trace with respect to any basis of orthogonal unit vectors containing $|\psi\rangle$. Then $\mathrm{Tr}(|\psi\rangle\langle\psi|Q\rangle) = \langle\psi|\psi\rangle\langle\psi|Q|\psi\rangle = \langle\psi|Q|\psi\rangle$, because $|\psi\rangle$ is a unit vector and so $\langle\psi|\psi\rangle = 1$.

Different mixtures can be represented by the same mixed state density operator. For example, the mixture of pure states $|0\rangle$, $|1\rangle$ with equal weights is represented by the same mixed state density operator as the mixture of pure states $|+\rangle$, $|-\rangle$ with equal weights:

$$\rho = \frac{1}{2}|0\rangle\langle 0| + \frac{1}{2}|1\rangle\langle 1| = \frac{1}{2}|+\rangle\langle +| + \frac{1}{2}|-\rangle\langle -|.$$

You can see this by writing out the ket and bra vectors $|+\rangle$, $\langle +|$ and $|-\rangle$, $\langle -|$ explicitly in terms of the ket and bra vectors $|0\rangle$, $\langle 0|$ and $|1\rangle$, $\langle 1|$ and multiplying out the expressions:

$$\frac{1}{2}\left(\frac{1}{\sqrt{2}}\left(|0\rangle + |1\rangle\right) \cdot \frac{1}{\sqrt{2}}\left(\langle 0| + \langle 1|\right)\right) + \frac{1}{2}\left(\frac{1}{\sqrt{2}}\left(|0\rangle - |1\rangle\right) \cdot \frac{1}{\sqrt{2}}\left(\langle 0| - \langle 1|\right)\right).$$

The $|0\rangle\langle 1|$ and $|1\rangle\langle 0|$ terms cancel out because they appear with $+$ signs in the expression you get from the first product and with $-$ signs in the expression you get from the second product. The remaining terms add up to

$$\frac{1}{2}\left(\frac{1}{2}|0\rangle\langle 0| + \frac{1}{2}|1\rangle\langle 1|\right) + \frac{1}{2}\left(\frac{1}{2}|0\rangle\langle 0| + \frac{1}{2}|1\rangle\langle 1|\right) = \frac{1}{2}|0\rangle\langle 0| + \frac{1}{2}|1\rangle\langle 1|.$$

Since the identity operator $I$ can be expressed as $|0\rangle\langle 0| + |1\rangle\langle 1|$ or as $|+\rangle\langle +| + |-\rangle\langle -|$, the mixed state of these two mixtures is represented by the density operator $\rho = I/2$. In fact, any completely random mixture of orthogonal qubit pure states is represented by the mixed state density operator $I/2$, which also represents mixtures of nonorthogonal qubit pure states.

Measurement probabilities for a mixture of states are given by the Born rule via the density operator $\rho$ and don't depend on the representation of $\rho$ as a specific mixture of pure states. So the fact that different mixtures of pure states are represented by the same density operator means that such mixtures are indistinguishable by any measurements. This is also a feature of mixtures with non-equal weights. You could also mix nonorthogonal pure states, and the indistinguishability will apply to these states as well.

This might seem odd, because you could prepare an equal weight mixture of linearly polarized photons in states $|0\rangle$, $|1\rangle$ by sending photons through a beamsplitter with its axis of polarization in a certain direction $z$, or you could prepare an equal weight mixture of diagonally polarized photons in states $|+\rangle$, $|-\rangle$ by passing photons through a beamsplitter with its axis of polarization rotated 45° to the $z$ direction, and these two mixtures would seem to be quite different.

But consider what happens if you select a photon at random from the mixture represented by the mixed state $\frac{1}{2}|+\rangle\langle +| + \frac{1}{2}|-\rangle\langle -|$ and send it through a beamsplitter with its axis of polarization in the $z$ direction. If you select a photon at random from the mixture, it will either be in the state $|+\rangle$ with probability $1/2$ or in the state $|-\rangle$ with probability $1/2$. If a photon in the state $|+\rangle$ goes through the beamsplitter, it will end up in the state $|0\rangle$ with probability $1/2$ or in the state $|1\rangle$ with probability $1/2$, and similarly for the state $|-\rangle$. So the probability that a photon selected at random from the mixture and sent through the beamsplitter ends up in the state $|0\rangle$ is $\frac{1}{2}\frac{1}{2}+\frac{1}{2}\frac{1}{2}=\frac{1}{2}$, and similarly the probability that the photon ends up in the state $|1\rangle$ is $\frac{1}{2}$. That's the same as the probabilities for the mixture $\frac{1}{2}|0\rangle\langle 0|+\frac{1}{2}|1\rangle\langle 1|$. The same thing will happen for any polarization measurements, so the two mixtures are really indistinguishable.

Finally, suppose you have an entangled state of two qubits, $A$ and $B$, like

$$|\psi\rangle = c_1|0\rangle_A|0\rangle_B + c_2|1\rangle_A|1\rangle_B,$$

where $c_1$ and $c_2$ are real number coefficients whose squares, representing probabilities, add up to 1. What's the quantum state of qubit $B$ alone?

It should be a state $\rho_B$ such that

$$\langle\psi|I_A Q_B|\psi\rangle = \mathrm{Tr}(\rho_B Q_B),$$

where $I_A$ is the identity operator for qubit $A$ and $Q_B$ is any observable of qubit $B$ alone. This expression is $I_A Q_B$ sandwiched between $\langle\psi| = c_1\,{}_A\langle 0|\,{}_B\langle 0| + c_2\,{}_A\langle 1|\,{}_B\langle 1|$ and $|\psi\rangle = c_1|0\rangle_A|0\rangle_B + c_2|1\rangle_A|1\rangle_B$. (I've written bra vectors as ${}_A\langle 0|, {}_B\langle 0|$ and ket vectors as $|0\rangle_A, |0\rangle_B$.)

If you work this out, you get

$$c_1^2\,{}_A\langle 0|\,{}_B\langle 0|I_A Q_B|0\rangle_A|0\rangle_B + c_2^2\,{}_A\langle 1|\,{}_B\langle 1|I_A Q_B|1\rangle_A|1\rangle_B$$
$$+ c_1c_2\,({}_A\langle 0|\,{}_B\langle 0|I_A Q_B|1\rangle_A|1\rangle_B + {}_A\langle 1|\,{}_B\langle 1|I_A Q_B|0\rangle_A|0\rangle_B).$$

The two terms with coefficients $c_1c_2$ are

$${}_A\langle 0|\,{}_B\langle 0|I_A Q_B|1\rangle_A|1\rangle_B = {}_A\langle 0|I_A|1\rangle_A \cdot {}_B\langle 0|Q_B|1\rangle_B = {}_A\langle 0|1\rangle_A \cdot {}_B\langle 0|Q_B|1\rangle_B = 0$$

and

$${}_A\langle 1|\,{}_B\langle 1|I_A Q_B|0\rangle_A|0\rangle_B = {}_A\langle 1|I_A|0\rangle_A \cdot {}_B\langle 1|Q_B|0\rangle_B = {}_A\langle 1|0\rangle_A \cdot {}_B\langle 1|Q_B|0\rangle_B = 0.$$

Both these terms are zero because $|0\rangle$ and $|1\rangle$ are orthogonal, so ${}_A\langle 0|1\rangle_A = {}_A\langle 1|0\rangle_A = 0$.

The other two terms are equal to

$$c_1^2\,{}_A\langle 0|I_A|0\rangle_A \cdot {}_B\langle 0|Q_B|0\rangle_B + c_2^2\,{}_A\langle 1|I_A|1\rangle_A \cdot {}_B\langle 1|Q_B|1\rangle_B$$
$$= c_1^2\,{}_A\langle 0|0\rangle_A \cdot {}_B\langle 0|Q_B|0\rangle_B + c_2^2\,{}_A\langle 1|1\rangle_A \cdot {}_B\langle 1|Q_B|1\rangle_B,$$

and this is just

$$c_1^2\,{}_B\langle 0|Q_B|0\rangle_B + c_2^2\,{}_B\langle 1|Q_B|1\rangle_B$$

because ${}_A\langle 0|0\rangle_A = {}_A\langle 1|1\rangle_A = 1$. So

$$\rho_B = c_1^2\,|0\rangle\langle 0| + c_2^2\,|1\rangle\langle 1|,$$

where I've left out the subscripts $B$.

This is a mixed state, called the "reduced state" of the subsystem $B$, representing a mixture of the states $|0\rangle$ and $|1\rangle$ with weights $p_1 = c_1^2$ and $p_2 = c_2^2$. Even though the two-qubit composite state is an entangled pure state, the state of a subsystem is a mixed state. This will always be the case, for an entangled pure state of any composite system, not necessarily two qubits, unless the state of the composite system is a product state.

## The bottom line

- A quantum system can be in a pure state that is a linear superposition of other pure states, or you can have a whole collection of quantum systems, called a mixture, prepared in different states that occur in the mixture with different probabilities or "weights."
- There's a neat way of expressing the expectation value of an observable in a mixture, or the probabilities of the possible values of an observable, in terms of an operator called the "density operator" of the mixture.
- The density operator plays the same role for mixtures as the state vector does for pure states. A quantum system selected at random from a mixture is said to be in a "mixed state" represented by the density operator of the mixture.
- Different mixtures can be represented by the same mixed state density operator, and such mixtures are indistinguishable by any measurements.
- If you have two (or more) quantum systems in an entangled pure state, the state of a subsystem is in general a mixed state, called the "reduced state" of the subsystem.

# INDEX